ABHANDLUNGEN DER SÄCHSISCHEN AKADEMIE
DER WISSENSCHAFTEN ZU LEIPZIG

Mathematisch-naturwissenschaftliche Klasse

Band 53 · Heft 1

3. INTERNATIONALES SYMPOSIUM ÜBER PROBLEME DER PASSIVITÄT VON METALLEN

Technische Universität Dresden

29. Mai bis 31. Mai 1975

Mit 150 Abbildungen und 27 Tabellen

AKADEMIE-VERLAG · BERLIN

1979

Vorträge, gehalten vom 29. bis 31. Mai 1975
an der Technischen Universität Dresden, Sektion Chemie,
anläßlich des 70. Geburtstages des Präsidenten
der Sächsischen Akademie der Wissenschaften zu Leipzig,
Professor Dr.-Ing. habil. Dr. h. c. mult. KURT SCHWABE

Manuskript eingereicht am 27. September 1976
Druckfertig erklärt am 14. Oktober 1978

Erschienen im Akademie-Verlag, 108 Berlin, Leipziger Straße 3—4

Lizenznummer: 202 · 100/557/79
Gesamtherstellung: VEB Druckhaus „Maxim Gorki“, 74 Altenburg
Bestellnummer: 762 270 4 (2025/53/1) · LSV 3015
Printed in GDR
DDR 42,— M

Sehr verehrter Herr Schwabe!

Wer Sie näher kennt, weiß um Ihre unüberwindbare Abneigung gegen jegliche Art von Feierlichkeit, die mit Ihrer eigenen Person zusammenhängt. Wenn sich Ihnen nun dieser Sammelband präsentiert, dann meine ich, Sie sollten ihn als das nehmen, was er ist: als Bekundung der Freundschaft, Dankbarkeit und Verehrung aus dem großen Kreis Ihrer Schüler, Mitarbeiter und Kollegen, für die die Begegnung mit Ihnen wichtig geworden ist.

Dieser Band, den vorzulegen ich namens der Sächsischen Akademie der Wissenschaften zu Leipzig als deren auswärtiges Mitglied und als Ihr Fachkollege die Ehre habe, vereint die Vorträge, die auf dem 3. Internationalen Symposium über Probleme der Passivität von Metallen vom 29. bis 31. Mai 1975 an der Technischen Universität Dresden aus Anlaß Ihres 70. Geburtstages gehalten worden sind. Sie behandeln aktuelle Themen der Passivierung der Metalle, also jenes Kernproblems der Korrosion und der Überlegungen zur Herstellung neuer korrosionsresistenter Legierungen und Schutzmittel, das Ihr eigenstes Forschungsgebiet ist und auf dem Sie selbst durch tiefgreifende Studien wesentliche Erkenntnisse gewonnen und fruchtbare Impulse vermittelt haben.

Allerdings ist Ihr Arbeitsfeld noch um sehr Vieles breiter. Davon zeugen über 350 wissenschaftliche Abhandlungen und Monographien, in denen Sie sich neben dem Problem der Passivität und Korrosion der Azidität, Polarographie, der elektrochemischen Analysentechnik sowie Fragen der Brennstoffelemente zuwenden. All dies weist Sie aus als einen Forscher von größter fachlicher Kompetenz und Originalität.

Daneben haben Sie sich den Forderungen, die die Öffentlichkeit an Sie stellte, nicht entzogen. Als Hochschullehrer, als Dresdener Rector magnificus, als Präsident der Sächsischen Akademie der Wissenschaften zu Leipzig und als Vizepräsident der Akademie der Wissenschaften der DDR haben Sie Rat gegeben, Arbeit übernommen und wissenschaftspolitisch gewirkt, und schließlich haben Sie das Forschungsinstitut Meinsberg, das nach wie vor unter Ihrer Leitung steht, aufgebaut und entscheidend geprägt.

Empfangen Sie also diesen Band, in dem nach altem Brauch die verschiedensten Arbeitsergebnisse zum Problem der Metallpassivität behandelt sind, als Zeugnis dankbaren Respektes. Mögen Sie Ihr reiches Lebenswerk in vielen künftigen Jahren der Gesundheit und des fruchtbaren Schaffens vollenden!

Moskau, den 27. September 1976 *Viktor I. Spitzin*

INHALT

M. POURBAIX

Some Practical Applications of Potential-pH Diagrams 9

1. Localized Corrosion of Copper and of Copper Alloys 9
2. Localized Corrosion of Iron and Steels 21
3. Conclusions 25
References 26

I. EPELBOIN, C. GABRIELLI, M. KEDDAM

Messungen der FARADAY-Impedanz zur Verifizierung eines Modells der Auflösung und Passivierung des Eisens in molarer Schwefelsäure 29

1. Meßanordnung und experimentelle Ergebnisse 29
   1.1. Problemstellung 30
   1.2. Probleme der Regelung der Grenzflächenpolarisation 31
   1.3. Anwendung auf die Aufnahme der Strom-Spannungs-Charakteristik des Eisens in Schwefelsäure 34
   1.4. Messung der Impedanz der Grenzfläche 36
   1.5. Anwendung auf die Messung der Impedanz der Grenzfläche Eisen—molare Schwefelsäure 37
2. Interpretation des anodischen Verhaltens des Eisens in schwefelsaurem Milieu 37
   2.1. Mathematische Formulierung des Problems 37
   2.2. Simulation des Modells 40
   2.2.1. Modell *A* 40
   2.2.2. Modell *B* 44
3. Diskussion 47
Literatur 47

E. STUPNIŠEK-LISAC, M. KARŠULIN

Über die intermediären Reaktionen bei der Passivierung des Eisens in Salpetersäure 49

1. Versuchsanordnung 49
2. Versuchsergebnisse 50
3. Diskussion 54
Literatur 58

M. PRAZÁK, Trinh Xuan SEN

The Kinetics of Redoc Reactions on Some Passive Metals and Alloys 59

1. Introduction 59
2. Experimental 59
3. Results and Discussion 61

4. The Effect of the OH⁻-Ion Concentration 64
5. Extraction of Components of Passive Films on Alloys 64
6. The Possibility of Evaluation of Artificial Passive Films 65
7. Conclusions 65
References 66

W. FORKER

Zum Aufbau und Verhalten von passivierenden Oxidschichten 67

Chr. WEISSMANTEL, P. MORZECK, G. REISSE

Passivierungs- und Korrosionsschutz mittels plasma- und ioneninduzierter Prozesse 73

1. Einführung 73
1.1. Stoßprozesse auf einem Target (Ion-sputtering) 74
1.2. Stoßprozesse im Plasma (Plasma-induced-deposition) 74
1.3. Stoßprozesse im Substrat und in der wachsenden Schicht (Ion assisted film formation) 74
2. Anlagen und Verfahren 75
2.1. Plasmazerstäubung 75
2.2. Ionenstrahlzerstäubung 76
2.3. Ionengestützte Beschichtung (IAFF) 78
2.4. Plasmainduzierte Schichtabscheidung aus der Gasphase (PID) 79
3. Herstellung und Eigenschaften spezieller Schichten 79
3.1. Aluminiumoxid 80
3.2. Siliziumnitrid 80
3.3. Polytetrafluoräthylen 83
4. Schlußfolgerungen 84
Literatur 84

M. SMIAŁOWSKI

Einfluß von Katalysatorgiften auf die Wasserstoffdurchlässigkeit von kathodisch polarisierten Stahlmembranen 85

H.-J. ENGELL, K. BOHNENKAMP, E. RIECKE, E.-G. DAFFT, F. KUSTER

Wasserstoffaufnahme und Wasserstoff-Versprödung bei der Korrosion von Stahl 93

V. I. SPITZIN, M. M. KURTEPOW, R. L. BARU

Elektrochemische Eigenschaften und Korrosionscharakteristiken von metallischem Technetium und mit Technetium legiertem Stahl 95

V. I. SPITZIN, A. F. KUSINA, A. A. OBLOWA, M. I. GLINKINA, L. I. STEPOWAJA

Über den Mechanismus der elektrochemischen Reduktion des TcO_4^--Ions 101

I. ROSENFELD, L. FROLOWA, B. LARKIN

Chemische Passivierung von Stahl durch Inhibitoren 105

K. SCHWABE

Zur Problematik der Passivitätsforschung 117

J. M. Kolotyrkin

Einige Gesetzmäßigkeiten der selektiven Auflösung von Eisen- und Nickellegierungen 131

N. D. Tomaschow, G. P. Tschernowa, L. A. Tschigirinskaja, M. F. German

Der Einfluß der Legierungselemente auf Passivierung und Korrosionsbeständigkeit der Legierungen auf Eisen-Chrom-Basis 141

I. Garz

Der Einfluß der Kaltverformung auf die Passivierung von Nickel und das Lochfraßpotential von Fe—Ni-Einkristallen 149

1. Einführung 149
2. Versuchsdurchführung und Ergebnisse 150
3. Diskussion 153
   3.1. Die Passivierungsstromdichte 153
   3.2. Die Passivstromdichte 154
   3.3. Der Aktiv-Passiv-Übergang 154
   3.4. Das Lochfraßpotential von Eisen-Nickel-Einkristallen 155

Literatur 156

L. Franke, W. Forker

Über das transpassive Verhalten von Eisen und Stählen in Salzlösungen bei hohen Stromdichten 157

L. Kiss, M. L. Varsányi

Die Passivierung des Eisens in nichtwäßrigen Lösungsmitteln 167

M. Pourbaix

Centre Belge d'Etude de la Corrosion, Brüssel

SOME PRACTICAL APPLICATIONS OF POTENTIAL-pH DIAGRAMS

Summary

Potential-pH diagrams may be either theoretical equilibrium diagrams (drawn on the basis of thermodynamic data) or experimental diagrams (drawn as a result of electrochemical experiments). Both may be useful for elucidating complex corrosion phenomena. Examples are given concerning the corrosion and protection of copper and copper-alloys and of iron and steels.

1. Localized Corrosion of Copper and of Copper Alloys

In the research of these materials we first tried to form artificially a pit similar to the natural ones by simply immersing in running Brussels water a copper plate on which was sealed a thin plastic tube containing some CuCl (Fig. 1); but, after about a month of immersion, we observed with surprise just the contrary of what we expected: instead of a pit there was under the CuCl-containing

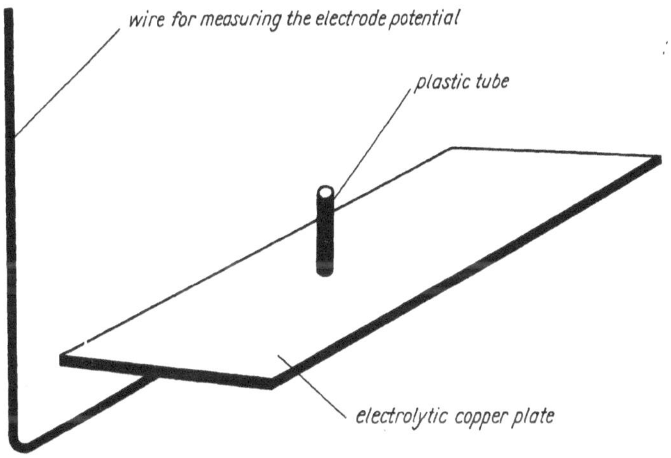

Fig. 1: Copper specimen used in experiments for attempting to produce artificial pits (first attempt).

tube a deposit of metallic copper ... We then used another experimental device (Fig. 2) consisting in a series of glass-tubes containing open sections of copper-tubes cut at 120°. Water was passed continuously through the glass-tubes at a velocity of 20 mm per second, and, as a function of time, we measured the values of the electrode potentials of the copper specimens and we observed visually their surface conditions. Tests were also made on thin copper-wires, where the development of a pit was detected by a rupture of the wire.

Fig. 3 shows a typical electrode potential-time curve for with 10% PO_4H_3 pickled electrolytic copper, in flowing Brussels water: during a first period of about ten days, the electrode potential remains at relatively low values (about −50 to −10 mV_{sce}) and the surface becomes covered with a reddish deposit of cuprous oxide Cu_2O; during a second period of about ten days, the potential increases steadily to about +40 mV_{sce}, and some greenish flakes (of basic carbonate malachite

$CuCO_3 \cdot Cu(OH)_2$) are progressively formed on the surface; during a third period periodic drops and rises of electrode potential occur.

We observed that illumination of the copper with a 200 watts tungsten flash lamp (which we performed occasionally for observing the surface-condition of the metal) does not affect its electrode potential during the first period shown at figure 8 (after one hour, at $-38\ mV_{sce}$) (Fig. 4), but that, during the third period (after 145 days, at $+77\ mV_{sce}$), it promotes very quickly, during a fraction of a second, a considerable potential drop of about 320 mV (until $-250\ mV_{sce}$), which is only temporary. We also observed that a stagnation of the water (Fig. 5) leads, during the first period, to

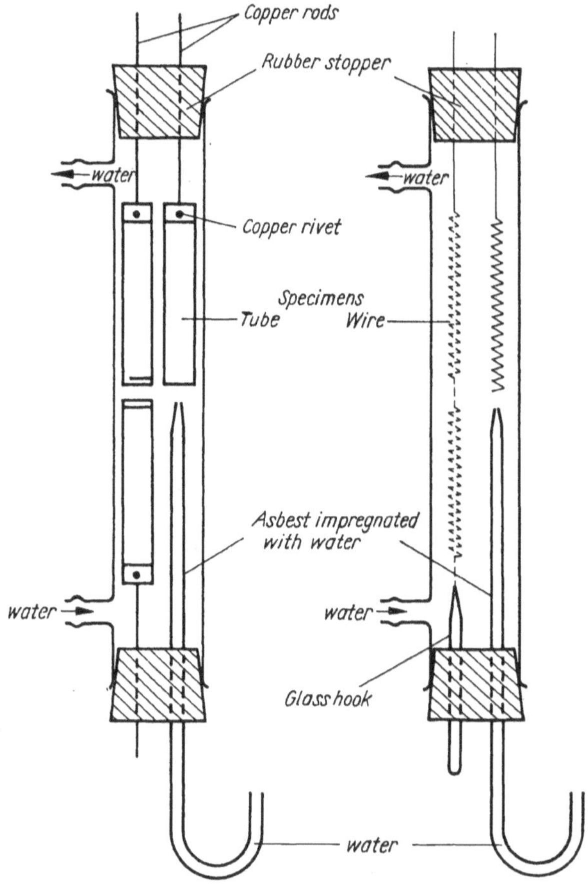

Fig. 2: Apparatus for corrosion testing of tube and wire specimens.

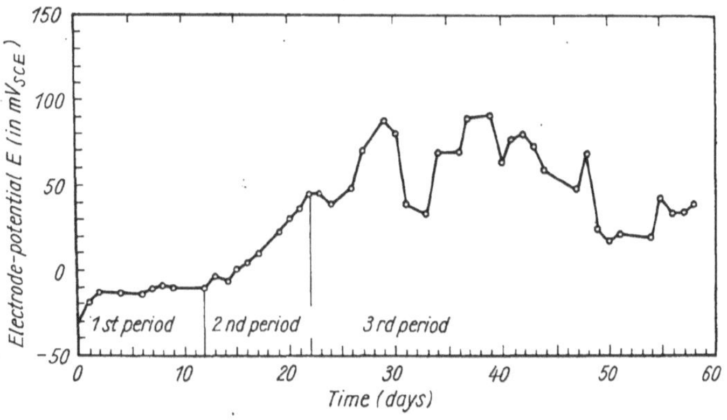

Fig. 3: Influence of immersion time on the electrode potential of electrolytic copper vertically immersed in Brussels water flowing continuously.

a slight increase (of about 20 mV) of the electrode potential, but that, during the third period, it promotes a slow and important decrease of the potential to the same value as does illumination ($-250\ mV_{sce}$), and also temporarily. I have to confess that these funny results were at that time quite ununderstandable to us, and that they even led some members of the team working with us to question the interest and the validity of electrode potential measurements. Besides this, we had performed numerous potentiokinetic and potentiostatic experiments, most of which were helpless an hopeless, until finally, after having tested during about 155 to 367 days 207 specimens of different composition (electrolytic copper, phosphorous deoxidized, arsenical, etc.) of different initial surface conditions (pickled, as coming from the factory, without or with "aquablasting", etc.), isolated or coupled with graphite or with different metals (gold, platinum, rhodium, silver, tin), without or with some additions (Fe^{++}, MnO_2, NaClO) to the Brussels water we were using for the tests we made an unexpected observation which brought back to us hope and confidence (Fig. 6 and 7).

Among these 207 specimens, 51 did show pitting and the 156 others did not. And all the 51 pitted specimens had reached, during the third period of the experiments conducted with the device shown

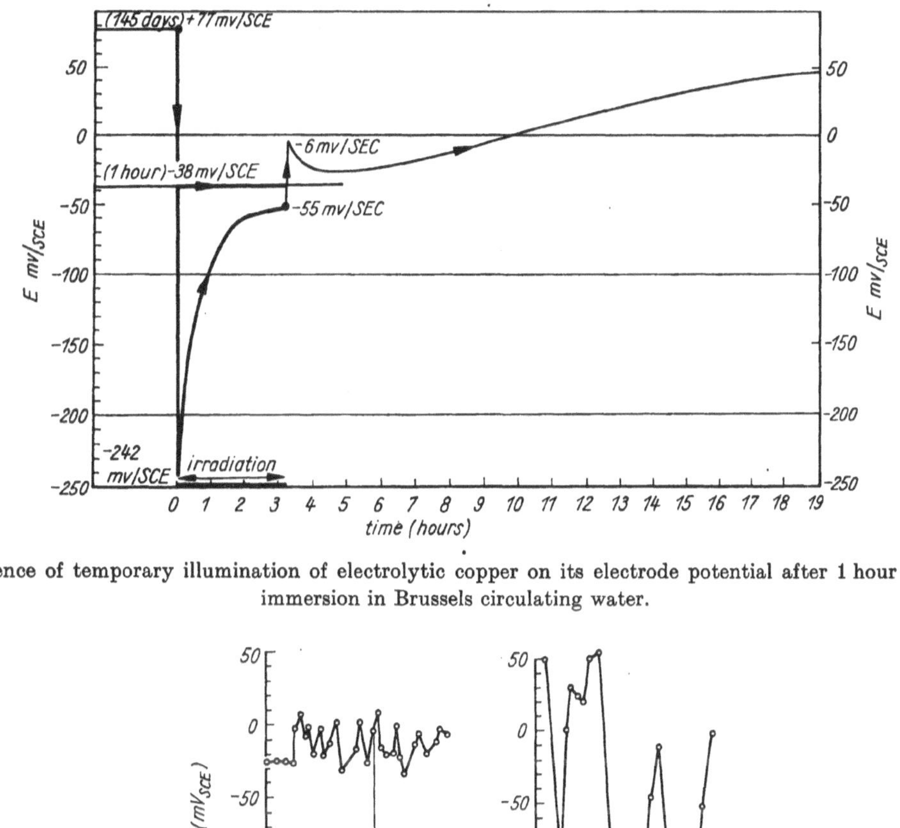

Fig. 4: Influence of temporary illumination of electrolytic copper on its electrode potential after 1 hour and 145 days immersion in Brussels circulating water.

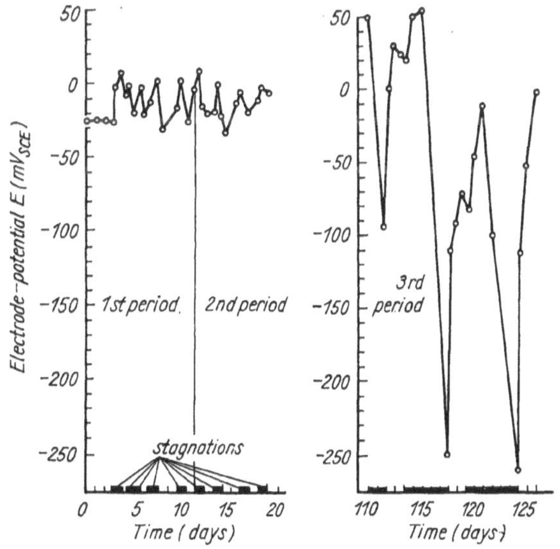

Fig. 5: Influence of water stagnation on the electrode potential of electrolytic copper immersed vertically.

at Fig. 2 an electrode potential higher (more positive) than a critical value, which was about $+100$ mV$_{sce}$ for wires and $+170$ mV$_{sce}$ for tubes, and these specimens were the only ones which reached such high values: all the 156 other specimens, which did not show pitting, had lower potentials. For trying to understand our experimental observations, we looked to cross-sections through copper pits and we used potential-pH equilibrium diagrams.

Fig. 8 shows a cross-section through a copper-pit in Brussels water. There is at the bottom of the pit some white cuprous chloride CuCl and a large amount of reddish cuprous oxide Cu_2O; the top of the pit is covered with green malachite $CuCO_3 \cdot Cu(OH)_2$ (and with some white calcium carbonate $CaCO_3$). Needless to say, the solution in the presence of which the copper-pit developes is not the bulk of the water, but a solution saturated in both CuCl and Cu_2O. Fig. 9 shows a further cross-section

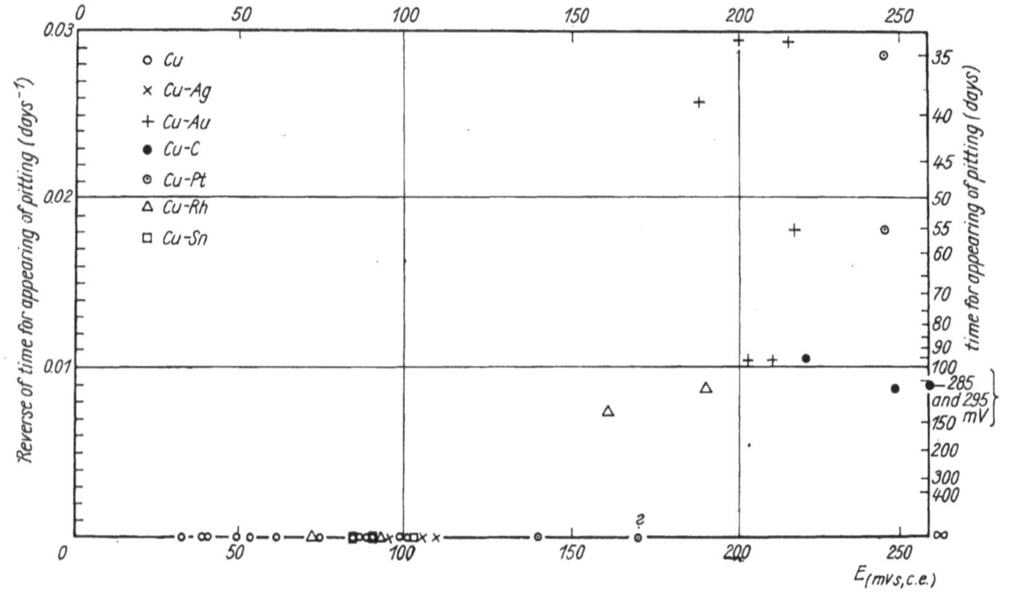

Fig. 6: Corrosimetric test on pickled copper tubes, in the presence of circulating Brussels water. Relationship between the maximum electrode potential and the time when corrosion pits appear.

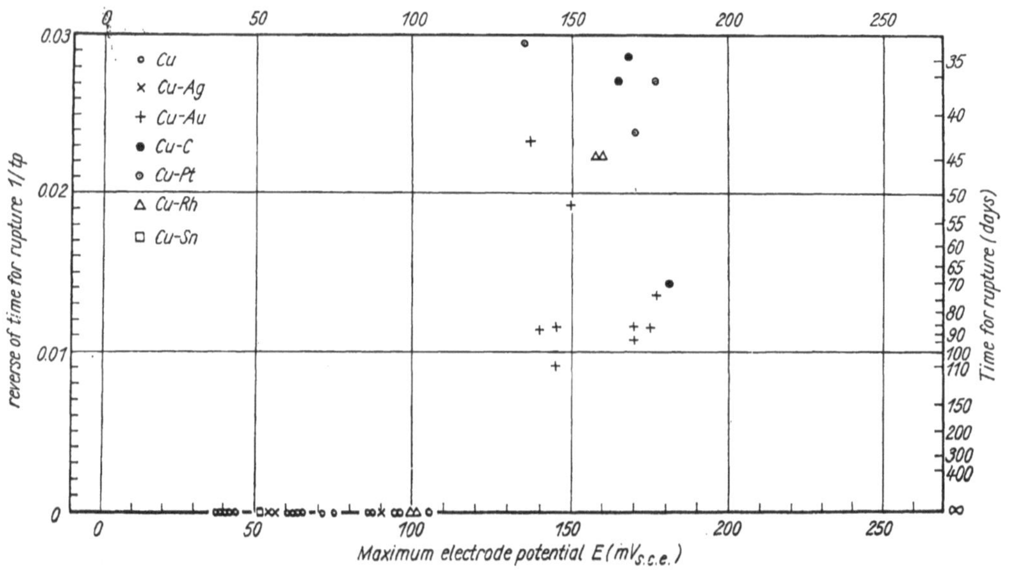

Fig. 7: Corrosimetric tests on electrolytic copper wires in the presence of Brussels water flowing continuously (11.5 cm · sec^{-1}). Relation between the maximum electrode potential and the time when rupture occurs. (Results after 250 days).

through a copper-pit similar to the one shown at Fig. 8, which clearly shows that most of the Cu_2O has disappeared from the cavity, and is thus a loose secondary corrosion product.

Fig. 10 shows a potential-pH equilibrium diagram for the binary system $Cu—H_2O$, at 25 °C, where may be seen notably the equilibrium conditions of Cu, Cu_2O and CuO; in the presence of pure water of pH 7, the stable oxidation product of copper is Cu_2O if no oxygen is present; it is CuO if oxygen

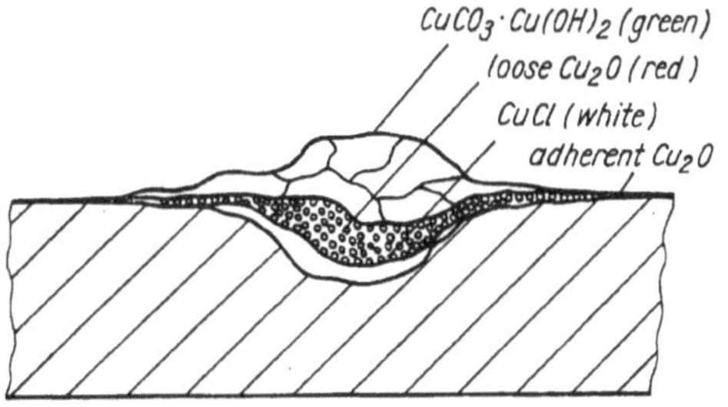

Fig. 8: Copper pit. Cross section showing the presence of red Cu_2O and white CuCl beneath a mushroom of green malachite.

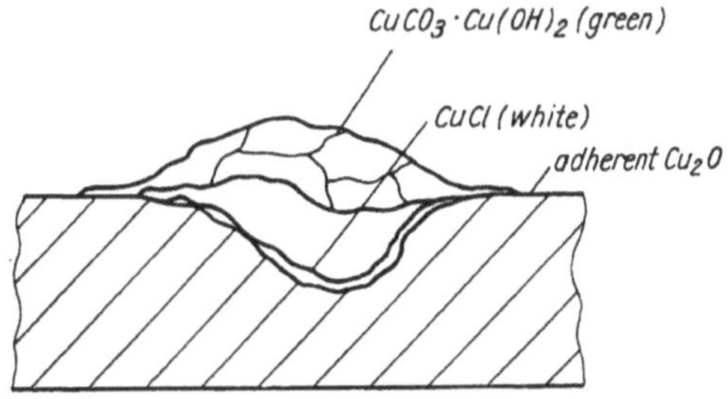

Fig. 9: Copper pit. Cross section of the pit shown in Fig. 8 after removal of the loose deposit of Cu_2O.

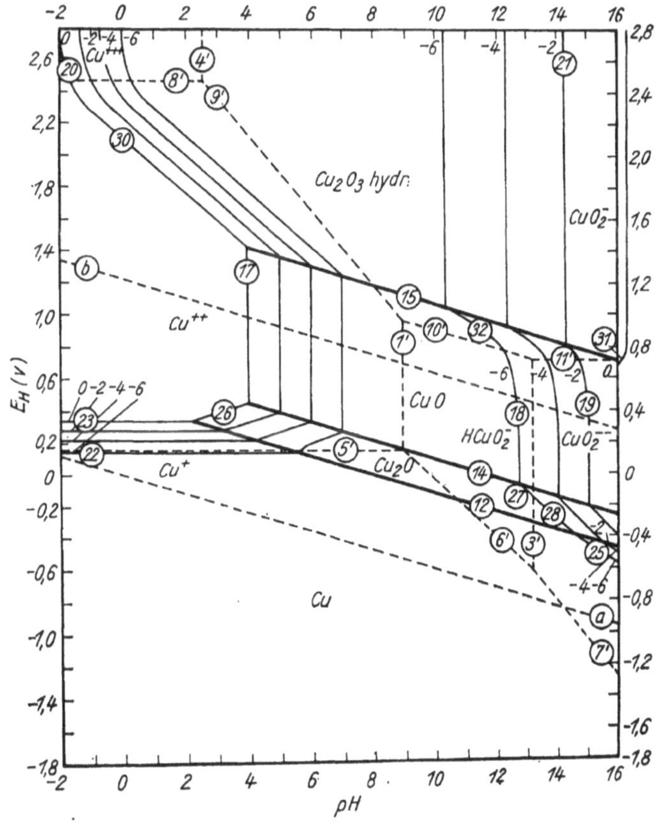

Fig. 10: Potential-pH equilibrium diagram for the copper-water system at 25 °C.

is present. Fig. 11 shows a potential-pH equilibrium diagram for the ternay system Cu—Cl—H_2O, valid for solutions containing 10^{-2} gr.ion Cl^- per liter, which is about the amount of Cl^- in a solution saturated in CuCl. This figure shows that CuCl is not stable in neutral solution of pH 7, where the stable form of univalent copper is Cu_2O; CuCl should thus, in such a solution, hydrolyse according to the reaction $2CuCl + H_2O \rightarrow Cu_2O + 2H^+ + 2Cl^-$. The only point of the diagram where Cu, Cu_2O and CuCl, which are coexisting at the bottom of a pit, are simultaneously stable, is the point of intersection of the three lines No. 12, 51 and 55. This point, whose coordinates are $E = +270$ mV_{ehs} and pH $= 3{,}5$, thus represents the equilibrium conditions at the bottom of the pit. For this point, the equilibrium concentrations in dissolved copper are, assuming that the solution is "perfect":

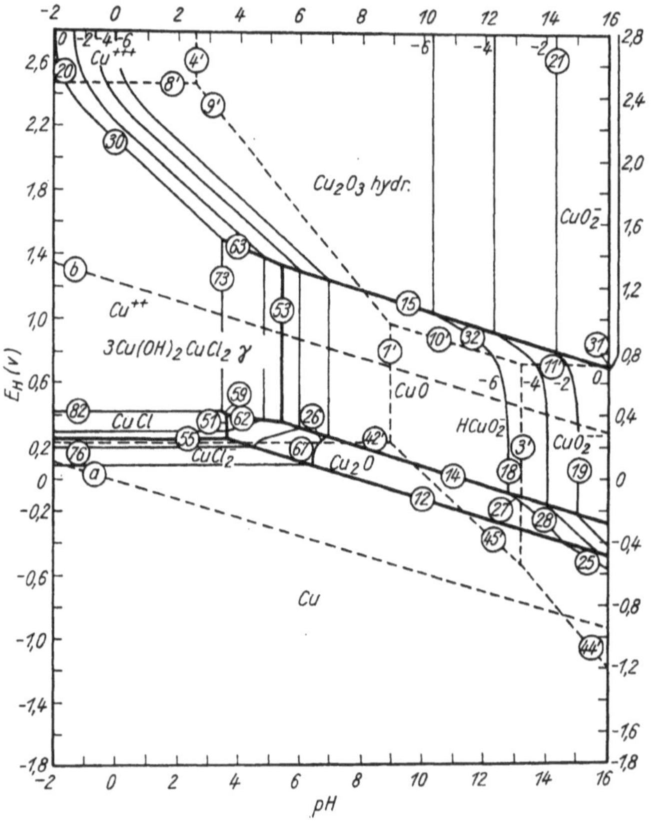

Fig. 11. Potential-pH equilibrium diagram for the 3-component system, Cu—Cl—H_2O, for solutions containing 355 ppm Cl^- (10^{-2} g.ion/liter).

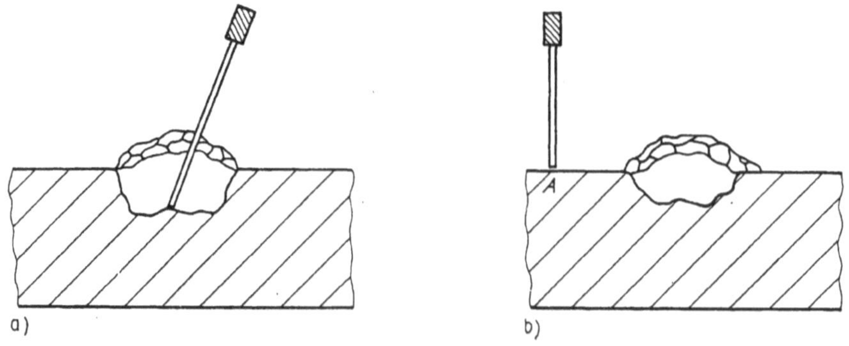

Fig. 12. The electrode potential of pitted copper

(*a*) as measured within the pit;

(*b*) as measured outside the pit, giving the "mixed potential".

$10^{-2.44}$ as Cu^{++}, $10^{-3.37}$ as $CuCl_2^-$ and $10^{-4.31}$ as Cu^+, i.e. a total of $10^{-2.43}$ (or 234 ppm) dissolved copper; the equilibrium concentrations in dissolved Cl are $10^{-2.17}$ as Cl^- and $10^{-3.07}$ as $CuCl_2^-$, i.e. a total of $10^{-2.12}$ at.gr/liter or 270 ppm dissolved Cl. Under these conditions of electrode potential and pH, the corrosion reaction of copper is reversible; it is thus easy to understand that any such pit will grow if the electrode potential inside the pit is higher (more positive) than $+270\ mV_{she}$ (i.e. $+20\ mV_{sce}$); on the contrary, such a pit will stifled, with redeposition of metallic copper, if this electrode potential is lower (more negative) than $+270\ mV_{she}$. And this is the very simple criterion which governs the problem of copper-pitting in waters containing some chloride: for avoiding the development of copper-pits in such waters, it is necessary and sufficient to prevent any increase of the electrode potential inside the cavities above $+270\ mV_{she}$ (or $+20\ mV_{sce}$). As electrode potentials inside cavities are usually not easily measurable, it is convenient to correlate them with the electrode potentials outside the cavities (Fig. 12) which, due to the existence of a diffusion potential, are generally about 80 to 150 mV higher (about +350 to $+420\ mV_{she}$ or +100 to $+170\ mV_{sce}$). And we thus have the explanation of the critical protection potential against copperpitting (about +100 to $170\ mV_{sce}$) which was observed experimentally.

Fig. 13 shows, together with the theoretical corrosion-immunity- and passivation conditions of copper in water at 25 °C, the equilibrium conditions inside a copper-pit ($E = +270\ mV_{she}$ and

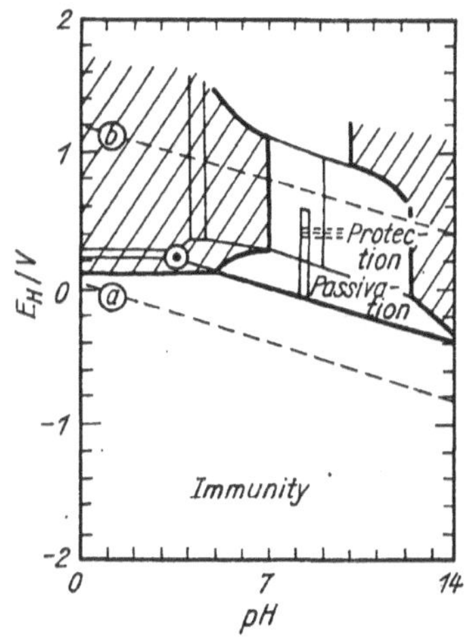

Fig. 13: Copper in Brussels water. pH and electrode potential conditions inside pits in equilibrium: pH = 3.5, $E = +270\ mV_{she}$. Possible electrode potentials outside pits at pH = 8: $E = 0$ to $+600\ mV_{she}$. Protection potential against localized corrosion: about +350 to $+420\ mV_{she}$.

pH = 3,5) and the possible extreme electrode potential values of copper in the presence of water of pH 8 (from 0 to about $+600\ mV_{she}$, according to the oxygen-content of the water and to the surface-condition of the metal). One sees that the protection potential against copper-pitting (about +350 to $+420\ mV_{she}$) lies between these two possible extreme values, and that copper-pits may thus in fact grow or be stifled.

This criterion was in harmony with all previously known facts, and has opened the way to new methods of protection and of control. The presence of carbon on the surface of copper which, thanks to H. S. Campbell [1 a] has been known since 1950 to promote pitting, increases the electrode potential of more or less perfectly passive copper up to about $+300\ mV_{sce}$, i.e. much above the critical value [2 b]. The addition of ferrous sulphate to sea-water, which has been shown in 1961 by T. W. Bostwick to prevent the pitting-corrosion of brass exchanger tubes [17], depresses the electrode potential below this critical value. Extensive exposure trials of different kinds of copper-tubes in different water supplies, with measurements of the electrode potentials, have been recently performed by F. J. Cornwell, J. Wildsmith and P. T. Gilbert [3]. As shown by Figs. 17 and 15, all the results of these tests are in full agreement with the above-mentioned criterion: whatever the metal and whatever the water, pitting only occurs (and always occurs) if the electrode potential is higher than about +100 to +170 mV_{sce} (i.e. +350 to +420 mV_{she}).

New precautions and new remedies have been suggested since 1962 as a result of this criterion, being understood that, if not only pitting corrosion but also any general corrosion should be avoided, the protection potential should be dropped from +270 to +100 mV_{she} (or from +20 to −150 mV_{sce}). When in presence of a corrosive water or of a corrosive atmosphere, copper should not be in contact with any substance whose zero-current electrode potential in the considered environment is higher than the critical one, such as carbon and gold for example; and this explains the recently observed deterioration of so many bronze sculptures which were covered with a porous gold deposit (the doors of the Baptistere in Firenze, the horses of Praxitele in Venice, etc.).

Silver and tin have lower electrode-potentials and are thus not harmful to copper, and this is why silvered and tinned copper ustensiles may be used for cooking. The copper hot-water tanks placed on the roof of many houses in London, whose corrosion has been for many years the source of great

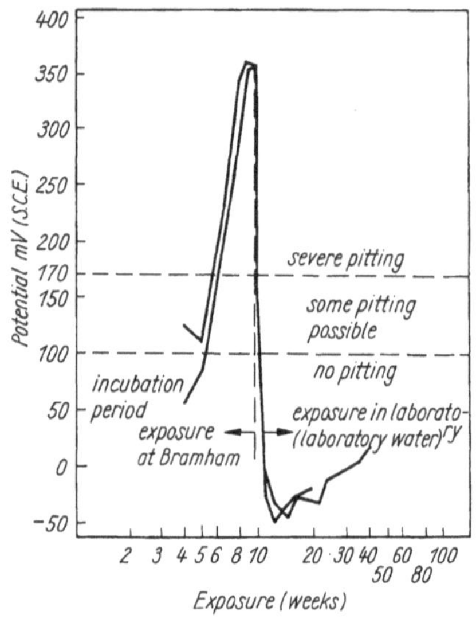

Fig. 14: Effect of transferring uncleaned copper tubes undergoing pitting at Bramham to the laboratory (Leeds water) [after CORNWELL et al.]

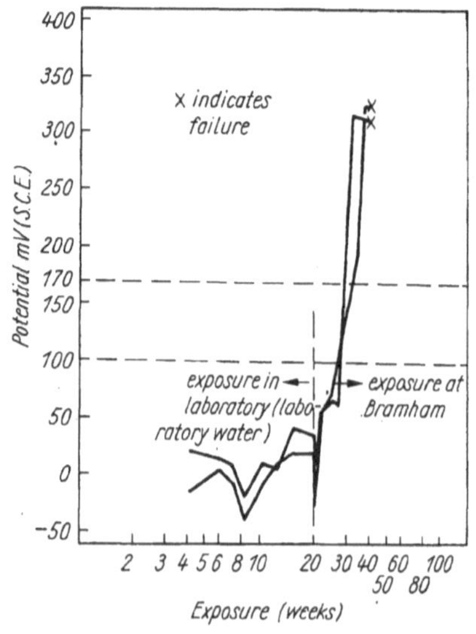

Fig. 15: Effect of transferring uncleaned copper tubes from the laboratory (Leeds water) to Bramham [after CORNWELL et al.]

concern, are since about 1966 protected cathodically after a recommendation of B.N.F.M.R.A., using temporarily so-called "protector rods" in aluminium-alloys, which depress the electrode potential below +100 mV_{she} [6]. A general method has been proposed for checking the pitting susceptibility of different sorts of copper and fo copper-alloys in different waters (Fig. 19) [2d, f, g], as well as for clarifying the mechanism of dealloying of copper-alloys, such as the denickelification of copper-nickels and the dezincification of brasses [4, 5]. However much work remains to be done before these problems might be considered as really mastered.

Now, coming back to the experiments which have been mentioned at the beginning of this talk, what is the mechanism of copper-pitting? What is the explanation of the "funny results" which were obtained concerning the influence of light and of stagnation on the electrode potential?

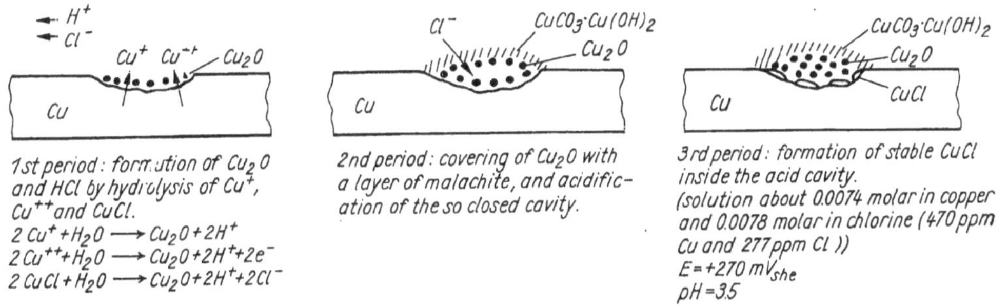

Fig. 16: Initiation and evolution of a copper corrosion-pit in the presence of water containing chloride.

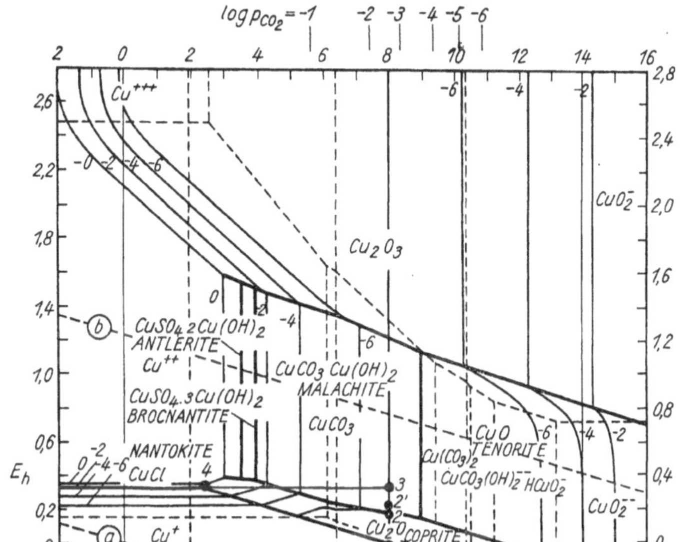

Fig. 17: Potential-pH equilibrium diagram for the 5 component system $Cu-Cl-CO_2-SO_3-H_2O$, at 25°C, for solutions containing 229 ppm CO_2 ($10^{-2.28}$ g.ion/liter), 46 ppm SO_3 ($10^{-3.24}$ g.ion/liter), and 22 ppm Cl ($10^{-3.21}$ g.ion/liter).

During the first stage of the experiment whose results are shown in Fig. 3, copper undergoes general corrosion in the bulk of the water (Fig. 16) with formation of Cu^+ and Cu^{++} ions and eventually solid CuCl which hydrolyse, for instance according to reaction $2\,CuCl + H_2O \rightarrow Cu_2O + 2\,H^+ + 2\,Cl^-$, with formation of solid Cu_2O which deposits on the metal and of hydrochloric acid HCl which is washed away by the water. According to Fig. 17, which is the equilibrium potential-pH diagram of the quinary system $Cu-Cl-CO_2-SO_3-H_2O$ for the contents of Cl, CO_2 and SO_3 existing in Brussels

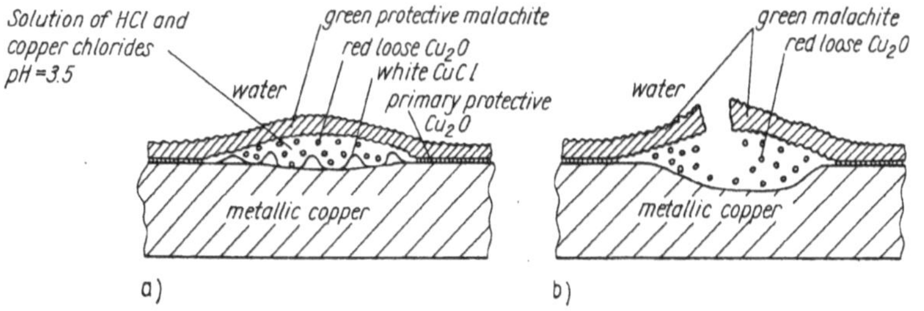

Fig. 18: Copper pit
(*a*) closed pit (in running water) $E = +20$ to $+77\ mV_{sce}$;
(*b*) open pit (in stagnant water) $E = -250\ mV_{sce}$.

water, malachite $CuO_3 \cdot Cu(OH)_2$ is the stable form of copper in the presence of aerated Brussels water of pH 8 (f.i. at point 3) and this is why, during the second stage of the experiment, Cu_2O becomes progressively covered with malachite which, being relatively protective [2e], forms a screen which progressively prevents the hydrochloric acid from being washed away by the bulk of the water. The pH inside the cavity thus decreases until, when the third stage is reached, the solution inside the cavity becomes saturated in CuCl which is being deposited on the metal; the solution inside the cavity has then a pH of about 3.5 and is then about 0.01 molar in CuCl; as previously said, the cavity increases in depth with dissolution of copper if the electrode potential is higher (more positive) than $+270\ mV_{she}$; on the contrary, the pit heals with redeposition of metallic copper if the electrode potential is lower (less positive) than $+270\ mV_{she}$.

As we have seen, both an illumination of the metal and a stagnation of the water lead, during the third stage of the experiment, to a temporary potential drop (which is sudden in the case of illumination and slow in the case of stagnation), and this to the same value: $-250\ mV_{sce}$ (i.e. $0\ mV_{she}$). As may be

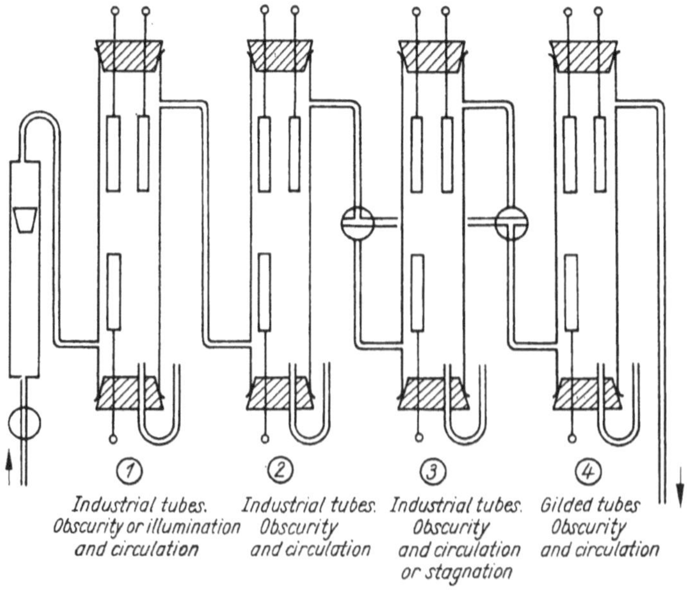

Fig. 19: Device for testing the pitting behaviour of copper in water

seen at Fig. 17 (point 1), this value 0 mV_{she} is, for the pH 8 of Brussels water, the equilibrium electrode potential of the system $Cu-Cu_2O$. It is thus likely that, during the third stage of the experiment, both an illumination of the copper and a stagnation of the water lead to a temporary contact of truly metallic copper with the bulk of the water. We think, without having had so far the opportunity to check this extensively, that:

— this sudden action of *illumination* of copper during the third stage of the experiment is due to an immediate photochemical action of light on CuCl with formation of metallic Cu (as the action of light on AgCl with formation of Ag in photography). Such an effect of light should thus indicate the presence of CuCl on the surface of the copper, and might thus detect an early stage of pit formation.

— this slow action of *stagnation* of the water is due to a progressive opening of some active cavities (Fig. 18), probably as a result of a chemical attack of the external malachite layer by the internal chlorhydric acid, with formation of "chimneys" through which gaseous CO_2 escapes from the cavity and bulk of the water penetrates till the bottom

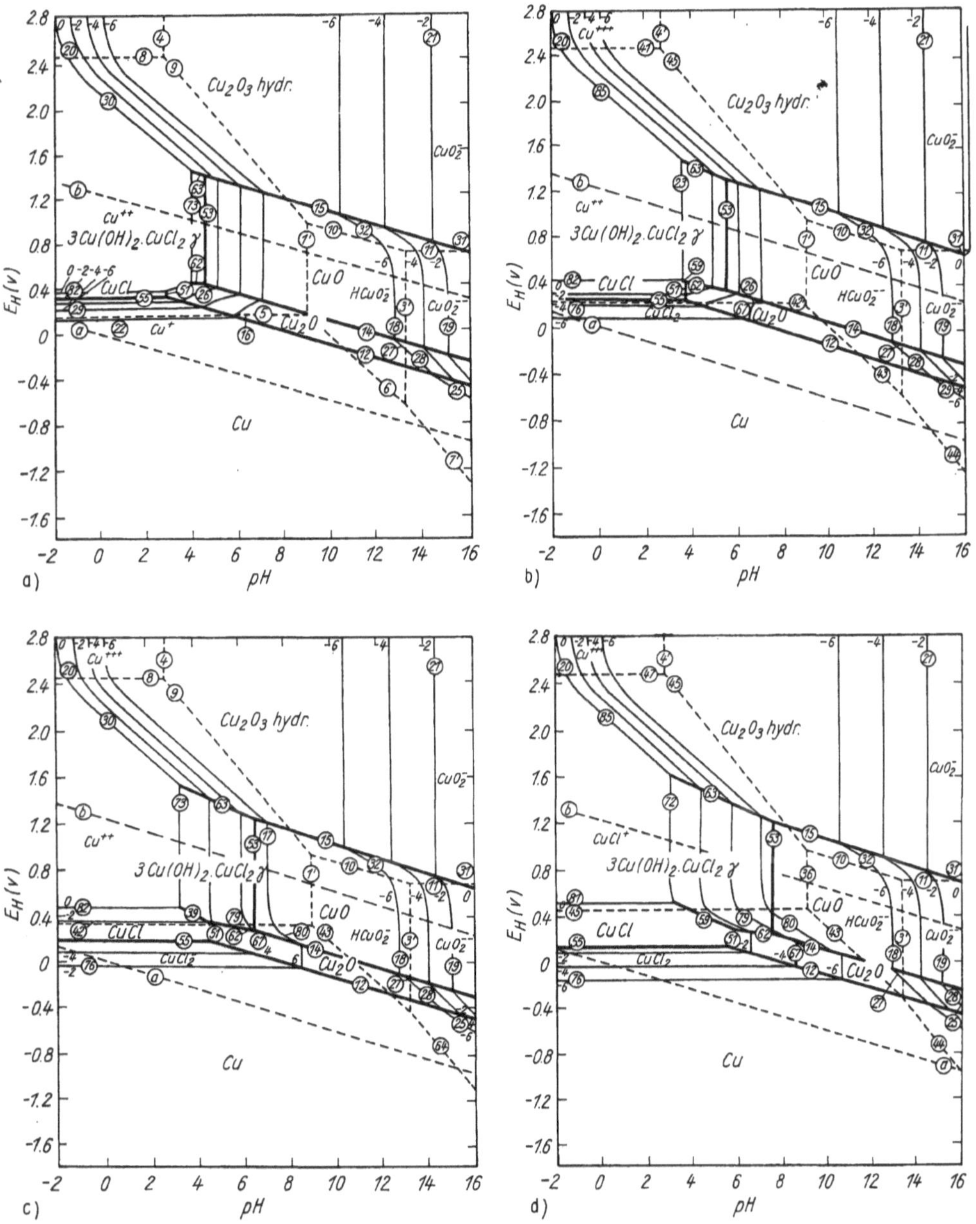

Fig. 20: Potential-pH equilibrium diagrams for the ternary system $Cu-Cl-H_2O$ at 25°C for (*a*) 10^{-3}, (*b*) 10^{-2}, (*c*) 10^{-1}, and (*d*) 1 g.ion Cl^-/liter (35, 355, 3550 and 35500 ppm).

of the cavity, where copper is really in a pickled metallic state. Such an effect of stagnation should thus detect and advanced stage of pit formation.

— both these actions are temporary because, due to the presence of oxygen in the bulk of the water, metallic copper becomes quickly oxidized.

As a result of all this, the testing device shown at Fig. 19 has been established. Water is passed through a series of four corrosimeters containing copper specimens, with recording of the electrode potentials of these specimens as a function of time. In the corrosimeters No. 1, 2 and 3, which are intended for checking the behaviour of different samples of copper and/or copper-alloys in a given water, water is passed in the darkness continuously (No. 2), in the darkness with periods of stagnation (No. 3), and continuously with periods of illumination (No. 1). In corrosimeter No. 4, which is intended for appreciating the danger of pitting in a given water, the water to be tested is passed in the darkness continuously on pickled copper partially covered with gold, supposed to represent a "deleterious surface condition".

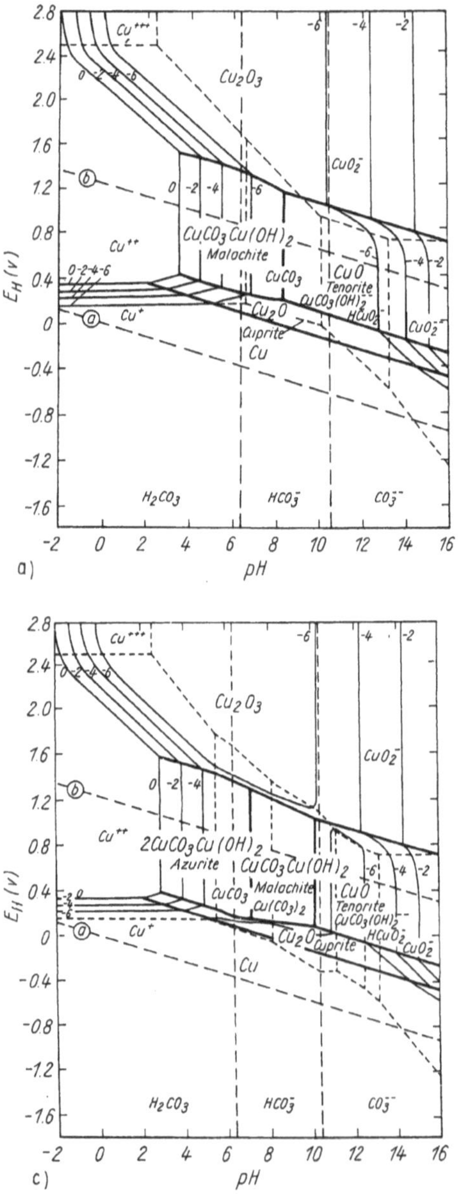

Fig. 21: Potential-pH equilibrium diagrams for the ternary system $Cu-CO_2-H_2O$ at 25°C for (*a*) 10^{-3}, (*b*) 10^{-2}, (*c*) 10^{-1}, and (*d*) 1 mole total dissolved CO_2 per liter (44, 440, 4400 and 44000 ppm).

Now, why is the pitting-corrosion of copper-tubes for cold water delivery not a problem in France, as it is in many other countries? Probably because France mostly uses hard-tubes which, contrarily to the soft ones, are not annealed in a reducing atmosphere and are free of the deleterious carbon film likely to be formed by the cracking of some oils.

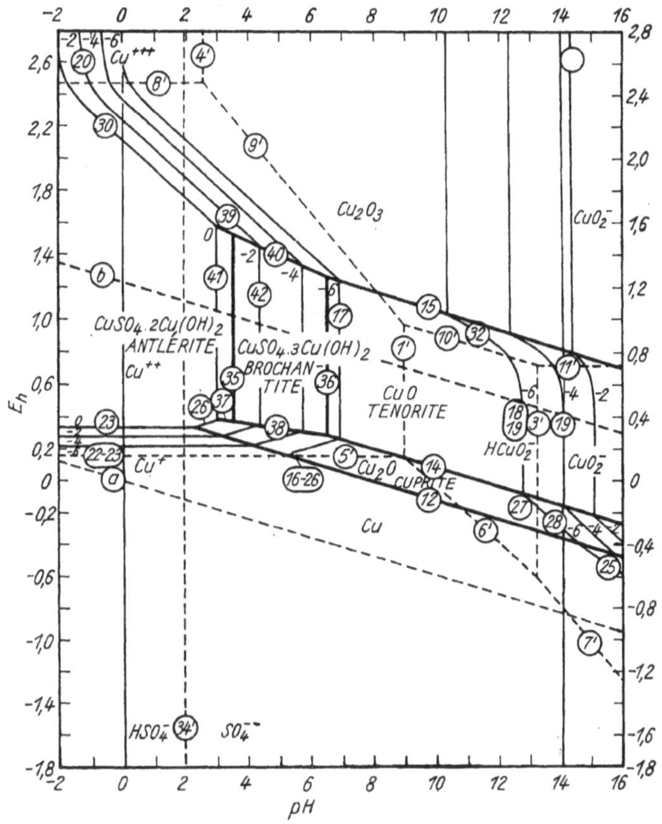

Fig. 22: Potential-pH equilibrium diagram for the ternary system $Cu-SO_3-H_2O$ at 25 °C for $10^{-3.24}$ mole dissolved SO_3 per liter (46 ppm).

Before closing this statement on copper, I wish to remind that, as has been emphasized by R. M. GARRELS [7], potential-pH equilibrium diagrams may be useful in geology. Fig. 20 indicates some stability conditions of the chlorides nantokite and γ paratacamite (which is the chilese copper mineral). Fig. 21 may be useful for checking the conditions of stability of the basic copper carbonates malachite and azurite. Fig. 22 may be useful for checking the conditions of stability of the basic copper sulphates brochantite and antlerite, as well as the conditions where patinas formed on copper roofing may be the black tenorite or the green brochantite.

2. Localized Corrosion of Iron and Steels

Fig. 23 shows a theoretical potential-pH diagram for the $Fe-H_2O$ system at 25 °C. Fig. 24 shows experimental potentiokinetic polarization curves for ion in the presence of chloride-free aqueous solutions of different pH's and Fig. 24 shows the corresponding experimental potential-pH conditions for immunity, general corrosion and passivity of iron [8]. Fig. 25*a* shows experimental potentiokinetic polarization curves for iron in the presence of solutions similar to those of Fig. 24*a*, but containing chloride; these curves show, besides passivation-potentials which are the same as in Fig. 24*a*, pitting-potentials and protection-potentials against localized corrosion, and make it possible to represent at Fig. 25*b* experimental potential-pH conditions for immunity, general corrosion, pitting corrosion,

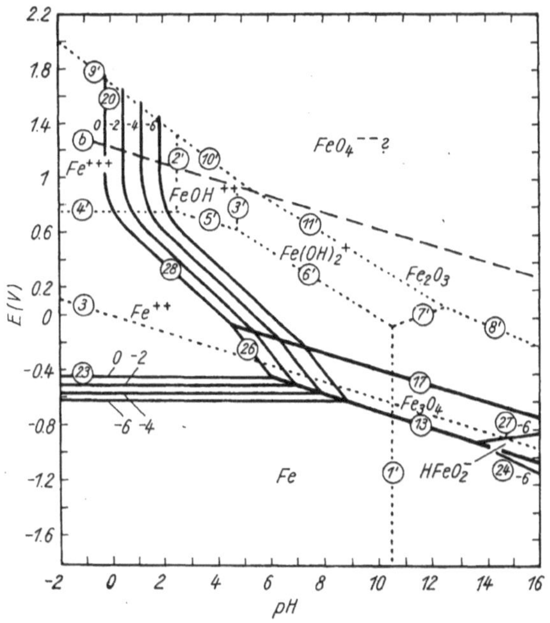

Fig. 23: Potential-pH equilibrium diagram for the iron-water system at 25 °C.

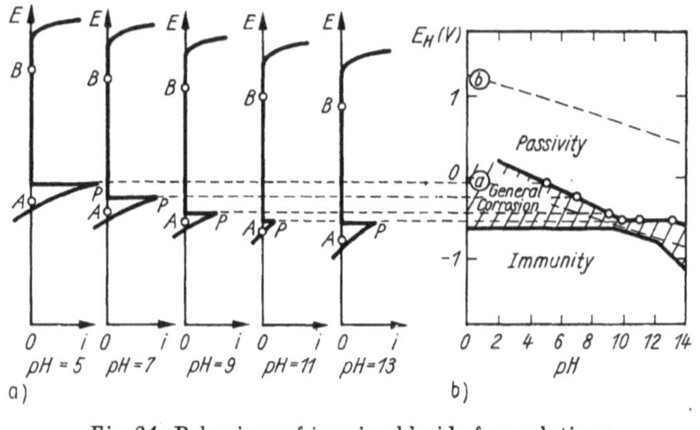

Fig. 24: Behaviour of iron in chloride-free solutions.

(*a*) polarization curves in the presence of solutions, for pH 5 to 13;

(*b*) experimental conditions of immunity, general corrosion, and passivity.

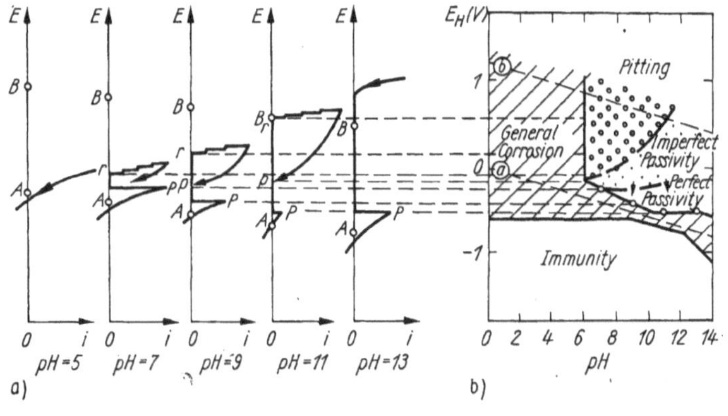

Fig. 25: Behaviour of iron in solutions containing chloride (355 ppm):

(*a*) polarization curves in the presence of solutions, for pH 5 to 13;

(*b*) experimental conditions of immunity, general corrosion, perfect and imperfect passivity, and pitting. Protection potential against localized corrosion.

imperfect passivity where no new pits are formed but where previously formed pits may grow, and perfect passivity where these pits do not grow.

Fig. 26 shows, together with the experimental potential-pH conditions of immunity, general corrosion, pitting corrosion, perfect and imperfect passivity shown at Fig. 25*b*, the potential and pH conditions inside an active oxygen-free corrosion pit or crevice, or other "occluded corrosion cell" O.C.C. as has been called by B. F. BROWN ($E = -325\ mV_{she}$ and pH = 3,8), and the possible extreme electrode potential values of iron in the presence of water of pH 8 (from about −550 to

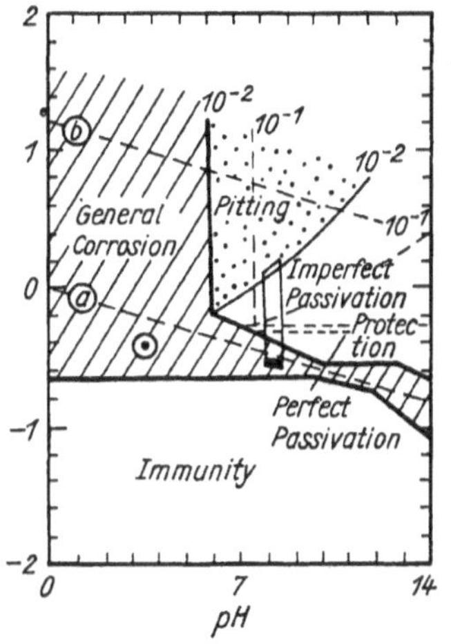

Fig. 26: Armco iron in solutions containing chloride (10^{-2} and 10^{-1} g.ion/liter). pH and electrode potential conditions inside oxygen-free living pits: pH = 3.8, $E = -325\ mV_{she}$. Possible electrode potentials outside pits at pH = 8: $E = -0.5$ to $0\ mV_{she}$ for 10^{-2} g.ion Cl^-/liter. Protection potential against localized corrosion: about $-300\ mV_{she}$.

$+100\ mV_{she}$ according to the oxygen content of the water). One sees in this figure that the protection-potential against localized corrosion of iron is, as it is for copper (Fig. 13), very close to the electrode potential inside an active O.C.C.

Fig. 27 shows a device which permits, by passing oxygen and nitrogen on respectively the external and the internal part of a duplex iron specimen, to produce on the internal part, as a result of differential aeration, a crevice under conditions which permit notably to measure the pH and the electrode potential on both the cathodic and the anodic area (which is an "occluded corrosion cell"). Fig. 28 shows, as a result of experiments performed with such a device by J. VAN MUYLDER et al. [9], the influence of a slow external cathodic polarization on the potential- and pH conditions inside an iron crevice. It is seen that this polarization progressively increases the pH inside the cavity and that,

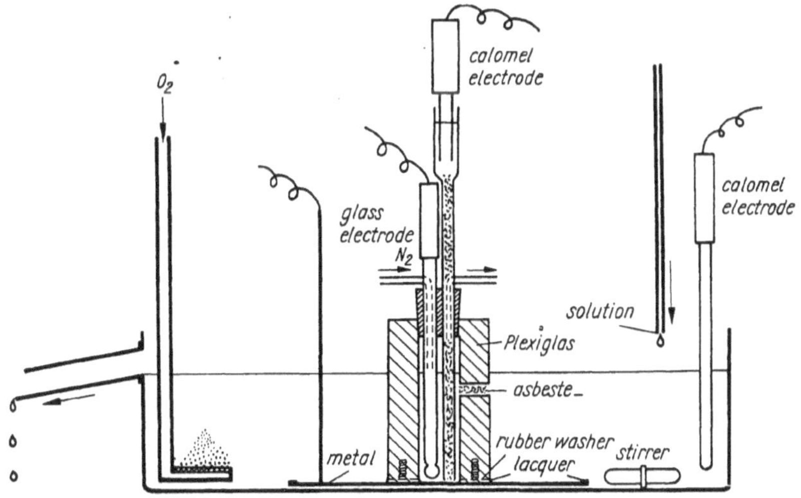

Fig. 27: Device for studying crevice-corrosion.

when the external potential reaches the "protection potential against localized corrosion", this solution is not acid anymore and the iron has become immune; however, there is still inside the cavity a hydrogen evolution whose affinity, which is measured by the vertical distance between the line relating to the crevice and the hydrogen equilibrium line *a* (about 100 mV), remains about the same until the solution becomes neutral, and very much increases when the solution has become definitely alkaline. When measuring, during a rapid external cathodic polarization, the current flow between a large external area and a small internal cavity of pH 5.5. Antoine POURBAIX has observed the two polarization curves shown at Fig. 29 [10]. It may be seen that, when the external cathodic polarization is increased, the cavity progressively ceases to act as an anode, reaches at -0.43 mV$_{she}$ the state of zero-current, and afterwards acts as a cathode. When the protection potential -0.38 mV$_{ehs}$ is reached on the external area, the potential inside the cavity has become -0.44 mV$_{she}$; the cavity has then

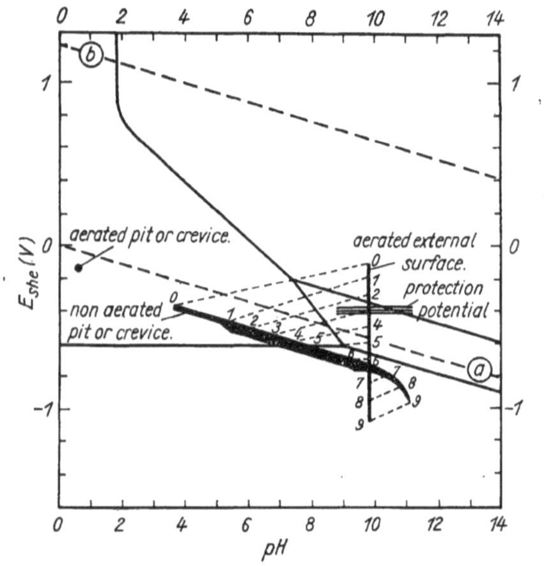

Fig. 28: Influence of a slow cathodic polarization on the potential- and pH conditions inside a crevice (mild steel with an aerated solution of 0.001 M NaOH and 0.001 M NaCl) [after J. VAN MUYLDER, L. SATHLER et al.].

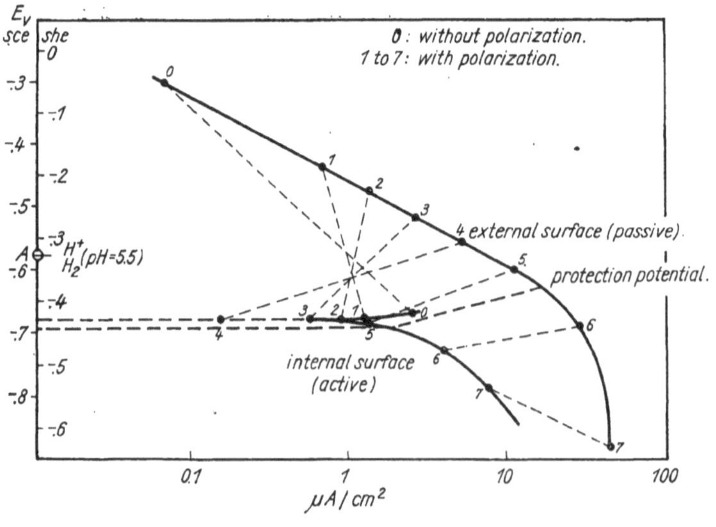

Fig. 29: Influence of a rapid external cathodic polarization on the electrochemical behaviour of mild steel outside and inside a crevice (external pH = 9.5; internal pH = 5.5) [after Antoine POURBAIX].

become immune and the cathodic current density, which is about 2 $\mu A \cdot cm^{-2}$, totally corresponds to a hydrogen evolution. Fig. 30 shows the results obtained by B. F. BROWN in his famous work on the cathodic polarization of steel-cracks, when using his freezing method [11]. One sees that BROWN's curve on cracks is very similar to VAN MUYLDER's curve on crevices, and this confirms the assumption that the chemistry and the electrochemistry of crevices and of cracks, as well as of any other form of O.C.C. (pits, corrosion in recesses, intergranular corrosion, etc.), is the same.

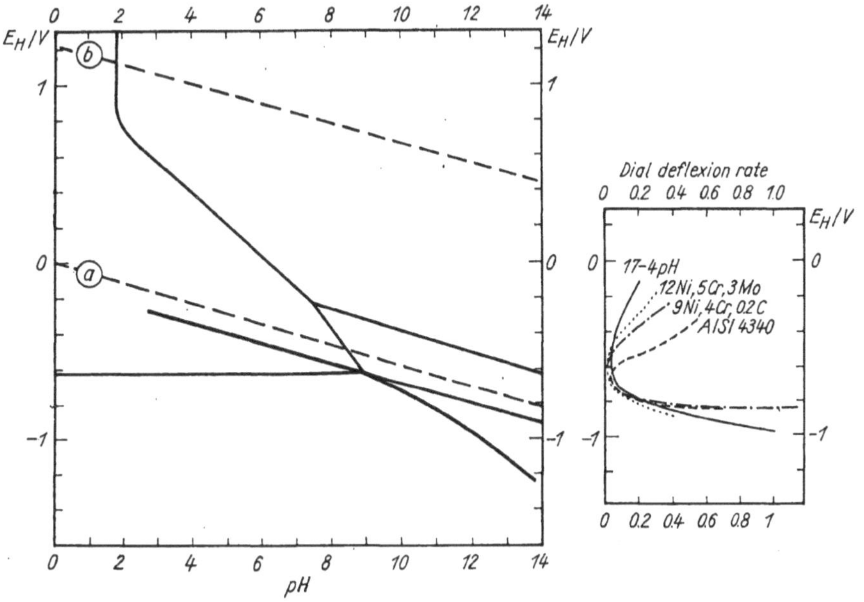

Fig. 30: Influence of a cathodic polarization on the electrochemical characteristics and on the propagation rate of a stress corrosion crack (chromium steel AISI 4340 in contact with NaCl 3.5% solution) [after B. F. BROWN].

As the representative point of the O.C.C. in both Figs. 28 and 30 is below the "hydrogen line" *a*, hydrogen is being evolved inside both crevices and cracks, even if the bulk of the solution is alkaline. This means that cathodic protection below the "protection potential against localized corrosion" (usually about -200 to -400 mV_{she}), which may often prevent any localized corrosion of iron and of steels, may be fully effective against deterioration of the material only if this material is not susceptible to hydrogen embrittlement. Needless to say, this is of particular importance for stress corrosion cracking.

3. Conclusions

Coming at the end of this rapid survey, I wish to raise the attention on the fact that, in both the cases of copper and of iron in the presence of solutions containing chloride which we have been considering in the present lecture, the solution existing inside the O.C.C. is in fact a solution of a chloride of the considered metal (CuCl or $FeCl_2 \cdot 4H_2O$) which is thermodynamically unstable in the bulk of the water and thus hydrolyses, with formation of a stable oxide (Cu_2O or Fe_3O_4), with decrease of the pH and increase of the chloride concentration until the solution becomes saturated in this chloride (about 0.01 molar and pH 3.5 for CuCl, and about 4.6 molar and pH 3.8 for $FeCl_2 \cdot 4H_2O$). The system then ceases to be univariant and becomes invariant.

Such solutions may be prepared synthetically by simply saturating oxygenfree water with this chloride, in the presence of the metal in powder-form. This makes it possible to prepare very easily a solution similar to the solution really existing inside a crevice or a crack, or along a corroding grain-boundary, or in any other O.C.C., and this in any desirable quantity (100 cm^3 or 1000 cm^3, or more) which might be suitable for any analysis and electrochemical experiment. This has been done for copper in 1962 by J. VAN MUYLDER [8], and for iron in 1973 by C. T. FUJII [12] and in 1974 by L. SATHLER [13]. Both C. T. FUJII and L. SATHLER have obtained for the O.C.C. of iron in chloride solutions, at 25 °C, a very concentrated solution respectively 4.2 and 4.6 molar in FeCl, and of pH 3.8. According to L. SATHLER, isolated iron corrodes in this solution with a velocity of 1.2 $\mu A \cdot cm^{-2}$ (13 $\mu m \cdot year^{-1}$), and passivates at $-185\ mV_{she}$ with a passivation current of $2.5\ mA \cdot cm^{-2}$ ($27\ mm \cdot year^{-1}$). If oxygen penetrates into this solution, the pH decreases from 3.8 to about 0.6, and the zero-current electrode potential of iron increases from $-325\ mV_{she}$ to $-125\ mV_{she}$ (Fig. 28), where it corrodes with a velocity of about $0.7\ mA \cdot cm^{-2}$ ($8\ mm \cdot year^{-1}$), and may not become passive.

Finally, I think that it is important to realize fully that, in any case where localized corrosion is to be considered, such corrosion does not develope in the bulk of the solution and at the electrode potentials existing on the external surfaces, but inside an O.C.C. where the composition of the solution and the electrode potential of the metal may be very different from what they are outside these cavities. For instance, many problems related to intergranular corrosion might be better understood by systematic research related to the electrochemical behaviour of substances existing along these grain-boundaries in these solutions.

Besides this, as stated recently [14], it is likely that, by adding to copper, to iron, or to other metals, alloying elements which, when dissolving in the solutions existing inside the O.C.C. of these metals, would render these solutions less aggressive and/or lead to the deposition of a less soluble salt or element under the there existing conditions of pH and of electrode potential, etc., it would be possible to explain better than presently the beneficial influence of some alloying elements to non-ferrous metlas (As and Fe in copper alloys? Sb and Bi in aluminium alloys?) and to some steels (Mo?), and to set up new and better alloys as far as their resistance to localized corrosion is concerned.

This last phrase has been written at a moment when no prove of the exactitude of this statement was available, as far as I know. Shortly after, T. KODAMA and J. R. AMBROSE [15], coworkers of J. KRUGER at the National Bureau of Standards NBS, have presented at the present meeting of the Electrochemical Society an outstanding paper (No. 75) entitled "Effect of molybdate ion on the repassivation kinetics of iron in sodium chloride solutions". According to this paper, and in full agreement with the present theory, molybdenum would dissolve in an iron O.C.C. as molybdate ions and would form with the dissolved ions insoluble ferrous molybdate $FeMoO_4$. It is most likely that this is the fundamental reason of the beneficial effect of molybdenum on the pitting- and crevice behaviour of alloy steels (such as the AISI 316); the just mentioned concept for setting new and better alloys might thus perhaps be correct.

As a conclusion, I wish to emphasize that, any time one has to study localized corrosion, and notably crevice corrosion, intergranular corrosion, and stress corrosion cracking, which have a so great practical importance, there is a great interest in the measurement of electrode potential- and pH values outside and inside the cavities, and to interprete the so-obtained values on the basis of both electrochemical thermodynamics and electrochemical kinetics. Such researches would preferably include a thorough study of the chemistry and of the electrochemistry of the solutions existing inside the occluded corrosion cells and of the anodic and cathodic behaviour of the constituents of the considered metal or alloy in these solutions [16].

References

[1] H. S. CAMPBELL,

(a) "Pitting corrosion of copper water pipes caused by films of carbonaceous material during manufacture", J. Inst. Metals, **77**, 345—356 (1950);

(b) "The influence of the composition of supply waters, and especially of traces of natural inhibitor, on pitting corrosion of copper water pipes", Proc. Soc. Water Treatment and Examination, **3**, 100—115 (1954);

(c) "Pitting corrosion of copper water pipes", Proc. 2nd International Congress Metallic Corrosion, New York (1963); National Association of Corrosion Engineers, Houston, Texas (1966) p. 243;

(d) "Pitting corrosion of copper and its alloys. A review", Proc. U. R. EVANS Conference on Localized Corrosion, Williamsburg (1971). Publ. NACE, Houston (1974), pp. 625—638.

[2] M. POURBAIX, J. VAN MUYLDER, P. VAN LAER, N. DE ZOUBOV and A. POURBAIX, CEBELCOR's Technical Reports (*RT*):

(a) Concerning electrode potentials of copper in Brussels tap water. Influence of light and water-circulation (*RT*. 125—1965);

(b) Relation between electrode potential and corrosion conditions of electrolytic copper in Brussels water. Influence of surface state, water circulation, water treatment and contact between copper, graphite and platinum (*RT*. 126—1965);
(c) Electrochemical characteristics of copper pitting corrosion in water and chloride solutions (*RT*. 127—1965; *E*. 61—1967);
(d) Accelerated test for checking the pitting-susceptibility of copper in the presence of water. Influence of copper contact with other metals or graphite (*RT*. 128—1965; *E*. 62—1967);
(e) Concerning the behaviour of copper in sodium bicarbonate solutions (*RT*. 133—1965);
(f) Test for the susceptibility of copper to pitting corrosion in presence of cold water (*RT*. 161—1969);
(g) Test for the susceptibility of copper to pitting corrosion in presence of cold water (*RT*. 201—1971).

[3] F. J. Cornwell, G. Wildsmith and P. T. Gilbert,
"Pitting corrosion in copper tubes in cold water service", British Corrosion Journal, 8, 202—209 (1973).

[4] Ph. A. Parrish,
"Use of experimentally determined Pourbaix diagrams to elucidate the role of iron in the passive behavior of copper-rich alloys containing nickel", Thesis University of Florida (1970).

[5] E. D. Verink and M. Pourbaix,
"Use of potentiokinetic methods at successively increasing and decreasing electrode potentials in developing alloys for saline exposure", Rapports Techniques CEBELCOR **117**, *RT*. 191 (1971).

[6] H. S. Campbell,
"Pitting corrosion of copper and its alloys. A review" (see ref. 1d).

[7] R. M. Garrels,
(a) "Mineral equilibria at low temperature and pressure", Publ. Harper, New York (1960);
(b) (in collaboration with C. L. Christ), "Solutions, Minerals and equilibria", Publ. Harper, New York (1965).

[8] M. Pourbaix,
(a) "Corrosion passivity and passivation of iron; the role of pH and of potential" (part Thesis Brussels 1945). Publication CEBELCOR F. 21 (1951);
(b) "Research in corrosion. Results of recent work." Visits to the U.S.A., Rapports Techniques CEBELCOR, **109**, *RT*. 157 (1969).
(c) Lectures on Electrochemical Corrosion, Publ. Plenum Press (1973).

[9] M. Pourbaix,
"Fundamentals of cathodic protection, and applications", Rapports Techniques CEBELCOR, **111**, *RT*. 166 (1969).

[10] A. Pourbaix,
"Study of localized corrosion of steel in chloride-containing solutions", Rapports Techniques CEBELCOR, **118**, *RT*. 198 (1971).

[11] B. F. Brown,
(a) (in collaboration with C. T. Fujii and E. P. Dahlberg), "Methods for studying the solution-chemistry within stress corrosion cracks", J. Electrochem. Soc. **116**, 218—219 (1969);
(b) "The role of the occluded corrosion cell in stress corrosion cracking of high strength steels", Rapports Techniques CEBELCOR, **112**, *RT*. 170 (1970).

[12] C. T. Fujii,
"Electrochemical and chemical aspects of localized corrosion", Rapports Techniques CEBELCOR, **123**, *RT*. 213 (1974).

[13] L. Sathler and J. van Muylder,
"On the chemical and electrochemical nature of the solutions inside occluded corrosion cells", Rapports Techniques CEBELCOR, **126**, *RT*. 224 (1975).

[14] M. Pourbaix,
(a) Joint USA-Japan Seminar on "Passivity and its Breakdown on Iron and Iron-base alloys", Honolulu, 10—12 March 1975. Discussion (to be published by the Japan Corrosion Society and by the NACE);
(b) "Inhibition of localized corrosion processes", Preprint 4th European Symposium on Corrosion Inhibitors, pp. 674—691, Ferrara 15—19 September 1975.

[15] T. Kodama and J. R. Ambrose,
"Effect of molybdate ion on the repassivation kinetics of iron in sodium chloride solutions", 148th Meeting of the Electrochemical Society, Dallas, October 5—9, 1975. Paper No. 75.

[16] M. Pourbaix,
Electrochemical aspects of stress corrosion cracking,
(a) Rapports Techniques CEBELCOR, **118**, *RT*. 199 (1971) (in French);
b) Proc. NATO Science Committee Research Evaluation Conference "The theory of stress corrosion cracking in alloys", Ed. J. C. Scully, pp. 17—63 (1971) (in English).

[17] T. W. Bostwick,
"Reducing corrosion of power plant condenser tubing with ferrous sulphate", Corrosion, **17**, 12 (1961); See as Technical Bulletin No. 7 of Yorkshire Imperial Metals Ltd (revised 1970).

I. Epelboin, C. Gabrielli und M. Keddam

C.N.R.S. Physique des liquide et électrochimie, Paris

MESSUNGEN DER Faraday-IMPEDANZ ZUR VERIFIZIERUNG EINES MODELLS DER AUFLÖSUNG UND PASSIVIERUNG DES EISENS IN MOLARER SCHWEFELSÄURE

Das anodische Verhalten des Eisens in saurem Milieu ist der Gegenstand einer großen Zahl von Arbeiten [1]. Die Aufmerksamkeit der Forscher ist dabei hauptsächlich auf den Mechanismus der anodischen Auflösung und der Passivierung gerichtet, auf die zwei Phänomene, die für die Korrosion des Metalls entscheidend sind.

Auf dem Gebiet der anodischen Auflösung sind die neuesten Arbeiten in sehr guter Übereinstimmung, sowohl in Hinsicht auf experimentelle Befunde als auch in Hinsicht auf den zu ihrer Erklärung vorgeschlagenen Mechanismus [2—7]. Potentiostatische Techniken und Messung der Impedanz bei sehr tiefen Frequenzen haben eine quantitative Prüfung [7, 8] von Reaktionsmodellen ermöglicht, bei denen die intermediäre Bildung der adsorbierten Verbindung FeOH beim Übergang des Metalls in die Lösung angenommen wird.

Dagegen konnte die Untersuchung der Passivierungsprozesse, mindestens in stark sauren Lösungen, infolge von Oszillations- und Hystereseerscheinungen nicht mit diesen Techniken ausgeführt werden. Diese Erscheinungen werden übrigens allgemein als charakteristisch für das Verhalten dieser Elektroden im Lauf der Passivierung angesehen [9, 10]. Die Untersuchung des passiven Zustands, der oft mit dem Auftreten einer Oxidschicht verbunden ist, blieb daher das einzige Mittel, um die Gesamtheit der Prozesse zu durchleuchten, die beim Übergang von der Aktivität zur Passivität des Eisens auftreten.

Mangelhafte Kontrolle des Stofftransports und Fehlanpassung der potentiostatischen Regelung beim elektrochemischen System Eisen—Schwefelsäure haben zu irrtümlichen Interpretationen der Passivierungserscheinungen geführt.

Es ist der Zweck dieses Artikels, zu zeigen, daß jüngste Fortschritte auf dem Gebiet der rotierenden Elektroden und der Regelung der Polarisation der Grenzfläche es ermöglicht haben, den Übergang des Eisens vom aktiven in den passiven Zustand in saurem Milieu eingehend zu untersuchen.

Die Einrichtung einer besonderen Meßanordnung erlaubt jetzt, den stationären Bereich und die Impedanz der Grenzfläche sogar im Verlauf des Übergangs aktiv—passiv [11, 12] zu untersuchen. Das anodische Verhalten des Systems Fe/H_2SO_4 1 M wird mit Hilfe eines Modells interpretiert, das die Kopplung zwischen der Diffusion und der reversiblen Bildung mehrerer adsorbierter Zwischenzustände berücksichtigt.

1. Meßanordnung und experimentelle Ergebnisse

Elektrochemische Systeme können in zwei Etappen untersucht werden. Zuerst bringt die Untersuchung des stationären Bereichs die im allgemeinen nichtlineare Änderung des Gleichstroms als Funktion des Potentials (Auftragung der I-U-Charakteristik). Obwohl diese Methode sehr nützlich ist, läßt sie nur die Identifizierung des langsamsten der durch den Stromfluß ausgelösten Prozesse zu. Sie reicht daher nicht aus, eine detaillierte Beschreibung der räumlichen und zeitlichen Verzahnung der Elementarprozesse an der Grenzfläche aufzustellen.

Bekanntlich kann in jedem Punkt der Kurve, die das im ganzen nicht-lineare Verhalten eines Systems repräsentiert, das lineare Verhalten dieses Systems vollständig durch seine Impedanz beschrieben werden [11, 13]. Zur Deutung der Elektrodenimpedanz als Funktion der Frequenz sind je nach Frequenz sehr verschiedene Aspekte der Eigenschaften der Grenzfläche zu berücksichtigen. Die experimentelle Bestimmung der Impedanz in jedem Punkt der Polarisation (I-U-Charakteristik) führt auf Probleme, die sowohl die Aufrechterhaltung der elektrochemischen Bedingungen während der Messung als auch die eigentliche Messung der Impedanz über einen breiten Frequenzbereich betreffen.

Die Regeleinrichtung für die Grenzflächenpolarisation ist daher nicht nur für Untersuchungen unter stationären Bedingungen, sondern auch für die Analyse der Elektrodenimpedanz wesentlich.

Außerdem ist eine geeignete Kontrolle des Stofftransportes wichtig, sobald hohe Stromdichten erreicht werden, wie es beim anodischen Verhalten des Eisens in sauren, etwa molaren Lösungen der Fall ist. Die Verwendung ruhender Elektroden führt infolge natürlicher Konvektion zu wenig reproduzierbaren Ergebnissen. Deshalb verwenden wir für unsere Untersuchungen durchweg die rotierende Scheibenelektrode, um die hydrodynamischen Erscheinungen zu beherrschen und um die heterogenen Phänomene leicht von den Transportphänomenen trennen zu können.

1.1. Problemstellung

Die ersten Regeleinrichtungen, die galvanostatisch mit eingeprägtem Strom arbeiteten, haben die Untersuchung elektrochemischer Systeme außerhalb des Gleichgewichts ermöglicht. Man ist aber sehr bald auf Metall-Elektrolyt-Grenzflächen gestoßen, die eine Strom-Spannungs-Charakteristik mit negativem Anstieg besitzen. So beobachtet man im Fall eines passivierbaren Metalls einen Sprung der Spannung V (Abb. 1*a*), wenn man den Strom I auf einen bestimmten Wert erhöht, der oft kritischer Strom I_c genannt wird [14]. Man findet einen anderen Sprung, in umgekehrter Richtung, wenn man den Strom vermindert.

Ein beträchtlicher Fortschritt wurde durch die Einführung der ersten Potentiostaten erzielt, die dem Experimentator bis dahin unzugängliche Bereiche öffneten. Man verfügte so über ein Instrument, das zur Lösung experimenteller Probleme in der Elektrochemie gut geeignet war. Jedoch im Fall einer Eisenelektrode, die in eine starke Säure taucht, bleiben bisweilen Oszillationen und sprunghafte Änderungen sogar bei der Verwendung eines Potentiostaten bestehen [9, 10, 15].

Legt man im Fall des Systems Eisen—molare Schwefelsäure mit Hilfe eines Potentiostaten ein zunehmend anodisches Potential an und beginnt beim Korrosionspotential, so löst sich das Eisen auf (*a—b*) bis ein Potential erreicht wird, wo die Geschwindigkeit der Auflösungsreaktion durch Diffusion begrenzt wird, was sich in einem Plateau des Stromes äußert (*b—d*). Charakterisiert wird indessen diese Strom-Spannungskurve durch die abrupte Änderung des Stroms beim Aktivierungspotential U_A und beim Passivierungspotential U_P. Wenn das angelegte Potential vergrößert wird, fällt der Strom beim Passivierungspotential U_P tatsächlich abrupt ab (*d—e*), wobei er fast Null wird, und das Metall löst sich praktisch nicht mehr auf; man sagt dann, die Elektrode ist passiv. Wenn das angelegte Potential vermindert wird, steigt der Strom beim Aktivierungspotential U_A abrupt an, bis zum Punkt *b*. Man sieht, daß diese abrupten aperiodischen Übergänge eine Hystereschleife (*b—d—e—f*) zwischen den Potentialen U_A und U_P bilden.

Derartige experimentelle Beobachtungen haben zu der Annahme geführt, daß sich das Eisen in molarer Schwefelsäure bei bestimmten Werten des Stromes oder der Spannung abrupt passiviert oder reaktiviert. Daher hat man denken können, daß das Eisen vom aktiven in den passiven Zustand und vice versa durch eine schnelle Blockade oder Deblockade der Oberfläche übergeht, ohne daß irgend ein stabiler Zwischenzustand existieren würde [9]. Lange Zeit ist deshalb versucht worden, den Prozeß der Passivierung durch die Begriffe des kritischen Stroms und des Passivierungspotentials zu charakterisieren.

Es ist jedoch vorstellbar [12, 16], daß die beobachteten sprunghaften Änderungen in Wirklichkeit nicht für die Passivierung charakteristisch sind, sondern durch die Regelung verursacht werden. Diese wäre dann ungeeignet, eventuelle Zwischenzustände zwischen aktivem und passivem Zustand nachzuweisen.

Eine solche Hypothese setzt also voraus, daß eine Klasse von elektrochemischen Systemen mit mehrfachen stationären Zuständen existieren kann, von denen einige selbst bei Anwendung einer

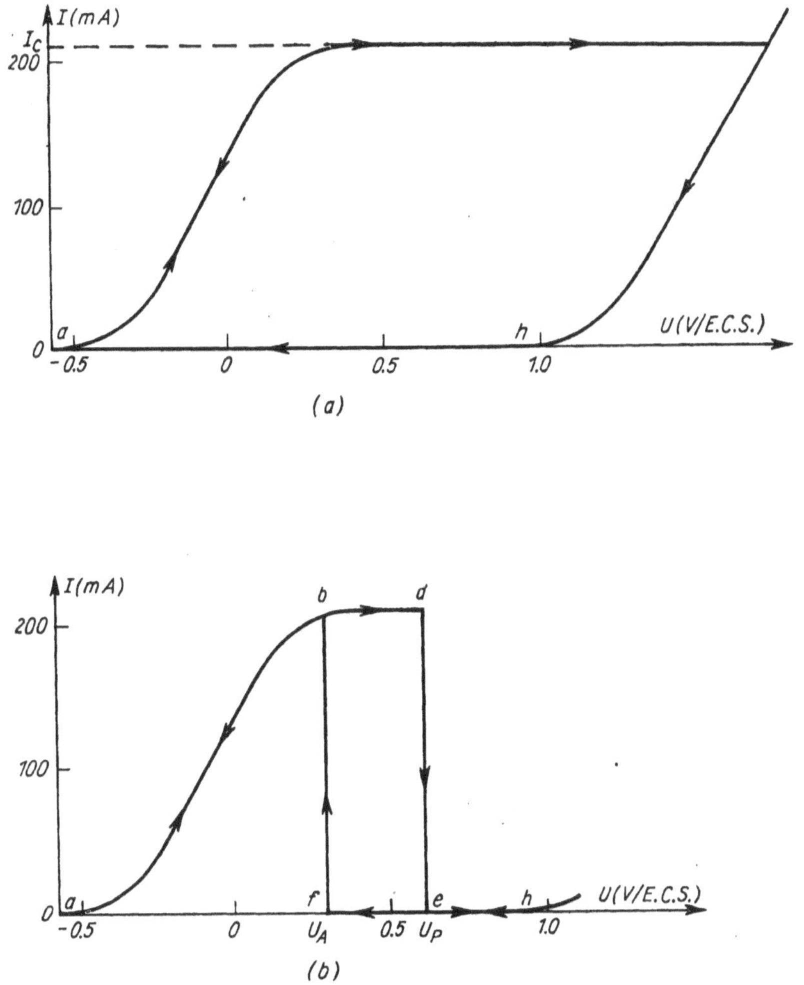

Abb. 1: Strom-Spannungskurven $I-U$ der Grenzfläche Eisen—molare Schwefelsäure, mit wachsenden, dann sinkenden Potentialen aufgenommen mit folgenden Reglern:

(a) galvanostatisch;
(b) potentiostatisch

I_c — kritischer Strom; U_p — Passivierungspotential; U_A — Aktivierungs- oder FLADE-Potential.

potentiostatischen Regelung nicht erreichbar sind. Eine potentiostatische Regelung würde daher für eine vollständige Untersuchung dieser Systeme und insbesondere der Grenzfläche Eisen—Schwefelsäure ungeeignet sein. Die Regelung der Polarisation der Grenzfläche muß zur Lösung dieses Problems von einer anderen Seite in Angriff genommen werden. Ganz allgemein scheint ein derartiges Verhalten zwangsläufig mit den Phänomen verknüpft zu sein, die auftreten, wenn ein offenes System fern vom Gleichgewichtszustand gehalten wird. Diese Phänomene bringen also direkt die Wechselwirkungen des Systems mit seiner Umgebung ins Spiel, die durch einen bestimmten, der untersuchten Grenzfläche auferlegten Zwang charakterisiert werden.

1.2. Probleme der Regelung der Grenzflächenpolarisation

Ein System entfernt sich von seinem Ruhezustand unter der Wirkung eines ihm auferlegten Zwanges. Dieser wird in der Elektrochemie gewöhnlich mit Hilfe einer Spannungsquelle E mit dem Innenwiderstand R ausgeübt, was etwa durch das Schema in Abb. 2 a dargestellt wird. Der der Grenzfläche eingeprägte Punkt der Polarisation wird bestimmt, indem man in der I-U-Ebene die I-U-Charakteristik der betrachteten Grenzfläche $f(I, U) = 0$ und die I-U-Charakteristik der Regelung $U = E - RI$, die auch Arbeitsgerade genannt wird, aufträgt. Die gesuchte Lösung wird durch den Schnittpunkt der beiden Kurven gegeben [11, 17].

Wir interessieren uns hier für den Fall, wo die Zahl der Schnittpunkte größer als 1 sein kann (Abb. 2 *b*). Zur Aufnahme der Strom-Spannungs-Charakteristik des betrachteten Systems vergrößert man im allgemeinen E und läßt den Widerstand R konstant. Angefangen vom Ausgangszustand 0 durchläuft man eine Folge von stabilen stationären Zuständen, deren Aufzeichnung die I-U-Kurve ergibt. Zuerst durchläuft man die Strecke $0A$, dann, wenn die Widerstandsgerade Tangente an die betrachtete Charakteristik wird (Punkt A), springt der Punkt der angelegten Polarisation nach A', der eine andere Lösung des Problems darstellt. Darauf durchläuft man $A'Q$. Wenn man E vermindert, beschreibt man die Strecke QB, der Punkt der Polarisation springt nach B', worauf $B'0$ durchlaufen wird. So beobachtet man 2 Sprünge $A - A'$ und $B - B'$, die eine Hystereseschleife $A - A' - B - B'$

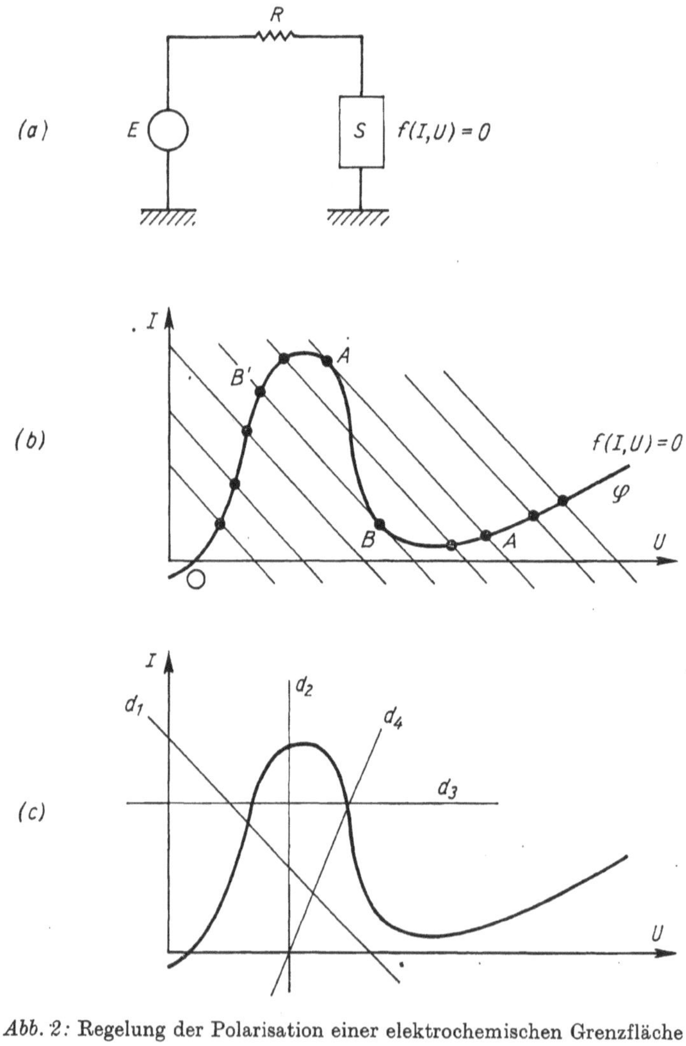

Abb. 2: Regelung der Polarisation einer elektrochemischen Grenzfläche

(*a*) Schema einer Regeleinrichtung mit Spannungsquelle E und Innenwiderstand R, belastet mit dem nichtlinearen System S der Strom-Spannungs-Charakteristik $f(I, U) = 0$;

(*b*) Bestimmung des Punktes der Polarisation des Systems S und der Regelung mit der Widerstandsgeraden $E = U + RI$ (in der Abbildung für verschiedene Werte von E dargestellt), $A - B$ mit dem gewählten Widerstand R unerreichbares Gebiet;

(*c*) Beispiele für Widerstandsgeraden, die die potentiometrische (d_1), potentiostatische (d_2), galvanostatische (d_3) und Regelung mit negativem Innenwiderstand (d_4) definieren.

bilden. Die durch den Bereich AB charakterisierten Zustände des Systems sind daher dem Experimentator nicht erreichbar, wenn er die betrachtete Regelung (E, R) anwendet. Diese Erscheinungen des Springens des Polarisationspunktes, die zu Hystereschleifen führen, könnte man durch Verwendung des Begriffs der Stabilität stationärer Zustände besser begründen [11]. Es ist offensichtlich, daß dieses spezielle Verhalten in ganzer Strenge nicht allein von der Form der Strom-Spannungs-Charakteristik, sondern auch vom dynamischen Verhalten her untersucht werden muß. Für das Folgende wird jedoch die obige Beschreibung ausreichend sein.

Die Widerstandsgerade kann 4 Lagen haben, die 4 Arten der Regelung entsprechend den Werten von R definieren (Abb. 2 *c*).

$\alpha - d_1$, wenn $R > 0$: potentiometrische Regelung,
$\beta - d_2$, wenn $R = 0$: potentiostatische Regelung,
$\gamma - d_3$, wenn $R \rightarrow \infty$: galvanostatische Regelung,
$\delta - d_4$, wenn $R < 0$: Regelung mit negativem Innenwiderstand.

Wenn man a priori verschiedene, mögliche Formen der I-U-Charakteristik betrachtet, so kann man jetzt das zu erwartende experimentelle Ergebnis je nach der verwendeten Art der Regelung voraussagen (Abb. 3).

Die monotone Charakteristik der Abb. 3 *a* kann mit Hilfe der Regelung α, β oder γ erhalten werden. Die S-förmige Charakteristik der Abb. 3 *b*, die man in gewissen Fällen der Elektrokristallisation

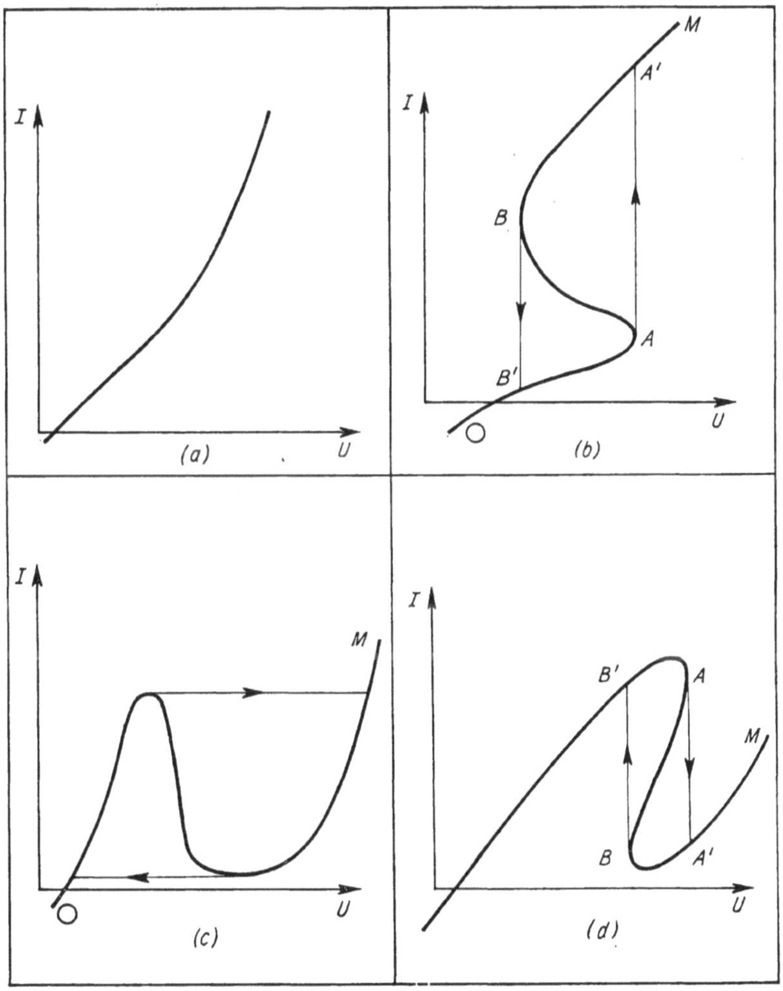

Abb. 3: Beispiele von nichtlinearen Strom-Spannungs-Charakteristiken des Typs:
(*a*) monoton: $I = \phi_1(U)$ oder $U = \phi_2(I)$; (*c*) festlegbar in der Spannung: $I = h(U)$;
(*b*) festlegbar im Strom: $U = g(I)$; (*d*) mehrdeutig: $f(I, U) = 0$.

antrifft (z. B. dendritisches Wachstum von Zn [18]) ist durch den Strom eindeutig bestimmt, d. h. man kann alle Zustände, die durch diese *I-U*-Kurve charakterisiert werden, mit Hilfe eines Galvanostaten erreichen. Die klassische Charakteristik eines passivierbaren Metalls (Abb. 3 *c*) ist durch das Potential eindeutig bestimmt, die Verwendung eines Galvanostaten führt zu der Hystereseschleife $A-A'-B-B'$. Schließlich ist der ganze Z-förmige Bereich der Abb. 3 *d* weder durch den Strom noch durch das Potential eindeutig festlegbar. Ein Potentiostat, der die übliche Regeleinrichtung darstellt und die Untersuchung des größten Gebiets zuläßt, führt nichtsdestoweniger zu einer Hystereseschleife $A-A'-B-B'$.

Der Vergleich der mit einem Galvanostaten (Abb. 3 *c*) und mit einem Potentiostaten (Abb. 3 *d*) erhaltenen Ergebnisse und der in Abb. 1 gegebenen experimentellen Resultate läßt Zweifel an der oft in der Literatur [9] zitierten Interpretation der Passivierung als ein Phänomen der sprunghaften Änderung aufkommen. So ist zu sehen, daß infolge schlechter Wahl der angewendeten Regelung abrupte Sprünge des Stroms oder des Potentials sogar dann beobachtbar sind, wenn die untersuchte Charakteristik stetig ist und wenn der Wechsel des Zustands aktiv—passiv über eine Folge von stationären Zwischenzuständen abläuft.

Nun ist aber offensichtlich, daß man leicht den Abschnitt *AB* der Abb. 3 *d* erreichen kann, wenn eine Widerstandsrichtung realisierbar ist, die eine Arbeitsgerade wie d_4 (Abb. 2 *c*) aufweist. So müßte man, wenn ein stetiger Übergang zwischen dem aktiven und dem passiven Zustand des Eisens in Schwefelsäure so wie (*d—f*) in Abb. 1 *b* existiert, dieses Gebiet mit Hilfe einer solchen Anordnung erreichen können.

1.3. *Anwendung auf die Aufnahme der Strom-Spannungs-Charakteristik des Eisens in Schwefelsäure*

Um diese Hypothese zu verifizieren, haben wir einen Regler mit negativem komplexen Innenwiderstand gebaut [11, 12]. Das Schema ist in Abb. 4 *a* dargestellt. Es ist die in der Elektronik klassische Schaltung eines Wandlers mit negativer Impedanz auf der Basis von Operationsverstärkern *A*. Zur Aufnahme der Strom-Spannungskurve wird das Potential *U* zwischen Bezugs- und geerdeter Arbeitselektrode und der Strom *I* durch die Meßzelle an den Klemmen des Widerstands R_2 gemessen. Die auf

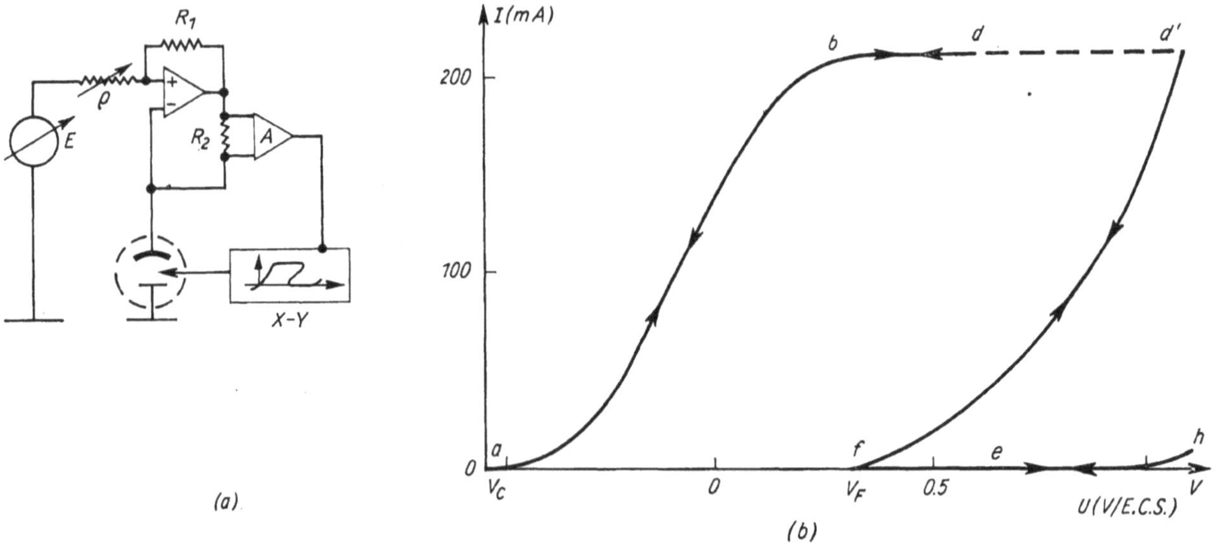

Abb. 4: Strom-Spannungs-Charakteristik der Grenzfläche Eisen—molare Schwefelsäure, aufgenommen mit einer Scheibenelektrode, die von dem ebenen Teil eines Eisenzylinders (reines Eisen, JOHNSON-MATTHEY) von 5 mm Durchmesser gebildet wird, Rotationsgeschwindigkeit 750 U/min^{-1}:

(*a*) Regelung mit negativem Innenwiderstand $\left(R_{\text{int}} = -\varrho \frac{R_2}{R_1}\right)$;

(*b*) Aufzeichnung der Strom-Spannungs-Charakteristik.

diese Weise z. B. mit einem X-Y-Schreiber registrierte Kurve ist allein durch das elektrochemische Verhalten des Eisens in Schwefelsäure charakterisiert. In Abb. 4 *b* haben wir die Strom-Spannungskurve aufgetragen, die mit der Anordnung der Abb. 4 *a* aufgenommen wurde. Wir finden die potentiostatische Kurve *a—b—d* und *f—h* (Abb. 1 *b*) wieder, aber die abrupten Stromänderungen sind verschwunden und statt dessen erhalten wir einen kontinuierlichen Übergang über eine Folge stationärer Zustände *d'—f* zwischen aktivem und passivem Zustand, der nicht von der Richtung des Durchlaufs abhängt. Die Strecke *d—d'* stellt einen Bereich dar, wo Stromschwankungen um einen Mittelwert auftreten, der hier punktiert wiedergegeben ist. Damit haben wir drei Ströme für ein und dasselbe vorgegebene Potential nachweisen und erstmals einen kontinuierlichen Übergang zwischen aktivem und passivem Bereich des Eisens in molarer Schwefelsäure messen können. Dieses Verhalten erklärt die mit einem Potentiostaten beobachteten Erscheinungen. So wird das System Potentiostat—Zelle bei *d* und *f* infolge von Schwankungen des Stroms oder infolge der Form der Strom-Spannungskurve instabil und die potentiostatische Regelung bricht zusammen.

Um den Einfluß des Stofftransports zu untersuchen, haben wir die Strom-Spannungs-Charakteristik des Eisens in molarer Schwefelsäure bei verschiedenen Rotationsgeschwindigkeiten der Elektrode aufgenommen (Abb. 5), [19]. Es ist festzustellen, daß die Höhe der Stromstufe sich wie $\Omega^{1/2}$ ändert, was mit der Theorie von LEVICH für den Stofftransport durch konvektive Diffusion übereinstimmt, die für konzentrierte Lösungen angepaßt wurde [20, 21].

Andererseits ist anzumerken, daß im Übergangsbereich zwischen aktivem und passivem Zustand die Auflösung des Eisens nicht gleichmäßig auf der Elektrodenoberfläche vonstatten geht. Die Auflösung, die anderweits untersucht wurde [12, 19], ist auf einem konzentrischen Ring der Scheibenelektrode lokalisiert.

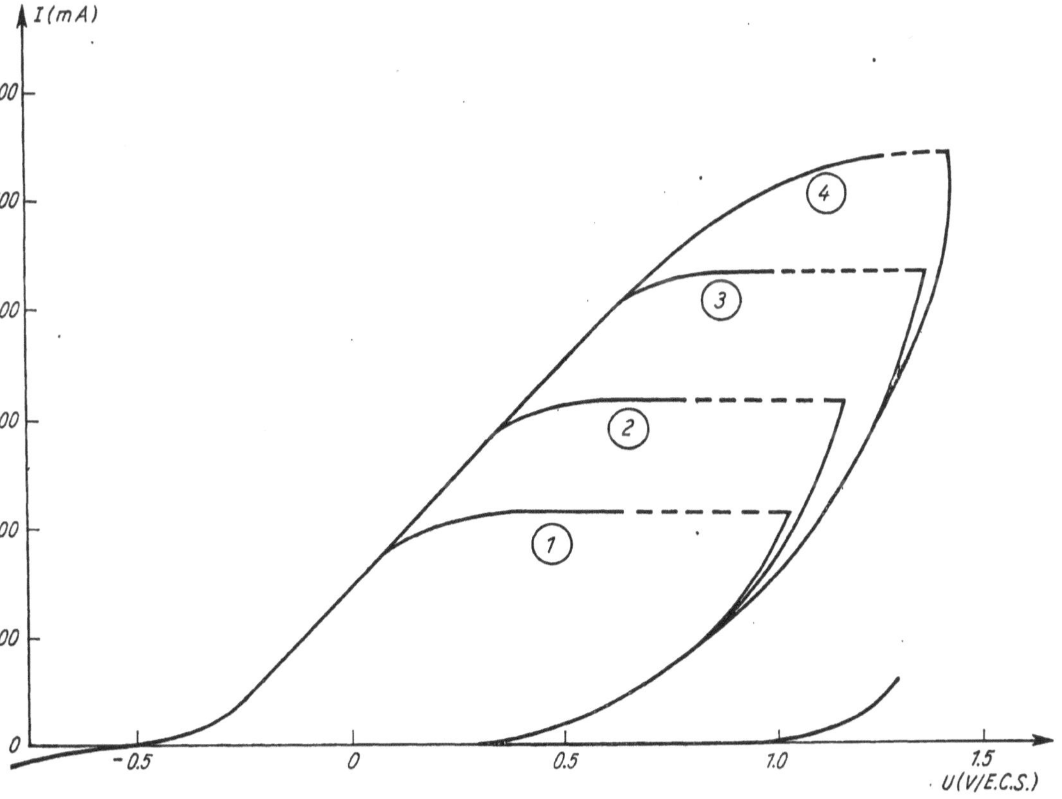

Abb. 5: Strom-Spannungs-Charakteristik des Eisens in Schwefelsäure (gleiche Bedingungen wie in Abb. 4) für verschiedene Rotationsgeschwindigkeiten der Elektrode.

1 — 750 U min⁻¹; *2* — 1600 U min⁻¹; *3* — 3000 U min⁻¹; *4* — 4500 U min⁻¹.

1.4. Messung der Impedanz der Grenzfläche

Die Messung der Grenzflächenimpedanz hat sich nach und nach zu einer wirksamen Untersuchungsmethode für elektrochemische Prozesse entwickelt [22]. In Hinsicht auf das Studium heterogener Reaktionen und auf den Stofftransport wurden die entscheidendsten Fortschritte erzielt, seitdem man begann, Messungen bei sehr tiefen Frequenzen [11, 23, 24] auszuführen. Inzwischen hat die Ausdehnung der Messungen auf diesen Frequenzbereich schwierige technische Probleme in zweierlei Hinsicht ausgelöst.

Einerseits muß die sinusförmige Störung ohne Trennung in Wechselstromkreis und Gleichspannungspolarisation überlagert werden; denn man kann die zwei Kreise nicht mehr unabhängig voneinander machen, wenn die Meßfrequenz nach Null geht. Dieses Problem ist jetzt dank der Anwendung von Operationsverstärkern in integrierter Bauweise gelöst, mit denen alle gewünschten Regelungen realisiert werden können [11, 12].

Andererseits reichen die klassischen Geräte der Impedanzmessung: Brücken, Impedanzmesser ... nicht mehr für die Messung bei sehr tiefen Frequenzen aus. Im Fall von Elektrolyten, wie die hier von uns verwendeten sauren wäßrigen Lösungen, ist es im allgemeinen nötig, Impedanzmessungen zwischen 10^{-3} Hz und 50 kHz auszuführen. Außerdem ist es unerläßlich, Meßsignale kleiner Amplitude (einige mV Scheitel—Scheitel) zu verwenden, da elektrochemische Systeme nichtlineare Systeme sind. Für den gegebenen Frequenzumfang und die Größe der Signale haben wir zu Versuchsdurchführung eine Anordnung zur Aufnahme der Übergangsfunktion mittels Korrelationsanalyse ausgesucht. Die Impedanzmeßanordnung wird daher in 3 verschiedene Baugruppen gegliedert. Zunächst die Polarisationseinrichtung ①, die ein Potentiostat, ein Galvanostat oder ein Regler mit negativem Innenwiderstand sein kann, wie es Abb. 6 illustriert; weiterhin eine Apparatur ② zur Erfassung und Formierung der Signale, die Differenzverstärker und Sollspannungsgeber für die Gleichspannung enthält. Schließlich die Anordnung zur Messung der Übertragungsfunktion ③, die die eigentliche Impedanzmessung ausführt.

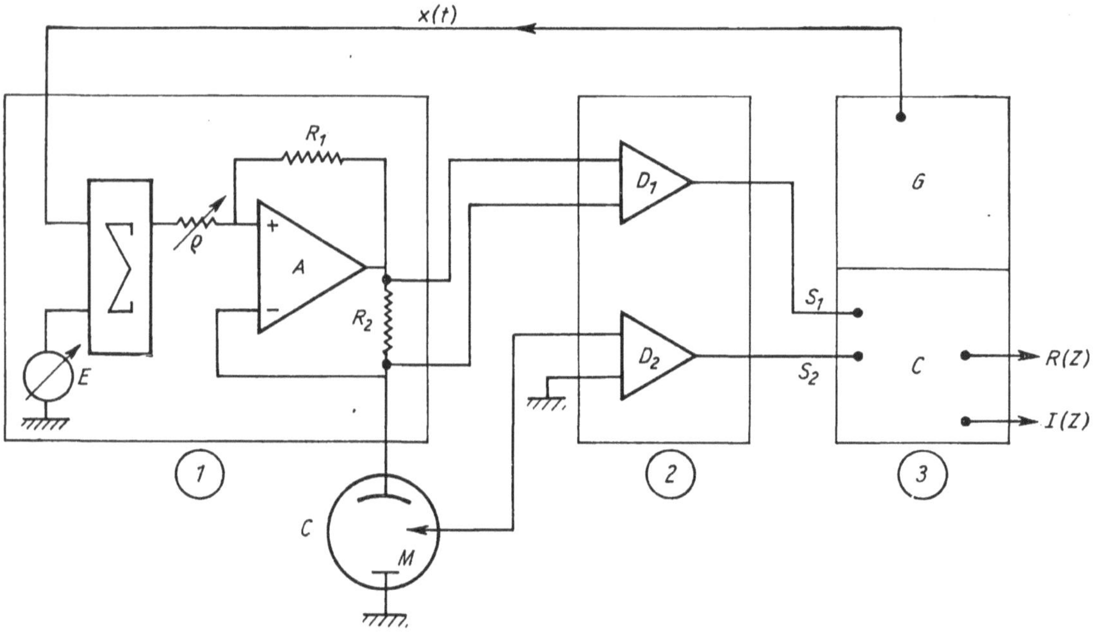

Abb. 6: Meßanordnung für die Impedanz einer Elektrode

① Regelung der Polarisation der Elektrode mit negativem Innenwiderstand (siehe Abb. 4)

② Formung der Meßsignale durch die Differenzverstärker D_1 und D_2

③ Messung der Impedanz mit einer Anordnung zur Bestimmung der Übergangsfunktion, die den Realteil $R(Z)$ und den Imaginärteil $I(Z)$ der Impedanz Z registriert. G — Generator, C — Korrelator, C — Zelle, M — Arbeitselektrode.

In Abb. 6 zeigen wir die experimentelle Anordnung, die von uns verwendet wurde. Die einzelnen Elemente, Regler mit komplexem Innenwiderstand A, Summator Z, Differenzverstärker D_1 und D_2, wurden hinsichtlich Durchlaßbreite und Brumm auf ihre Eignung in einem derartigen Meßgerät ③ besonders untersucht. Der Verstärker D_1 greift an dem Widerstand R_2 den Strom durch die Zelle C ab, und D_2 mißt das Potential der Arbeitselektrode, die an Masse liegt, gegen eine Bezugselektrode. Die beiden Verstärker sind mit einem Sollspannungsgeber ausgerüstet, der die im Meßsignal vorhandene Gleichspannungskomponente eliminiert, die Anordnung zur Messung der Übertragungsfunktion mißt die Übergangsadmittanz der Zelle C, die direkt die Impedanz der Arbeitselektrode M liefert. Es ist zu bemerken, daß der Meßplatz zur Bestimmung der Übergangsfunktion: Solartron Typ 1172, der ein Meßsignal $x(t)$ abgibt und die Übergangsfunktion zwischen zwei beliebigen Punkten S_1 und S_2 des zu untersuchenden Kreises mißt, hierfür besonders geeignet ist. So wird die Stabilität der Anordnung Regler—Zelle, im Gegensatz zum Fall eines Potentiostaten, auf Kosten der Bandbreite der Regelung erreicht, was die Übertragung des Signals $x(t)$ ohne Phasenverschiebung auf die Arbeitselektrode nicht zuläßt. Die Verwendung einer klassischen Anordnung zur Aufnahme der Übergangsfunktion (z. B. Solartron JM 1600) nötigt demnach zur sukzessiven Messung der Übergangsfunktion von S_1 und S_2 bezüglich des Ausgangs des Generators $x(t)$, um daraus erst die Übergangsfunktion zwischen S_1 und S_2 berechnen zu können, die mit der Impedanz der Arbeitselektrode verknüpft ist [11]. Die hier in einem Arbeitsgang ausgeführte Messung liefert einen beträchtlichen Zeitgewinn, der bei diesen Untersuchungen sehr nützlich ist, weil sich der Zustand der Elektrodenoberfläche schnell verändert.

1.5. Anwendung auf die Messung der Impedanz der Grenzfläche Eisen—molare Schwefelsäure

Mit dieser Anordnung konnten wir die Impedanz über die ganze Strom-Spannungs-Charakteristik der Eisenelektrode in molarer Schwefelsäure aufnehmen. Die gemessenen Impedanzen haben wir in Abb. 7 in der komplexen Ebene aufgetragen. (Die imaginären, kapazitiven Partien sind im 1. Quadranten zu finden, wie es in der Elektrochemie üblich ist.)

2. Interpretation des anodischen Verhaltens des Eisens in schwefelsaurem Milieu

Abrupte Übergänge und Hysterese, die das Verhalten der Grenzfläche Eisen—Schwefelsäure unter potentiostatischen Bedingungen charakterisieren, sind auch für andere Metalle und andere Lösungen angegeben worden, die hohe anodische Stromdichten aufweisen ($1\ A \cdot cm^{-2}$) [25]. Da die oben angeführten Ergebnisse den Einfluß des Materietransports deutlich zeigen, haben wir ein Modell der Kopplung zwischen den elektrochemischen Reaktionen und der Diffusion gesucht, das für die Interpretation dieser Verhaltensweise geeignet ist.

2.1. Mathematische Formulierung des Problems

Wir haben auf einem anderen Weg gezeigt [26], daß ein einfaches Modell, das auf der Kopplung zwischen Aktivierungs-Passivierungs-Reaktion und der Diffusion der passivierenden Spezies beruht, zu einem elektrochemischen System mit mehreren stationären Zuständen für ein vorgegebenes Potential führt. Jedoch kann dieses Modell das experimentell beobachtete Verhalten nur näherungsweise beschreiben; denn die anodische Auflösung und die Passivierung sind nur durch globale Reaktionen erfaßt. Wir wollen hier ein realistisches Modell vorführen, das die Tatsache in Rechnung stellt, daß sowohl die Auflösung in zwei Etappen erfolgt, wobei ein adsorbierter Zwischenzustand eine Rolle spielt [27], als auch die Passivierung in zwei aufeinanderfolgenden Schritten abläuft. Dieser Reaktionsmechanismus, bei dem das Hydroxylion eine aktivierende und dann passivierende Rolle spielt, ist schon im Fall des Eisens [28] betrachtet worden, konnte aber durch Impedanzmessungen mit den

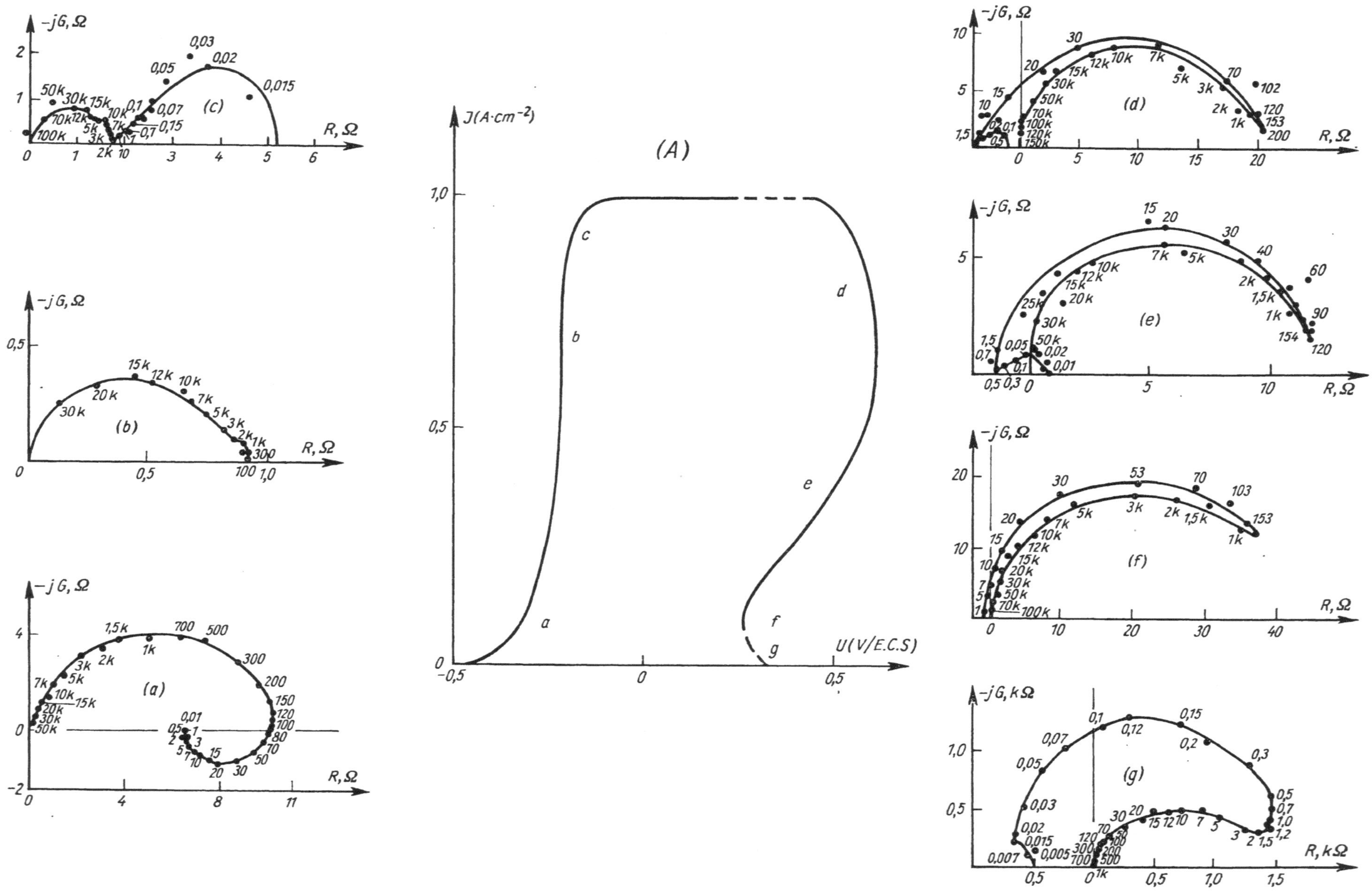
(A)
J (A·cm⁻²)
U (V/E.C.S)
a
b
c
d
e
f
g
(a)
(b)
(c)
(d)
(e)
(f)
(g)
−jG, Ω
R, Ω
−jG, kΩ
R, kΩ

experimentellen Erfahrungen nur bei der Auflösung [7, 23] verglichen werden. Er kann dargestellt werden durch

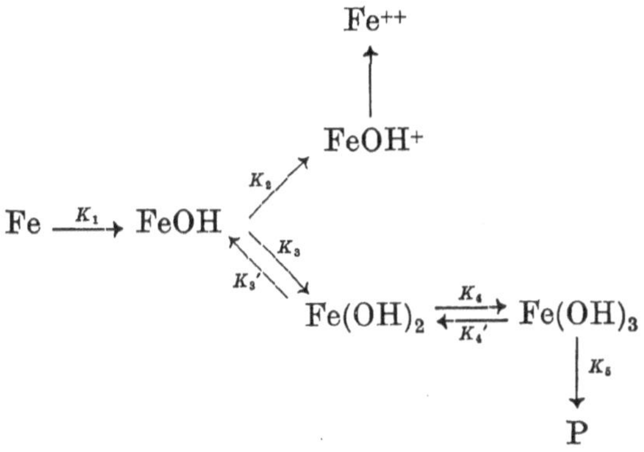

Dieses Reaktionsschema kann durch folgende heterogene Reaktionen ausgedrückt werden:

$$
(A)\quad
\begin{array}{lll}
Fe + OH^- & \underset{K_2}{\xrightarrow{K_1}} FeOH + e, & (I) \\
FeOH & \xrightarrow[K_3]{} FeOH^+ + e, & (II) \\
FeOH + OH^- & \underset{K_3'}{\rightleftharpoons} Fe(OH)^2 + e, & (III) \\
Fe(OH)_2 + OH^- & \underset{K_4'}{\overset{K_4}{\rightleftharpoons}} Fe(OH)_3 + e, & (IV) \\
Fe(OH)_3 & \xrightarrow{K_5} P & (V)
\end{array}
$$

Die Reaktionen *I* und *II* geben die Auflösung des Eisens wieder, wobei die adsorbierte Zwischenform FeOH [7, 8] ins Spiel kommt; die Reaktionen *III* und *IV* beziehen sich auf die zwei aufeinanderfolgenden Etappen der Passivierung und die Reaktion *V* auf eine chemische Reaktion, die die Passivschicht zerstört.

Wenn θ_1, θ_2 und θ_3 die entsprechenden Bedeckungsgrade der Adsorbate FeOH, $Fe(OH)_2$ und $Fe(OH)_3$, β_1, β_2 und β_3 deren maximale Oberflächenkonzentrationen und c die Aktivität von OH^- an der Grenzfläche sind, führen die Bilanzen von FeOH, von $Fe(OH)_2$, von $Fe(OH)_3$ und der Elektronen zu

$$\beta_1 \frac{d\theta_1}{dt} = K_1 c(1 - \theta_1 - \theta_2 - \theta_3) - K_2\theta_1 - K_3 c\theta_1 + K_3'\theta_2,$$

$$\beta_2 \frac{d\theta_2}{dt} = K_3 c\theta_1 - K_3'\theta_2 - K_4 c\theta_2 + K_4'\theta_3,$$

$$\beta_3 \frac{d\theta_3}{dt} = K_4\theta_2 - K_4'\theta_3 - K_5\theta_3,$$

$$\frac{J}{F} = K_1 c(1 - \theta_1 - \theta_2 - \theta_3) + K_2\theta_1 + K_3 c\theta_1 - K_3'\theta_2 + K_4 c\theta_2 - K_4'\theta_3.$$

Abb. 7: Impedanzdiagramme, Eisenelektrode in Schwefelsäure
(*A*) Strom-Spannungs-Charakteristik $\Omega = 750$ U/min^{-1}; Scheibenelektrode (∅ 5 mm) H_2SO_4 1 M (pH = 0);
(*a*) Auflösung mit kleinem Strom $I = 4$ mA; $\Omega = 2000$ U/min^{-1}; Scheibenelektrode pH = 0;
(*b*) Auflösung mit großem Strom $I = 50$ mA, $\Omega = 2000$ U/min^{-1}; Scheibenelektrode pH = 3;
(*c*) Auflösung nahe beim Diffusionsplateau $I = 50$ mA; $\Omega = 100$ U/min^{-1}; Scheibenelektrode pH = 3;
(*d*) Passivierung mit großem Strom — die Neigung der I—U-Kurve ist negativ: $I = 100$ mA; $\Omega = 750$ U/min^{-1}; Ringelektrode ($\varnothing_{\text{inn}} = 3$ mm) pH = 0.
(*e*) Passivierung mit großem Strom — die Neigung der I-U-Kurve ist negativ. $I = 60$ mA, $\Omega = 750$ U/min^{-1}; Ringelektrode ($\varnothing_{\text{inn}} = 3$ mm) pH = 0
(*f*) Passivierung mit kleinem Strom, $I = 15$ mA; $\Omega = 750$ U/min^{-1}; Ringelektrode ($\varnothing_{\text{inn}} = 4{,}5$ mm) pH = 0;
(*g*) Passivierung mit sehr kleinem Strom, $I = 60$ µA; $\Omega = 750$ U/min^{-1}; Scheibenelektrode pH = 0; $\varnothing_{\text{inn}}$: Innendurchmesser, der Außendurchmesser ist 5 mm.

Mit den geläufigen Annahmen [7, 29]:

a) LANGMUIR-Adsorptionsisotherme für die Adsorbate,

b) TAFELsche Kinetik für die Geschwindigkeitskonstanten der elektrochemischen Reaktionen, d. h.

$$K_i = K_{i0} \exp b_i V,$$

c) Diffusion von OH^- in einer Diffusionsschicht der Dicke δ_N nach dem NERNSTschen Modell, gegeben durch

$$c(x) = c - (c - c_0) \frac{x}{\delta_N} \quad \text{für} \quad 0 < x < \delta_N,$$

$$c(x) = c_0 \quad \text{für} \quad x \geq \delta_N.$$

2.2. *Simulation des Modells*

Um diese Untersuchung zu vereinfachen, soll sie in zwei Etappen vorgenommen werden.

Zunächst betrachten wir den Potentialbereich, in welchem die Reaktion *IV* eine vernachlässigbare Rolle spielt. Dazu genügt es, im allgemeinen Modell

$$K_4 = K_4' = K_5 = 0 \quad \text{und} \quad \theta_3 = 0$$

zu setzen; dieser Fall soll Modell *A* genannt werden.

Dann interessieren wir uns für den Potentialbereich, wo K_1 sehr groß ist, was zu $\theta_1 = 1 - \theta_2 - \theta_3$ führt. Das wird Modell *B* genannt. In diesem Modell werden nur kleine Ströme betrachtet, die Diffusion wird dann keine Rolle spielen.

2.2.1. Modell *A* [30]

Es ist definiert durch die Gleichungen

$$\beta_1 \frac{d\theta_1}{dt} = K_1 c(1 - \theta_1 - \theta_2) - K_2\theta_1 - K_3\theta_1 c + K_3'\theta_2, \tag{1}$$

$$\beta_2 \frac{d\theta_2}{dt} = K_3\theta_1 c - K_3'\theta_2, \tag{2}$$

$$J = F[K_1 c(1 - \theta_1 - \theta_2) + K_2\theta_1 + K_3 c\theta_1 - K_3'\theta_2]. \tag{3}$$

Der stationäre Strom *I* kann ausgedrückt werden entweder mit Hilfe der Geschwindigkeitskonstanten der Reaktionen (*I, II, III*)

$$J_1 = \frac{2FK_1K_2c}{K_1c(1 + kc) + K_2}; \quad k = \frac{K_3}{K_3'}, \tag{4}$$

die durch Eliminierung von θ_1 und θ_2 aus den Gln. (1), (2) und (3) unter stationären Bedingungen $\frac{d\theta_1}{dt} = \frac{d\theta_2}{dt} = 0$ erhalten werden; oder aber über den Diffusionsfluß von OH^-

$$J_2 = 2F \frac{D}{\delta_N} (c_0 - c). \tag{5}$$

Hier ist *D* der Diffusionskoeffizient von OH^-.

Die Identität $J_1 = J_2$ liefert eine Gleichung 3. Grades in c:

$$kc^3 + c^2(1 - kc_0) + c\left[K_2\left(\frac{\delta_N}{D} + \frac{1}{K_1}\right) - c_0\right] - \frac{K_2}{K_1}c_0 = 0. \tag{6}$$

Für angemessene Werte der Parameter hat diese Gleichung 3 Lösungen zwischen 0 und c_0, denen 3 Werte des Stromes für ein gegebenes Potential entsprechen. Das kann man bei genügend großen Werten von δ_N/D sehen.

Um auf strengere Weise das Verhalten des Modells zu charakterisieren, haben wir mit denselben Annahmen wie oben die FARADAY-Impedanz $Z_{F_1}(\omega)$ berechnet. So erhält man die Antworten δJ, δc, $\delta\theta_1$, $\delta\theta_2$ auf die Störung des Potentials $\delta V e^{j\omega t}$ der Art

$$\begin{aligned} j\omega\beta_1\,\delta\theta_1 = {} & \delta V[b_1K_1c(1 - \theta_1 - \theta_2) - b_2K_2\theta_1 - b_3K_3\theta_1c - b_3'K_3'\theta_2] \\ & + \delta\theta_1[-K_1c - K_2 - K_3c] + \delta\theta_2[-K_1c + K_3'] \\ & + \delta c[K_1(1 - \theta_1 - \theta_2) - K_3\theta_1], \end{aligned} \tag{7}$$

$$j\omega\beta_2\,\delta\theta_2 = \delta V[b_3K_3\theta_1c + b_3'K_3'\theta_2] + \delta\theta_1K_3c - K_3'\,\delta\theta_2 + K_3'\theta_1\,\delta c, \tag{8}$$

$$\begin{aligned} \delta J = {} & F\{\delta V[b_1K_1c(1 - \theta_1 - \theta_2) + b_2K_2\theta_1 + b_3K_3c\theta_1 + b_3'K_3'\theta_2] \\ & + \delta\theta_1[-K_1c + K_2 + K_3c] + \delta\theta_2[-K_1c - K_3'] \\ & + \delta c[K_1(1 - \theta_1 - \theta_2) + K_3\theta_1]\}. \end{aligned} \tag{9}$$

Außerdem kann man zeigen [11], daß δc und δJ durch

$$\frac{\delta c}{\delta J} = \frac{1}{2FE(\omega)} \tag{10}$$

verknüpft sind, wo

$$\frac{1}{E(\omega)} = -\frac{\delta_N}{D}\,\frac{th\sqrt{j\omega\delta_N^2/D}}{\sqrt{j\omega\delta_N^2/D}}.$$

Wir eliminieren $\delta\theta_1$, $\delta\theta_2$, δc aus den Gln. (7), (8), (9) und (10) und können die FARADAY-Impedanz $Z_{F_1}(\omega)$ berechnen mit

$$Z_{F_1}(\omega) = \frac{\delta V}{\delta J}$$

$$\begin{vmatrix} j\omega\beta_1 + K_1c + K_2 + K_3c & K_1c - K_3' & -\dfrac{K_1(K_2 - K_3c)}{2AE(\omega)} \\ -K_3c & j\omega\beta_2 + K_3' & -K_1K_3c/2AE(\omega) \\ K_1c - K_2 - K_3c & K_1c + K_3' & 1 - \dfrac{K_1(K_2 + K_3c)}{2AE(\omega)} \end{vmatrix}$$

$$Z_{F_1}(\omega) = \frac{1}{F} \tag{11}$$

$$\begin{vmatrix} j\omega\beta_1 + K_1c + K_2 + K_3c & K_1c - K_3' & K_1c[K_2(b_1 - b_2) - K_3c(b_3 + b_3')]/A \\ -K_3c & j\omega\beta_2 + K_3' & K_1K_2c^2(b_3 + b_3')/A \\ K_1c - K_2 - K_3c & K_1c + K_3' & K_1c[K_2(b_1 + b_2) + K_3c(b_3 + b_3')]/A \end{vmatrix}$$

wo $A = K_1c(1 + kc) + K_2$ ist.

Abb. 8 gibt ein Beispiel der Strom-Spannungskurve und der Impedanzen, die mit den voraufgehenden Gleichungen berechnet wurden. Aus Gl. (6) erhält man für jedes Potential die Werte für c, die man zur Bestimmung des stationären Stromes in Gl. (5) und zur Bestimmung der Impedanz in Gl. (11) einsetzt. Bei Betrachtung der glockenförmigen Kurve, die allgemein dem Reaktionsmechanismus (A) [11] entspricht und die man für $\delta_N/D = 0$ und durch Einsetzen von $c = c_0$ in Gl. (4) erhält, wird man beobachten, daß die Diffusion sich in einem Grenzstrom (Diffusionsstufe) und in der Existenz dreier verschiedener Ströme für ein festgelegtes Potential auswirkt.

Die verschiedenen Kurventypen für die vorausberechnete Änderung von $Z_{F_1}(\omega)$ mit der Kreisfrequenz ω in der komplexen Ebene sind in Abb. 8 b dargestellt, wobei die Doppelschichtkapazität berücksichtigt wurde. Beim Übergang vom Durchtrittswiderstand R_t zum Polarisationswiderstand R_p kann man die Zeitkonstanten finden, die zur Adsorption-Desorption von $FeOH_{ads}$ und von $Fe(OH)_2$ gehören und außerdem eine andere mit sehr charakteristischer Form. die der Impedanz der Diffusion von OH^- entspricht.

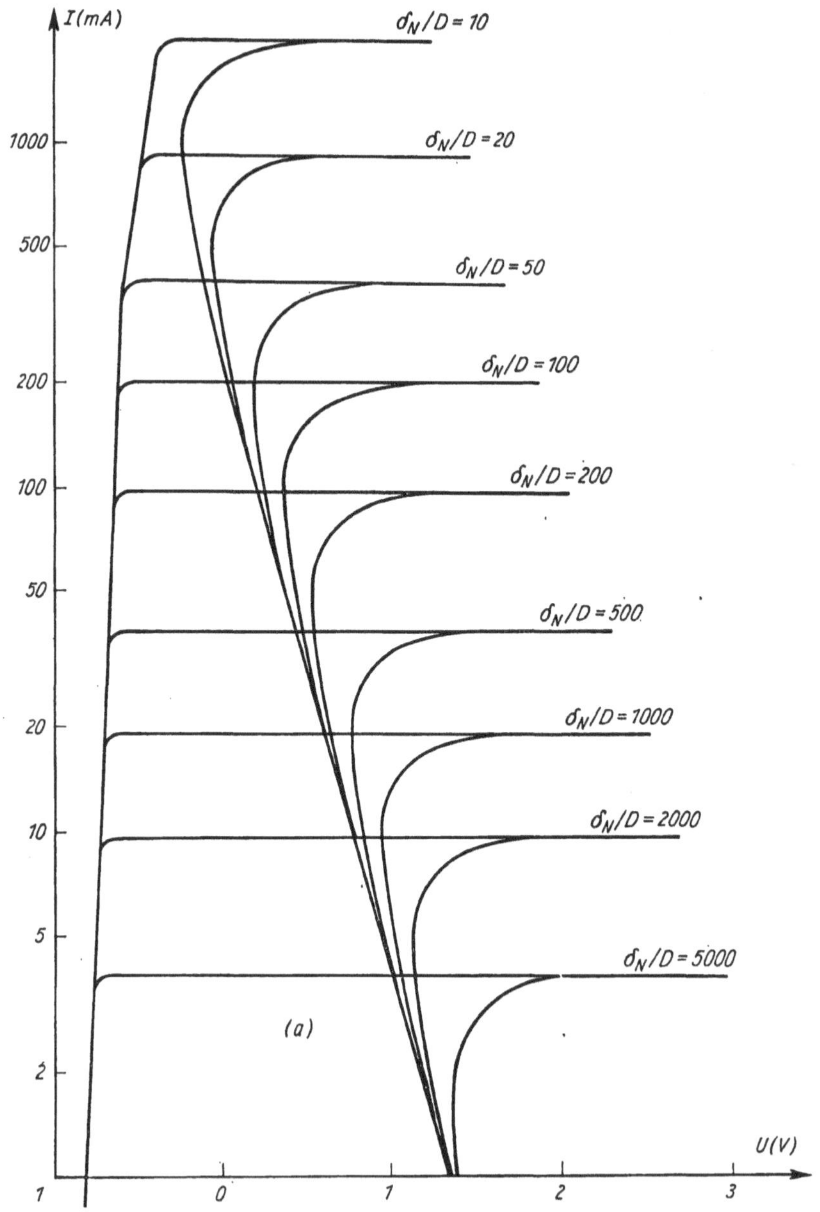

Abb. 8: Simulation des Verhaltens des Modells A mit $K_1 = 4 \cdot 10^8 \exp 36$ V; $K_2 = 10^{-3} \exp 10$ V; $K_3 = 2 \cdot 10^{-2} \exp 12$ V; $K_3' = 2 \cdot 10^{-8} \exp -5$ V; $\beta_1 = \beta_2 = 2 \cdot 10^{-9}$, $c_0 = 10^{-4}$. $\delta_N/D = 20$, $\delta_N{}^2/D = 1$. Kapazität der Doppelschicht: 25 μF.

(a) Strom-Spannungs-Kurven für verschiedene Werte des Verhältnisses δ_N/D;

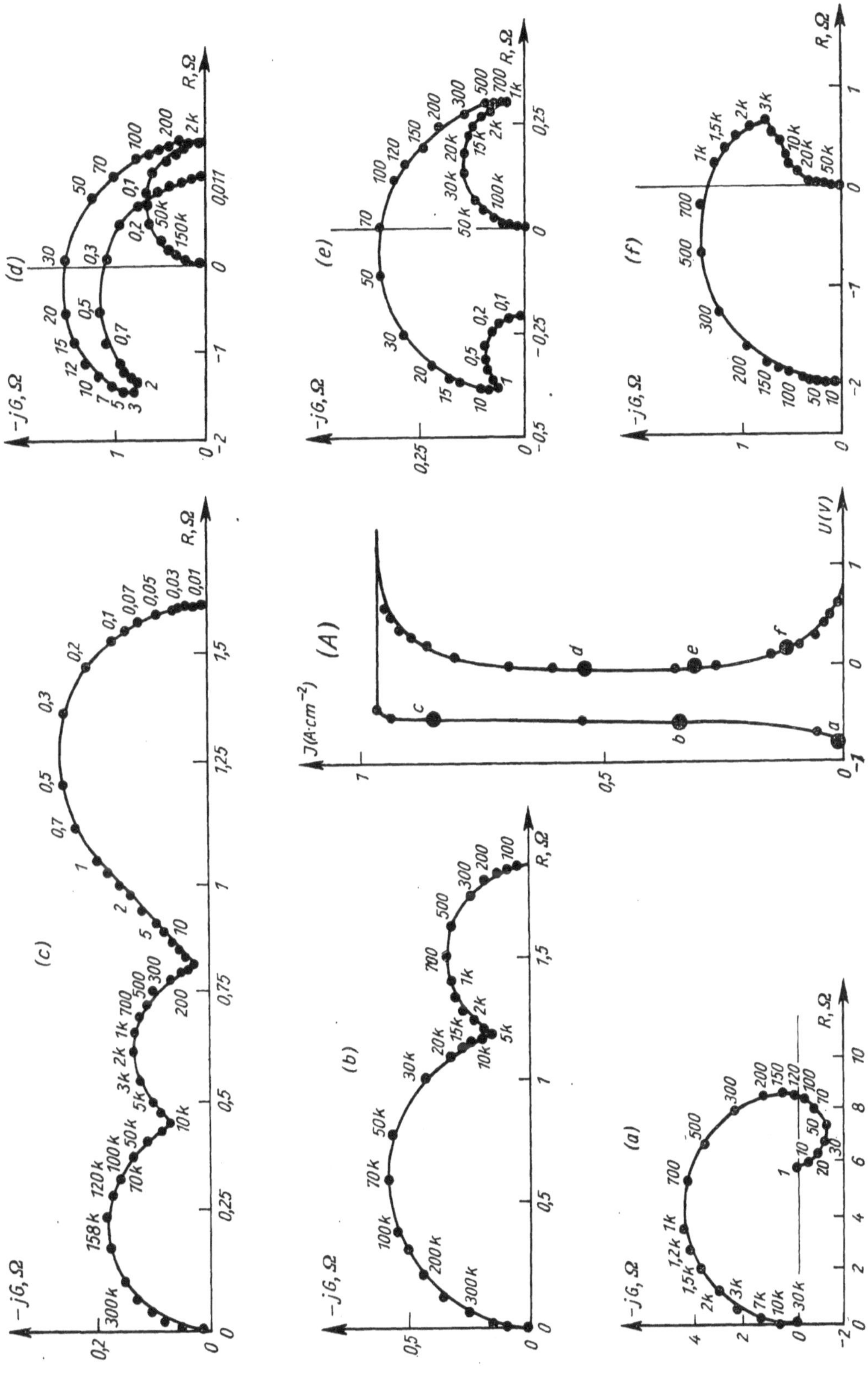

(b) Berechnete Impedanzdiagramme

(A) Strom-Spannungskurve (a—f) Impedanzdiagramme, berechnet für die Punkte der Polarisation, die den Punkten a bis f auf der Kurve A entsprechen.

Die Diagramme a und b entsprechen dem Mechanismus der Eisenauflösung in 2 Etappen, der seit 1968 nachgewiesen worden ist [7, 20, 23]. Das Diagramm f stellt gleichermaßen einen für Passivierungserscheinungen wohlbekannten Typ dar [24]. Die Diagramme c, d und e sind im Gegensatz dazu für die hier betrachtete Kopplung zwischen den heterogenen Reaktionen und der Diffusion charakteristisch. Die spezielle Lage der Diffusionsimpedanz gibt daher eine einfache Möglichkeit, die durch ein derartiges Modell beschriebenen elektrochemischen Prozesse zu untersuchen.

2.2.2. Modell B

Es ist definiert durch die Gleichungen

$$\beta_2 \frac{d\theta_2}{dt} = K_3(1 - \theta_2 - \theta_3) - K_3'\theta_2 - K_4\theta_2 + K_4'\theta_3, \tag{12}$$

$$\beta_3 \frac{d\theta_3}{dt} = K_4\theta_2 - K_4'\theta_3 - K_5\theta_3, \tag{13}$$

$$\frac{J}{F} = 2K_2(1 - \theta_2 - \theta_3) + 2K_3(1 - \theta_2 - \theta_3) - 2K_3'\theta_2 + K_4\theta_2 - K_4'\theta_3. \tag{14}$$

Führt man dieselben Rechnungen wie für Modell A aus, so erhält man für die stationäre Strom-Spannungskurve

$$J = F\,\frac{2K_2(K_3'K_4' + K_3'K_5 + K_4K_5) + 2K_3K_4K_5}{(K_3 + K_3' + K_4)(K_4' + K_5) + K_4(K_3 - K_4')} \tag{15}$$

und für die Faraday-Impedanz

$$Z_{F_2}(\omega) = \frac{1}{F}\,\frac{\begin{vmatrix} j\omega\beta_2 + K_3 + K_3' + K_4 & K_3 - K_4' \\ -K_4 & j\omega\beta_3 + K_4' + K_5 \end{vmatrix}}{\begin{vmatrix} j\omega\beta_2 + K_3 + K_3' + K_4' & K_3 - K_4' & A_1 \\ -K_4 & j\omega\beta_3 + K_4' + K_5 & A_2 \\ 2K_2 + 2K_3 + 2K_3' - K_4 & 2K_2 + 2K_3 + K_4' & A_3 \end{vmatrix}} \tag{16}$$

mit

$$A_2 = \frac{1}{D} K_3K_4[K_4'(b_4 + b_4') + b_4K_5],$$

$$A_1 = \frac{K_3}{D}[(b_3 + b_3')K_3'K_4' + K_5(b_3(K_3' + K_4) + b_3'K_3')] \quad -A_2,$$

$$A_3 = 2A_1 + 3A_2 + \frac{2b_2K_2}{D}[K_3'K_4' + K_5(K_3' + K_4)],$$

$$D = (K_3 + K_3' + K_4)(K_4' + K_5) + K_4(K_3 - K_4').$$

Abb. 9 gibt ein Beispiel für Strom-Spannungskurven und Impedanzen, die mit den obigen Gln. (15) und (16) berechnet wurden. Die verschiedenen Typen von zu erwartenden Diagrammen in der komplexen Ebene für die Änderung von $Z_{F_2}(\omega)$ mit der Kreisfrequenz sind unter Berücksichtigung der Kapazität der Doppelschicht parallel zur Faraday-Impedanz gegeben [7].

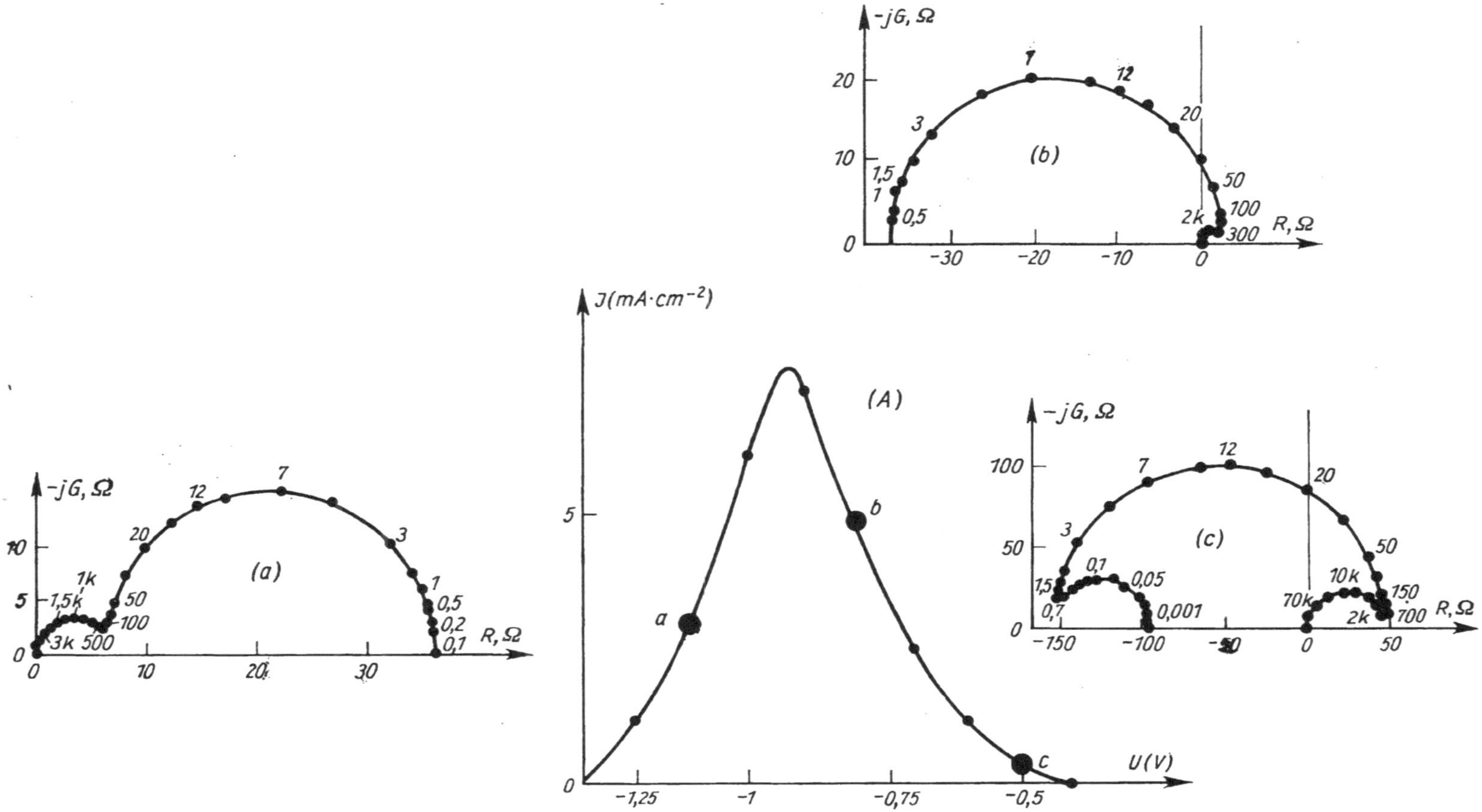

Abb. 9: Simulation des Verhaltens des Modells *B* mit $K_2 = 10^{-3} \exp 10V$; $K_3 = 2 \cdot 10^{-2} \exp 12V$; $K_3 = 2 \cdot 10^{-8} \exp -5V$; $K_4 = 10^{-5} \exp 31.6\ V$; $K_4' = 10^{-10} \exp -5\ V$; $K_5 = 10^{-11}$; $\beta_2 = \beta_3 = 2 \cdot 10^{-9}$; Kapazität der Doppelschicht 25 µF.
(*A*) Strom-Spannungs-Kurve; (*a*—*c*) Impedanzdiagramme, berechnet für 2 Punkte der Polarisation, die den Punkten *a* bis *c* auf der Kurve *A* entsprechen.

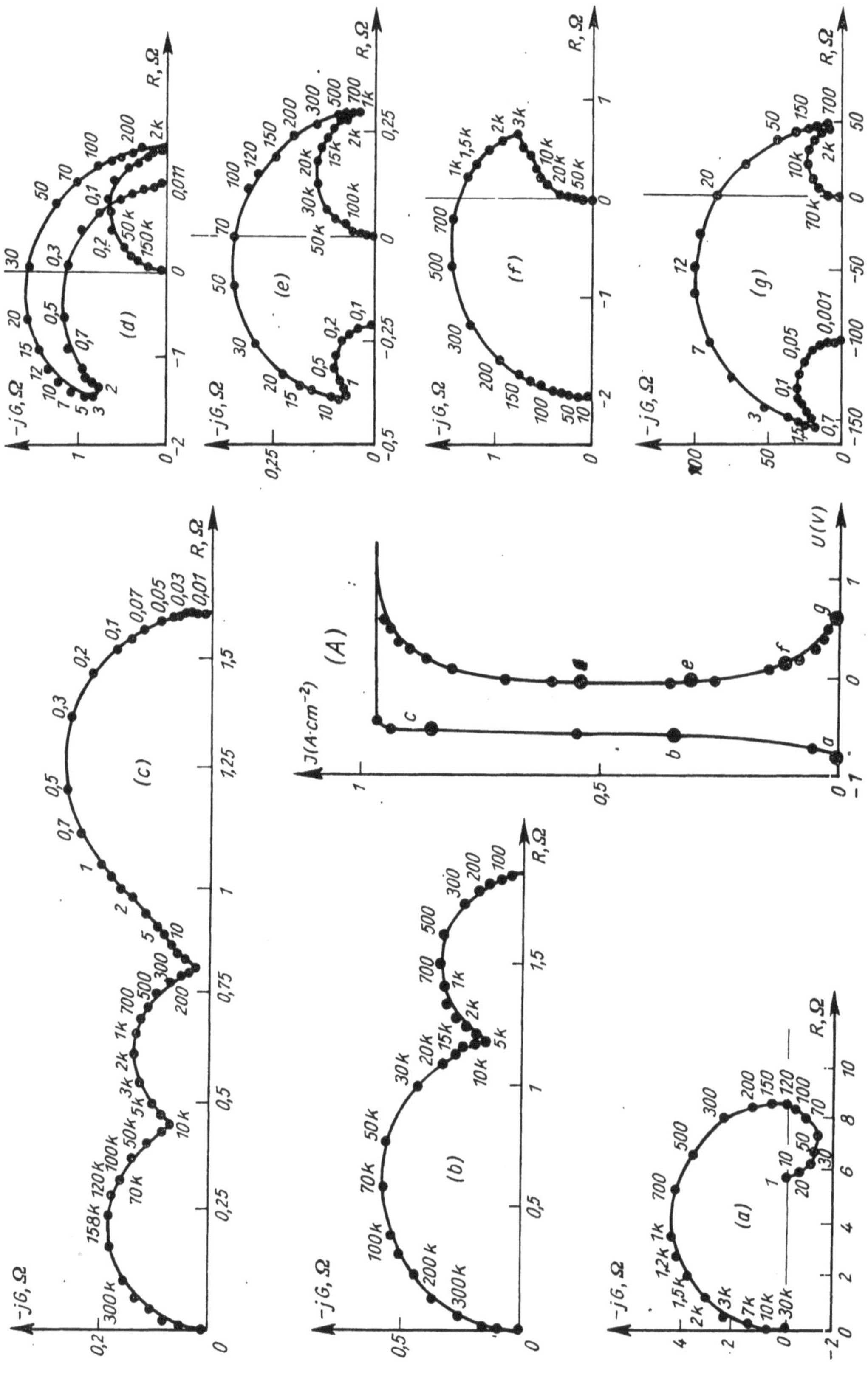

Abb. 10: Simulation des Verhaltens des Gesamtmodells (Zahlenwerte identisch mit Abb. 8 und 9).

3. Diskussion

Von den Abb. 8 und 9 ausgehend kann man das Gesamtverhalten des vorgeführten Modells konstruieren (Abb. 10). Der Gültigkeitsbereich der Näherungen, die zu Modell *A* führen, erstreckt sich vom Korrosionspotential bis zu den Punkten *a*, *b*, *c*, *d*, *e*, *f*. Die Gültigkeit des Modells *B* ist auf die Passivierung beschränkt (Punkt *b* und *c* in Abb. 9). Der Übergang erfolgt in der Nähe des Punktes *f* von Abb. 8 und von *b* in Abb. 9, in diesen Punkten werden übrigens identische Impedanzdiagramme erhalten.

Der Vergleich der Abb. 7 und 10 zeigt, daß das vorgeschlagene Modell eine angemessene Beschreibung des anodischen Verhaltens der Grenzfläche Eisen—molare Schwefelsäure liefert. Es ist jedoch zu bemerken, daß dieses Modell a priori in seiner Anwendung in zwei Punkten begrenzt ist. Zuerst in Hinsicht auf die Diffusion des Hydroxylions kann es die Größenordnung des beobachteten Grenzstroms [27] nicht erklären, obwohl es relativ gut die Form der Strom-Spannungs-Kurve und die Impedanzen wiedergibt. Zum anderen kann dieses Modell die Prozesse, die sich während der Passivität abspielen, nicht erklären, wenn auch die Betrachtung der zwei passivierenden Adsorbate dem Phänomen der Passivierung Rechnung trägt. Dafür wäre es erforderlich, zusätzliche Annahmen einzuführen.

Das hier vorgeschlagene Modell für die Auflösung und die Passivierung des Eisens konnte durch Messung der Faraday-Impedanz in einem großen Frequenzbereich verifiziert werden. Jedoch erfordern die oben erwähnten Verbesserungen dieses Modells die Verwendung schneller Methoden der Impedanzanalyse, die zur Zeit in unserem Laboratorium auf der Grundlage der Korrelation eines weißen Rauschens mit der Antwort auf dieses Signal eingerichtet werden [31].

Beschränkt man sich auf den Bereich, wo das vorgeschlagene Modell anwendbar ist, so erhält man Aufschlüsse in zwei Richtungen. Zuerst kann man bestätigen, daß die Diffusion der passivierenden Spezies für die Tatsache verantwortlich ist, daß das Eisen noch bei Potentialen aktiv ist, wo es passiv sein müßte. Das hat die Möglichkeit gegeben, das Wachstum einer Lochfraßstelle auf der Eisenelektrode zu interpretieren [32, 33]. Dann wird bekräftigt, daß einerseits [7, 23] bei der Auflösung eine adsorbierte Zwischenform FeOH ins Spiel kommt und daß andererseits die Passivierung in zwei konsekutiven Etappen abläuft. Es würde also zwei passivierende Adsorbate geben, die als $Fe(OH)_2$ und $Fe(OH)_3$ identifiziert werden können.

Schließlich möchten wir darauf aufmerksam machen, daß Kopplungen zwischen heterogenen Reaktionen der Passivierung einerseits und der Diffusion oder dem ohmschen Spannungsabfall [11] andererseits, die zu elektrochemischen Systemen mit mehrfachen stationären Zuständen führen können, inzwischen unter anderen Bedingungen gefunden wurden (Kohleelektrode in geschmolzenem Kryolith [34], Eisenelektrode in Salpetersäure [35], Kobaltelektrode in Schwefelsäure [36]).

Literatur

[1] K. Schwabe, H. D. Zushke und M. Time, Zashita Metallov, **10**, No 5, 491 (1974)
[2] U. Ebersbach, K. Schwabe und P. König, Electrochim. Acta, **14**, 773 (1969)
[3] G. Kreysa, U. Ebersbach und K. Schwabe, Electrochim. Acta, **16**, 1489 (1971)
[4] U. Ebersbach, G. Kreysa und K. Schwabe, Electrochim. Acta, **17**, 445 (1972)
[5] H. Rosswag, G. Eichkorn und W. J. Lorenz, Werkstoffe und Korrosion, **25**, No 2, 86 (1974)
[6] G. J. Bignold und M. Fleischmann, Electrochim. Acta, **19**, 363 (1974)
[7] I. Epelboin und M. Keddam, J. of Electrochem. Soc., **117**, No 8, 1052 (1970)
[8] I. Epelboin und M. Keddam, Electrochim. Acta, **17**, 177 (1972)
[9] K. J. Vetter, „*Electrochemische Kinetik*", Springer-Verlag, Berlin 1961
[10] U. F. Franck und R. Fitzhugh, Z. für Elektrochem., **65**, No 2, 156 (1961)
[11] C. Gabrielli, Thèse, Paris 1973. Métaux, Corrosion, Industrie (1973), pp. 171—223, 309—356
[12] I. Epelboin, C. Gabrielli, M. Keddam, J. C. Lestrade und H. Takenouti, J. of Electrochem. Soc., **119**, 1632 (1972)
[13] C. Gabrielli und M. Keddam, Electrochim. Acta, **19**, 355 (1974)
[14] M. Prazak und J. Spanily, Coll. Czechoslov. Chem. Comm., **26**, 2828 (1961)

[15] U. R. EVANS, "*The corrosion and oxydation of metals*", Butler and Tanner Ltd., London 1960
[16] J. OSTERWALD, Z. Elektrochem., **66**, 401 (1962)
[17] U. F. FRANCK, Z. Physik. Chem., **3**, 183 (1955)
[18] I. EPELBOIN, E. LEJAY und R. WIART, C.R. Acad. Sci. Paris, **274**C, 1780 (1972); I. EPELBOIN, M. KSOURI, E. LEJAY und R. WIART, 25th Meeting of the I.S.E., Ext. Abs. p. 143, Electrochim. Acta (à paraître)
[19] C. GABRIELLI, M. KEDDAM, J. C. LESTRADE und H. TAKENOUTI, C.R. Acad. Sci. Paris, **274**C, 123 (1972)
[20] M. DAGUENET und D. SCHUMANN, C.R. Acad. Sci. Paris, **260** gr **7**, 4731 (1965)
[21] M. DAGUENET, Thèse Paris 1965, Ed. du Ministère de l'Air
[22] Intermediates in Electrochemistry, Faraday Disc. of the Chemical Society, n⁰ 56 (1973) p. 264
[23] M. KEDDAM, Thèse Paris 1968
[24] M. L. BOYER, I. EPELBOIN und M. KEDDAM, Electrochim. Acta, **11**, 221 (1966)
[25] U. F. FRANCK, Corrosion, Anticorrosion, **7**, No 3, 83 (1959)
[26] C. GABRIELLI und M. KEDDAM, J. of Electroanal. Chem., **45**, 267 (1973)
[27] I. EPELBOIN, C. GABRIELLI, M. KEDDAM und H. TAKENOUTI, 25th Meeting of the I.S.E., Brighton (1974), Ext. Abstr. p. 354, Electrochim. Acta (à paraître)
[28] J. O'M. BOCKRIS, M. A. GENSHAW, V. BRUSIC und H. WROBLOWA, Electrochim. Acta, **16**, 1859 (1971)
[29] I. EPELBOIN und R. WIART, J. of Electrochem. Soc., **118**, No 10, 1577 (1971)
[30] I. EPELBOIN, C. GABRIELLI und M. KEDDAM, 74e Manif. de la Féd. Europ. de Corrosion, Budapest 1974, Comptes Rendus du Congrès p. 105
[31] G. BLANC, I. EPELBOIN, C. GABRIELLI und M. KEDDAM, J. of Electroanal. Chem. (à paraître)
[32] C. GABRIELLI, M. KEDDAM und H. TAKENOUTI, C.R. Acad. Sci. (Paris), **277**C, 743 (1973)
[33] I. EPELBOIN, C. GABRIELLI und M. KEDDAM, Corrosion Science, **15**, 155 (1975)
[34] A. KEROUANTON, C. GABRIELLI, M. KEDDAM und V. PLICHON, J. of Electroanal. Chem., **57**, 273 (1974)
[35] C. GABRIELLI, M. KEDDAM, E. LISAC und H. TAKENOUTI, Electrochim. Acta (sous presse)
[36] I. EPELBOIN, C. GABRIELLI und Ph. MOREL, Electrochim. Acta, **18**, 509 (1973)

E. Stupnišek-Lisac und M. Karšulin

Institut für physikalische Chemie, Universität Zagreb

ÜBER DIE INTERMEDIÄREN REAKTIONEN BEI DER PASSIVIERUNG DES EISENS IN SALPETERSÄURE

Zusammenfassung

Die Passivierung des Eisens mittels Salpetersäure erfolgt über Ferronitroso-Komplex, $Fe(NO)^{++}$, als Zwischenprodukt der Oxydation des Fe(II)-Ions. Da uns die intermediäre Ferronitroso-Bildung wesentlich für die Deutung des Mechanismus der Passivierung erschien, wurde die Kinetik der Fe(II)-Oxydation mittels Salpetersäure näher untersucht.

Die kinetischen Untersuchungen ergaben folgende Resultate:

1. Salpetersäure, die frei von Salpetrigsäure ist, kann Fe(II)-Ionen nicht oxydieren. Die vollkommene Eliminierung der Salpetrigsäure erfolgt durch Zusatz von p-Nitroanilin.

2. Salpetrigsäure beeinflußt die Oxydation von Ferro-Ionen autokatalytisch, wobei während eines Reaktionszyklus die Konzentration der Salpetrigsäure verdoppelt wird.

3. Auf Grund der kinetischen Befunde wurde ein (vereinfachter) Reaktionsmechanismus der Fe(II)-Oxydation angenommen. Die auf Grund dieses Reaktionsmechanismus aufgestellten Differentialgleichungen wurden programmiert und mittels einer Analog-Rechenmaschine gelöst. Die dabei erhaltenen Kurven entsprechen den experimentell erhaltenen Kurvenverläufen.

4. Es wurde ferner festgestellt, daß geringe Zusätze von Cl^- bzw. Ti^{+4}-Ion die Oxydation der Fe(II)-Ionen inhibieren. Daraus wird geschlossen, daß der Reaktionsmechanismus komplizierter ist als der oben angenommene, daß an der Reaktion wahrscheinlich auch Radikale (OH, HO_2) teilnehmen.

Über den Mechanismus der Passivierung sowie auch über die Ursache des passiven Zustandes des Eisens in Salpetersäure wurden bereits eine große Zahl von Ergebnissen veröffentlicht. Eine zusammenfassende Darstellung, die die diesbezüglichen Arbeiten bis 1930 umfaßt, findet man im Gmelin-Handbuch [1]. Über die bis etwa 1960 publizierten Arbeiten berichtet K. J. Vetter [2].

Vor längerer Zeit wurde mitgeteilt [3], daß beim Übergang des in Salpetersäure eingetauchten Eisens aus dem aktiven in den passiven Zustand intermediär das komplexe Ferronitroso-Ion, $[Fe(NO)]^{2+}$, gebildet wird. Diese Verbindung entsteht auch bei der Oxydation von Fe(II)-Salzlösungen mittels Salpetersäure [4] sowie beim direkten Einleiten von NO in Ferrosalzlösung. Eingehend wurden die Eigenschaften des komplexen $Fe(NO)^{2+}$-Ions von W. Manchot u. Mitarb. [5] beschrieben.

Da wir der intermediären Bildung von $Fe(NO)^{2+}$-Ion für die Deutung des Mechanismus der Passivierung des Eisens durch Salpetersäure eine wesentliche Bedeutung zuschreiben, untersuchten wir die Kinetik der Oxydation von Fe(II)-Ionen mittels Salpetersäure.

1. Versuchsanordnung

Die Fe(II)-Lösungen wurden durch Auflösen von Ferroammoniumsulfat (p. a. E. Merck) in zweifach destilliertem Wasser hergestellt. Lösungen von Fe(II)-Nitrat gaben analoge Versuchsergebnisse, sind jedoch wesentlich instabiler als das Mohrsche Salz. Die Konzentration der Fe(II)-Lösungen war bei allen Versuchen konstant und betrug $7{,}24 \cdot 10^{-4}$ g Ion Fe(II)/4 cm³ Reaktionslösung. Die Vorratslösung der Salpetersäure wurde durch Verdünnen mit zweifach destilliertem Wasser (2 Teile HNO_3 [p. a. E. Merck, $d \sim 1{,}4$] : 1 Teil H_2O) hergestellt und der Gehalt an Säure durch genaue Dichtemessung bestimmt. Die Konzentration der Salpetersäure wurde bei den Versuchen zwischen $1{,}1 \cdot 10^{-2}$ und $1{,}49 \cdot 10^{-2}$ g Mol HNO_3/4 cm³ Versuchslösung variiert. Zusätze von Cl^--, Ti^{4+}- bzw. Fe^{3+}-Ionen waren mit der Salpetersäure in die Reaktionslösung gebracht worden.

In die in einer optischen Küvette befindlichen Fe(II)-Lösung wurde mittels einer Injektionspipette unter Rühren die Salpetersäure rasch hinzugefügt und gleichzeitig mit der Messung begonnen. Abb. 1 zeigt die Versuchsanordnung, mit der die zeitlich veränderliche Lichtabsorption des intermediär gebildeten braungefärbten $Fe(NO)^{2+}$-Komplexes mittels einer Fotozelle gemessen wurde. Gleichzeitig mit der Lichtabsorption wurde auch die zeitliche Änderung der Potentiale einer in der Versuchslösung befindlichen Platinelektrode gegen eine ges. Kalomelelektrode (ges. KNO_3-Brücke) bestimmt. Der zeitliche Verlauf der Reaktion wurde mittels eines Zweifachschreibers registriert. Die Messungen waren gut reproduzierbar.

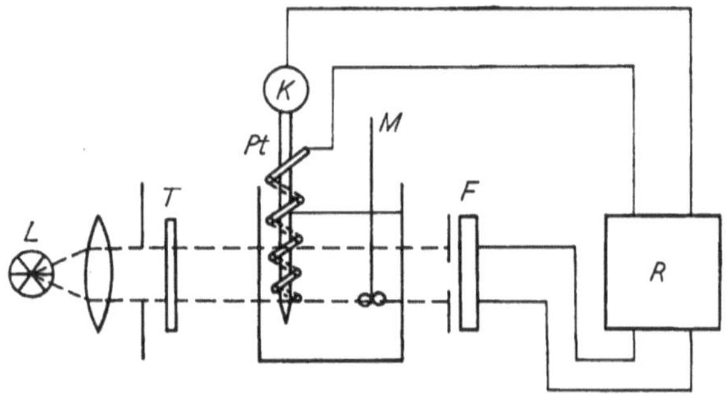

Abb. 1: Versuchsanordnung zur Bestimmung der Lichtabsorption von $Fe(NO)^{2+}$ und der Potentiale im System Fe^{2+}—HNO_3 —HNO_2 in Abhängigkeit von der Zeit.

F — Fotozelle; *K* — Kalomel-Elektrode; *L* — Lichtquelle; *M* — Rührer; Pt — Platin-Elektrode; *R* — Schreiber; *T* — Wärmefilter

2. Versuchsergebnisse

Abb. 2 zeigt den zeitlichen Verlauf der Oxydation von Fe(II)-Ionen in Abhängigkeit von der HNO_3-Konzentration, und zwar die Lichtabsorption des $Fe(NO)^{2+}$-Ions und die Potentiale der Pt-Elektrode. Die Reaktion beginnt mit einer Induktionsperiode, deren Dauer mit abnehmender HNO_3-Konzentration zunimmt. Entsprechend der Braunfärbung durch intermediär gebildetes $Fe(NO)^{2+}$ steigt die Absorptionskurve zunächst langsam, dann immer schneller an, um schließlich einen Maximalwert zu erreichen. Die Höhe und die zeitliche Lage des Maximums der Lichtabsorptionskurve hängen in charakteristischer Weise von der HNO_3-Konzentration ab. Nach Erreichung der maximal gebildeten Menge der $Fe(NO)^{2+}$-Verbindung erfolgt plötzlich deren oxydativer Zerfall, der durch den steilen

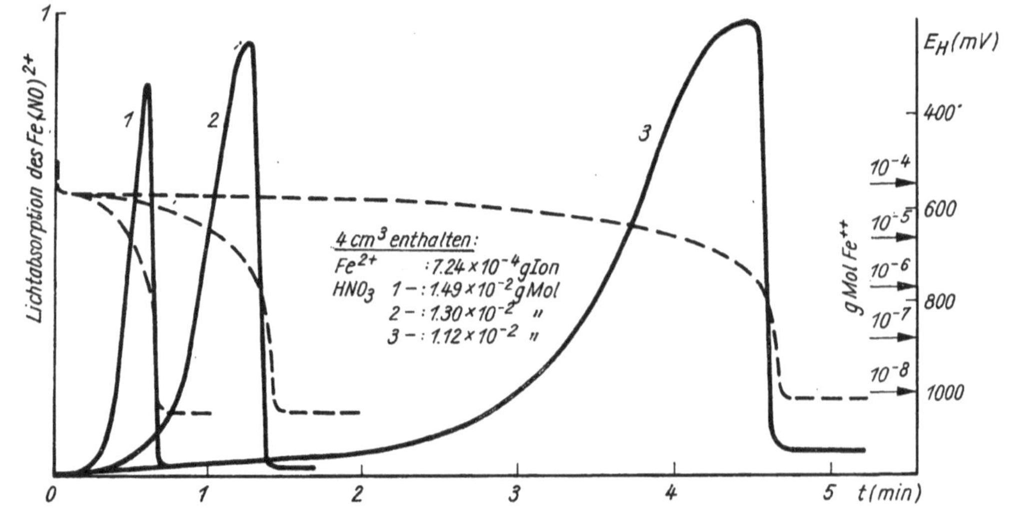

Abb. 2: Zeitlicher Verlauf der Oxydation von Fe^{2+} in Abhängigkeit von der HNO_3-Konzentration

——— Lichtabsorption durch $Fe(NO)^{2+}$, - - - - Potentialänderung ([mV] während der Reaktion

Abfall der Lichtabsorptionskurve angezeigt wird. Die Lichtabsorption erreicht dabei aber nicht den Nullwert, da die Lösung durch Fe(III)-Ionen schwach gelblich gefärbt ist.

Die von der Platinelektrode angezeigten Fe(II)/Fe(III)-Potentiale werden mit zunehmender $Fe(NO)^{2+}$-Bildung langsam positiver und steigen im Bereich des oxydativen Ferronitroso-Zerfalls plötzlich auf +1000 mV(NHE).

Ein Zusatz von Cl^--Ionen zur Reaktionslösung wirkt stark inhibierend auf die Oxydation von Ferro-Ionen durch Salpetersäure, wie dies aus Abb. 3 hervorgeht. Die der Reaktion vorgelegten Mengen an Chlor-Ionen entsprechen einer Größenordnung von nur 10^{-7} g Ion Cl^-/4 cm^3. Bei Zugabe von $18{,}15 \cdot 10^{-7}$ g Ion Cl^-/4 cm^3 erfolgte vollkommene Inhibierung der Reaktion (Abb. 3, Kurve *4*). Die reaktionshemmende Wirkung der Chlor-Ionen erfolgt möglicherweise durch Einfangen von NO, vielleicht als NOCl.

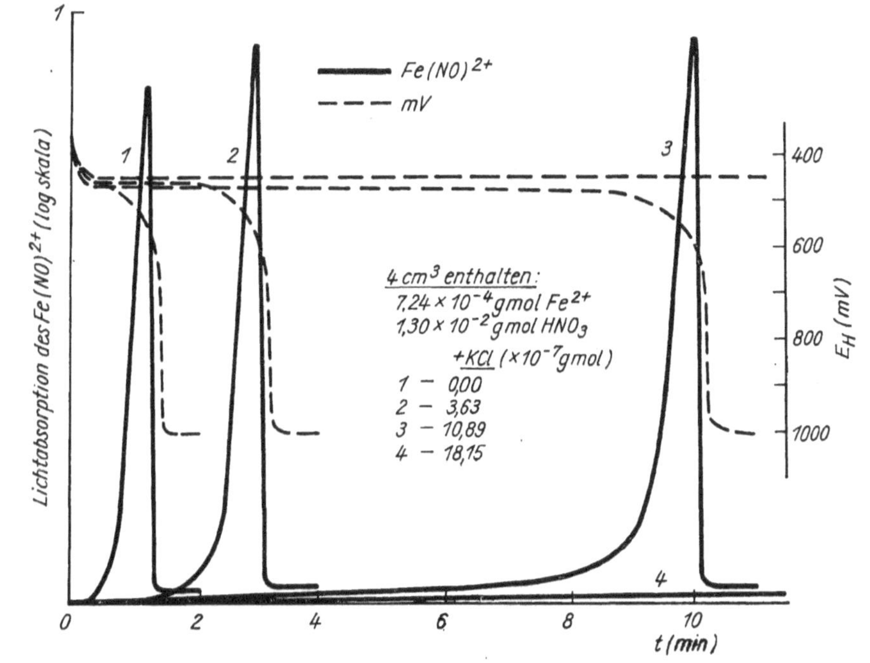

Abb. 3: Verlauf der Oxydation von Fe^{2+} mit HNO_3 bei Zusatz von Cl^--Ionen.

In Abb. 4 ist der reaktionshemmende Einfluß von Ti^{4+} zu erkennen. Die Ti-Menge ist hier in g $TiO_2 \cdot 10^{-5}$/4 cm^3 angegeben, und es ist ersichtlich, daß die Zugabe von $1{,}7 \cdot 10^{-5}$ g TiO_2/4 cm^3 die Reaktion vollkommen hemmt (Abb. 4, Kurve *5*). Der Verlauf der Kurve *5* zeigt eine geringe Zunahme der Lichtabsorption, die durch Gelbfärbung der Reaktionslösung verursacht wird, die möglicherweise die Bildung von Titanperoxihydrat anzeigt. In Abb. 4 ist bei den einzelnen Kurven auch der mittels eines Thermoelementes gemessene kleine Temperaturanstieg (ca. 1,7 °C) eingezeichnet. Diese geringe Temperaturerhöhung fällt in die Phase des oxydativen Zerfalls von Ferronitroso.

Der Einfluß von salpetriger Säure ist für die Reaktion wesentlich, und zwar nicht nur für den Verlauf, vielmehr auch für das Zustandekommen der Oxydation von Ferro-Ionen durch Salpetersäure überhaupt. Bereits E. Abel u. Mitarb. [6] haben bei ihren Untersuchungen des Gleichgewichtes und der Reaktionskinetik des Systems $3\,HNO_2 \rightleftharpoons H^+ + NO_3^- + 2NO + H_2O$ auf den autokatalytischen Einfluß der salpetrigen Säure hingewiesen. Es wurde betont, daß „zufällig" in der Salpetersäure vorhandene salpetrige Säure bei den Versuchen berücksichtigt werden muß [7].

Die autokatalytische Wirkung der salpetrigen Säure auf die Oxydation von Ferro-Ionen durch Salpetersäure wurde auch von E. Schröer [8] festgestellt. Nach diesem Autor wird die Oxydation von Ferro-Ionen überhaupt erst durch salpetrige Säure ermöglicht, und zwar über NO_2, $Fe^{2+} + NO_2 \rightarrow Fe^{3+} + NO_2^-$, wobei die salpetrige Säure weitere Ferro-Ionen in die Reaktion hineinzieht.

E. SCHRÖER arbeitete dabei mit Konzentrationen, bei denen keine sichtbare Ferronitroso-Bildung wahrzunehmen war, um Komplikationen des Reaktionsablaufes zu vermeiden.

Der Einfluß der salpetrigen Säure auf die Passivierung sowie auch die Aktivierung des Eisens in Salpetersäure wurde auch von K. J. VETTER [9] untersucht. Um den Einfluß der salpetrigen Säure auszuschalten, verwendete K. J. VETTER Harnstoff-Zusatz, der den Eintritt der Passivität verzögert.

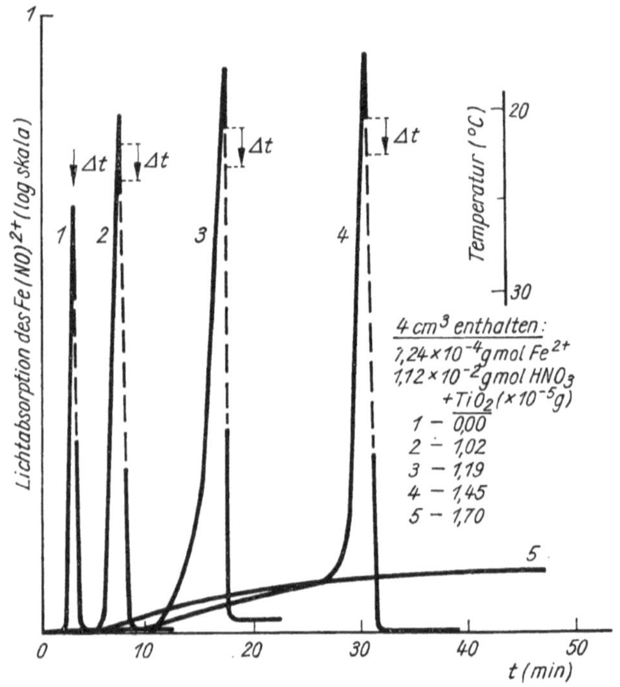

Abb. 4: Einfluß des Zusatzes von Ti^{4+} auf die Kinetik der Reaktion. (Die Temperaturänderung im Gebiet des oxydativen Zerfalls von $Fe(NO)^{2+}$ ist eingezeichnet.)

K. J. VETTER stellte jedoch fest, daß mittels Zusatz von Harnstoff die salpetrige Säure nicht vollkommen eliminiert werden kann, daß der Gehalt an salpetriger Säure nur bis auf 10^{-5} Mol/l erniedrigt wird.

Bei Versuchen des radiolytischen Zerfalls der Salpetersäure konnten A. R. KAZANIJAN u. Mitarb. [10] die anfangs in der Salpetersäure befindliche salpetrige Säure mittels Zusatz von p-Nitroanilin vollkommen eliminieren. Salpetrige Säure reagiert mit p-Nitroanilin, wobei primär das Diazonium-Ion gebildet wird, das mit p-Nitroanilin eine Diazoverbindung eingeht.

Nach E. E. GARCIA [11] reagiert salpetrige Säure mit Azulen-p-Nitroanilin, wobei ein blauer Diazo-Azulenfarbstoff gebildet wird, der die kolorimetrische Bestimmung auch sehr geringer Mengen von Salpetrigsäure (γ/l) ermöglicht. Mittels der von E. E. GARCIA angegebenen Methode bestimmten wir den Gehalt an salpetriger Säure in der von uns verwendeten Salpetersäure ($d \sim 1{,}4$) zu $8{,}5 \cdot 10^{-3}$ Mol HNO_2/l HNO_3. Die bei unseren Versuchen verwendeten Salpetersäure-Lösungen enthielten demnach folgende Mengen an salpetriger Säure:

1. $1{,}49 \cdot 10^{-2}$ g Mol HNO_3—$9{,}0 \cdot 10^{-6}$ g Mol HNO_2/4 cm³;
2. $1{,}30 \cdot 10^{-2}$ g Mol HNO_3—$7{,}7 \cdot 10^{-6}$ g Mol HNO_2/4 cm³;
3. $1{,}12 \cdot 10^{-2}$ g Mol HNO_3—$6{,}76 \cdot 10^{-6}$ g Mol HNO_2/4 cm³.

Wird zu den von uns untersuchten Reaktionslösungen eine der in ihnen enthaltenen salpetrigen Säure äquivalente Menge von p-Nitroanilin zugesetzt, so wird die Oxydation von Ferroion durch Salpetersäure vollkommen inhibiert. Bei Vorlage einer kleineren Menge von p-Nitroanilin, z. B. von $4{,}52 \cdot 10^{-6}$ g Mol p-Nitroanilin/4 cm³ zu einer $9{,}0 \cdot 10^{-6}$ g Mol HNO_2/4 cm³ enthaltenden Reaktionslösung, wird die Reaktionszeit um Stunden verlängert, wie dies in Abb. 5 wiedergegeben wird. Bei den angegebenen Konzentrationen dauert die Induktionsperiode etwa 2 Stunden, dann erfolgt die

Ferronitroso-Bildung und nach etwa 6 Stunden tritt der plötzliche oxydative Zerfall der Komplexverbindung ein. Der in Abb. 5 eingezeichnete Potentialverlauf entspricht der Ferronitroso-Kinetik.

Versuche, bei denen wir Harnstoff, zur teilweisen Eliminierung von salpetriger Säure, der Reaktionslösung zusetzten, waren sehr aufschlußreich bezüglich der Deutung des Reaktionsgeschehens.

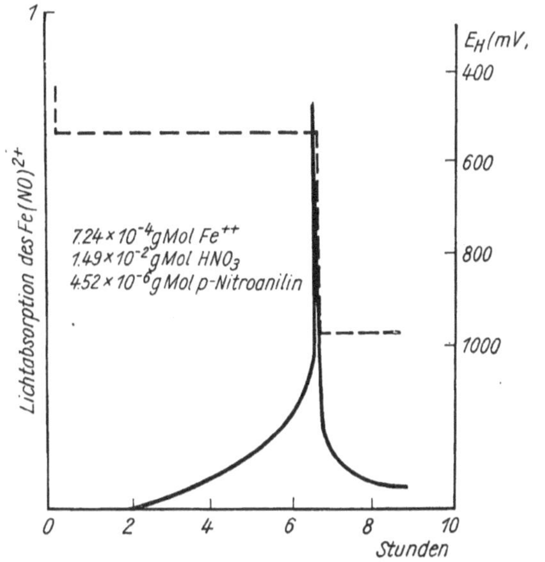

Abb. 5: Inhibierung der Reaktion durch Zusatz von p-Nitroanilin.

——— Lichtabsorption durch $Fe(NO)^{2+}$, - - - - Potentialverlauf an der Pt-Elektrode

Wird nämlich einer Reaktionslösung der Zusammensetzung $7{,}52 \cdot 10^{-4}$ g Ion $Fe^{2+}/4\ cm^3$, $1{,}34 \cdot 10^{-2}$ g Mol $HNO_3/4\ cm^3$ und $1{,}10^{-4}$ g Mol Harnstoff/4 cm^3 zugegeben, so erfolgt bei ruhender Lösung ein Reaktionsverlauf, wie ihn Abb. 6, Kurve *I*, wiedergibt. Wird jedoch die Lösung während der Reaktion gerührt, so tritt am steil abfallenden Ast der Lichtabsorptionskurve eine lebhafte Entwicklung feinstverteilter Gasblasen (N_2, CO_2) auf, die einen unregelmäßig oszillierenden Verlauf der Absorp-

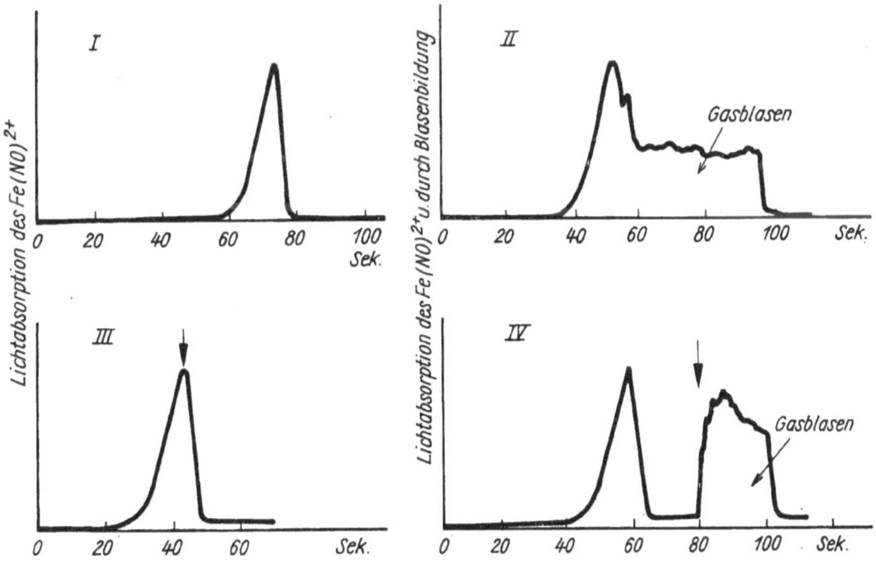

Abb. 6: Einfluß von Harnstoff auf die Oxydation von Fe^{2+} durch HNO_2 bzw. HNO_3.

Kurve I: ruhende Lösung; *Kurve II:* Rühren der Lösung; *Kurve III:* Rühren der Lösung bis zum Pfeil; *Kurve IV:* Ruhende Lösung bis zum Pfeil, dann Rühren tionskurve verursachen (Abb. 6, Kurve *II*). Wird zu Beginn der Oxydation gerührt, dann aber nach Erreichung der maximalen Lichtabsorption die Rührung unterbrochen (Pfeil auf Kurve *III*, Abb. 6), so erfolgt ein stetiger Abfall der Kurve. Wird schließlich bei ruhender Lösung, die einen normalen Verlauf der Lichtabsorptionskurve ergibt, etwa 20 Sekunden nach erfolgter Reaktion die Lösung gerührt (Pfeil auf Kurve *IV*, Abb. 6), so wird wieder lebhafte Gasentwicklung ausgelöst. Diese Versuche ergaben, daß entlang des steil abfallenden Astes der Lichtabsorptionskurve, entsprechend der plötzlichen Oxydation von Ferronitroso, salpetrige Säure gebildet wird. Bei ruhender Lösung erfolgt Übersättigung der Lösung an CO_2 und N_2, wohingegen durch Rühren die Übersättigung aufgehoben wird.

Nach Beendigung der Reaktion ergab die Titration der Lösung einen $KMnO_4$-Verbrauch, der etwa der Hälfte an Salpetrigsäure entsprach, verglichen mit der vorgelegten Menge an Ferro-Ion. In weiteren Versuchen soll die beim oxydativen Zerfall von Ferronitroso entstehende Salpetrigsäure quantitativ mittels p-Nitroanilin bestimmt werden.

3. Diskussion

Auf Grund der von uns erhaltenen Informationen bezüglich des Reaktionsablaufes der Oxydation von Ferro-Ion durch Salpeter- bzw. salpetriger Säure stellten wir folgendes vereinfachtes Reaktionsschema auf:

$$Fe^{++} + HNO_2 \xrightarrow[k_1]{HNO_3} Fe^{+++} + NO_3^- + NO + H_2O, \qquad (1-0)$$

$$Fe^{++} + NO \xrightarrow[k_2]{} Fe(NO)^{++}, \qquad (2-0)$$

$$Fe(NO)^{++} + HNO_3 \xrightarrow[k_3]{HNO_3} Fe^{+++} + NO_3^- + 2\,HNO_2. \qquad (3-0)$$

Die Reaktion (1 — 0) kann auch als Primärreaktion formuliert werden nach

$$[Fe(H_2O)_6]^{2+} + HONO \xrightarrow[k_1']{} [Fe(H_2O)_5OH]^{2+} + NO + H_2O \qquad (1-1)$$

und weiter zu

$$[Fe(H_2O)_5OH]^{2+} + HNO_3 \xrightarrow[k_1'']{} [Fe(H_2O)_6]^{3+} + NO_3^-. \qquad (1-2)$$

Dabei wurde berücksichtigt, daß die Ferro- bzw. Ferri-Ionen als Hexaaquokomplexe vorliegen. Den hydratisierten Ionen kommt dabei eine für den von uns untersuchten Reaktionsablauf wesentliche Rolle zu, wie dies von A. R. UBBELOHDE [12], N. URI [13] und RABINOWITSCH (zitiert nach N. URI [13]) aufgezeigt wurde. Wir versuchten das intermediär gebildete OH-Radikal mittels *ESR* nachzuweisen, was jedoch mit der uns zur Verfügung stehenden Apparatur nicht gelang.

Zur Reaktion (3 — 0) möchten wir bemerken, daß sie jedenfalls als Primärprozeß aufgefaßt werden kann, und zwar nach

$$[Fe(H_2O)_5NO]^{2+} + HNO_3 \xrightarrow[k_3']{H_2O} [Fe(H_2O)_6]^{3+} + NO_3^- + HNO \qquad (3-1)$$

und

$$HNO + HNO_3 \xrightarrow[k_3'']{} 2\,HNO_2. \qquad (3-2)$$

Auch hier gelang es uns nicht, die intermediäre Bildung der untersalpetrigen Säure nachzuweisen. Eine Abhängigkeit des Hydratationsgrades der Ferronitroso-Ionen von der Wasserstoffionenkonzentration wurde bereits von SCHLESINGER und von VALKENBURGH (zitiert nach ABBEGS Handbuch [14]) angenommen.

Wir versuchten für den von uns aufgestellten einfachen Reaktionsverlauf der Oxydation von Ferro-Ion mittels salpetriger bzw. Salpetersäure die entsprechenden Differentialgleichungen mittels eines Analogrechners zu simulieren. Es wurden folgende Gleichungen programmiert:

$$\frac{d[NO]}{dt} = k_1[Fe^{++}][HNO_2] - k_2[Fe^{++}][NO], \tag{1}$$

$$\frac{d[Fe(NO)^{++}]}{dt} = k_2[Fe^{++}][NO] - k_3[Fe(NO)^{++}], \tag{2}$$

$$\frac{d[HNO_2]}{dt} = -k_1[Fe^{++}][HNO_2] + k_3'c[Fe(NO)^{++}], \tag{3}$$

$$\frac{d[Fe^{++}]}{dt} = -k_1[Fe^{++}][HNO_2] - k_2[Fe^{++}][NO], \tag{4}$$

$$\frac{d[Fe^{+++}]}{dt} = k_1[Fe^{++}][HNO_2] + k_3[Fe(NO)^{++}]. \tag{5}$$

Entsprechend der bei unseren Versuchen verwendeten Konzentrationen wurden dem Analogrechner folgende Verhältnisse der Anfangskonzentrationen bzw. der Geschwindigkeitskonstanten vorgelegt:

$$b_{HNO_2} : a_{Fe^{++}} : c_{HNO_3} = 1 : 120 : 1860,$$

$$k_1 : k_2 : k_3 = 1 : 0{,}5 : 0{,}07,$$

wobei die Anfangskonzentration der Salpetersäure c in die Konstante k_3 einging zu $k_3 = k_3' \cdot c$. Der große Überschuß an Salpetersäure gestattet es, deren Konzentration als konstant zu behandeln.

Wurden dem Analogrechner folgende Werte für $k_3 = k_3' \cdot c$ vorgelegt: (1) = 0,050; (2) = 0,075; (3) = 0,100 und (4) = 0,200, so erhielt man einen „Reaktionsverlauf", wie ihn Abb. 7 wiedergibt (Kurven *1* bis *4*).

Die vom Rechner simulierte Teillösung des zeitlichen Verlaufs der Ferri- bzw. Ferro-Ionenkonzentration wurde für den Konzentrationsintervall: $7{,}24 \cdot 10^{-4}$ g Ion Fe^{2+} — 10^{-8} g Ion Fe^{3+} bzw. 10^{-8} g Ion Fe^{2+} — 7,24 g Ion Fe^{3+} in Potentiale der Ferro/Ferri-Elektrode umgerechnet (für k_3 der Kurve *1* und in Abb. 7 gestrichelt eingezeichnet). Es resultiert dabei ein zeitlicher Verlauf Ferro/Ferri-Potentiale, der dem im Versuch erhaltenen analog ist.

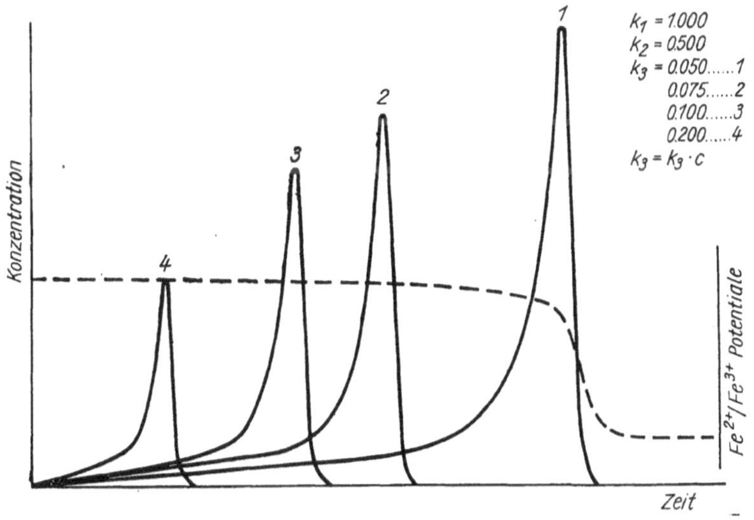

Abb. 7: Mit Analogrechner simulierter „Reaktionsverlauf" im System $Fe^{2+}-HNO_2-HNO_3$.

Die vollausgezogenen Kurven *1–4* entsprechen den „Geschwindigkeitskonstanten" k_3(*1–4*). Die gestrichelte Kurve – Potentiale Fe^{2+}/Fe^{3+} wurde aus der Teillösung für den zeitlichen Verlauf von Fe^{3+}/Fe^{2+} berechnet

Die mit dem Analogrechner erhaltenen Lösungen der auf Grund experimenteller Befunde aufgestellten Differentialgleichungen machen es demnach wahrscheinlich, daß der tatsächliche Reaktionsverlauf der Oxydation von Ferro-Ionen durch salpetrige Säure bzw. Salpetersäure etwa nach den von uns vorgeschlagenen Reaktionsgleichungen erfolgen könnte.

Eingangs wurde erwähnt, daß wir der intermediären Bildung von Ferronitroso-Ionen im Verlauf der Oxydation von Ferro-Ionen durch salpetrige Säure bzw. Salpetersäure als wesentlich betrachten für die Deutung des Passivierungsvorganges des Eisens in Salpetersäure. Auf Grund der von uns erhaltenen Resultate des Studiums der Kinetik der Ferro-Ionen-Oxydation mittels salpetriger Säure bzw. Salpetersäure kommen wir zu folgender Beschreibung der einzelnen Phasen des Passivierungsvorganges:

1. Das Eisen geht in Salpetersäure primär als Ferro-Ion in Lösung, wobei zum Teil atomarer, zum Teil molekularer Wasserstoff gebildet wird,

$$Fe + 2HNO_3 \rightarrow Fe^{++} + 2NO_3^- + 2H(H_2).$$

Dieser Vorgang ist elektrochemischer Natur, entweder als Funktion von Lokalelementen oder als Folge anodischer Polarisation.

2. Der nach 1. gebildete atomare Wasserstoff vermag die Salpetersäure zur salpetrigen Säure zu reduzieren, nach

$$2H + HNO_3 \rightarrow HNO_2 + H_2O.$$

3. Gebildetes Ferro-Ion wird durch salpetrige Säure und Salpetersäure unter intermediärer Bildung von Ferro-Nitroso zu Ferri-Ion oxydiert, bei gleichzeitiger autokatalytischer Bildung von salpetriger Säure. Die während des Passivierungsvorganges intermediär gebildete braune Ferro-Nitroso-Verbindung kann experimentell nachgewiesen werden. Zu diesem Zweck wurde spektralreines Eisen in Salpetersäure (55 Gew.-%) coulostatisch mit $6 \cdot 10^{-5}$ Coulomb/0,07 cm² aktiviert und anschließend wurde die Repassivierungskurve der Eisenpotentiale (gegen eine ges. Kalomelelektrode) mittels eines Speicheroszilloskops (Tektronix) registriert. Gleichzeitig wurde auch die während der Passivierung erfolgende zeitliche Änderung des von der Eisenelektrode reflektierten Lichtes oszilloskopisch aufgezeichnet. Abb. 8 gibt den Potentialverlauf (Kurve *1*) und die Änderung der Intensität des reflek-

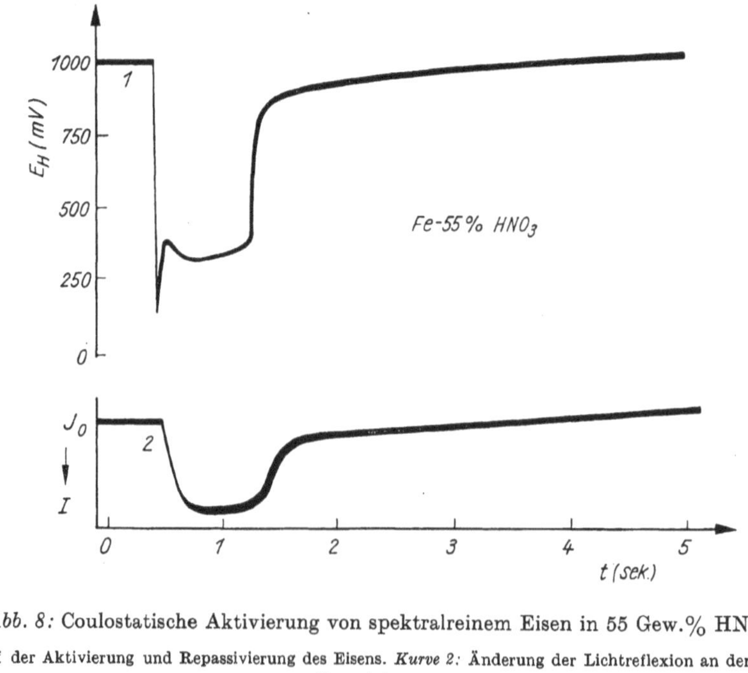

Abb. 8: Coulostatische Aktivierung von spektralreinem Eisen in 55 Gew.% HNO_3.

Kurve 1: Potentialverlauf der Aktivierung und Repassivierung des Eisens. *Kurve 2:* Änderung der Lichtreflexion an der Fe-Elektrode während der Repassivierung tierten Lichtes (Kurve 2) während der Repassivierung des Eisens wieder. Dem Potentialbereich zwischen +125 mV (E_h) und +380 mV (E_h) entspricht eine Abnahme der Intensität des reflektierten Lichtes, die durch intermediär gebildetes Ferro-Nitroso-Ion verursacht wird. Bemerkenswert ist dabei, daß sowohl die Potentiale des Eisens als auch die Lichtreflexion nach dem oxydativen Zerfall des Ferronitroso-Komplexes nur langsam die Werte, die dem passiven Eisen entsprechen, erreichen.

4. Die Versuche über die Kinetik im System Fe(II)-Ion $-HNO_2-HNO_3$ ergaben, daß infolge des oxydativen Zerfalls von Ferronitroso, neben salpetriger Säure, als Endprodukt der Reaktion Ferri-Ion gebildet wird. Es ist daher nicht unbegründet anzunehmen, daß der Potentialsprung, den die Repassivierungskurve des Eisens in Salpetersäure aufweist, ebenfalls der Bildung von Ferri-Ion als Endprodukt der Ferronitroso-Oxydation entspricht. Diese Annahme führt notwendigerweise zu Überlegungen bezüglich der Löslichkeit der während der Repassivierung gebildeten Ferri-Verbindung

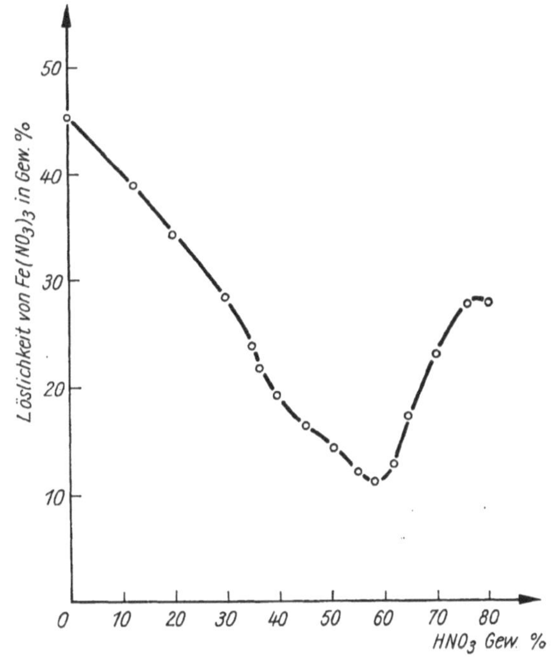

Abb. 9: Aus den Löslichkeitsbestimmungen des Systems $Fe(NO_3)_3-HNO_3-H_2O$ von G. MALQUORI bzw. ROBINSON und CAMERON.

Berechnete Löslichkeitskurve: Ordinaten Gew.% $Fe(NO_3)_3$/100 g Lösung, Abszissen: Gew.% HNO_3

Dazu liegen Untersuchungen von ROBINSON und CAMERON sowie auch von G. MALQUORI [15] über das Löslichkeitsverhalten des Ferrinitrates in wäßrigen Lösungen der Salpetersäure vor. Wir übertrugen die von G. MALQUORI in ternärer Darstellung gegebene Löslichkeitskurve des Systems Ferrinitrat—Salpetersäure—Wasser in das System Gew.% $Fe(NO_3)_3$ — Gew.% HNO_3, das in Abb. 9 wiedergegeben ist. Es ist ersichtlich, daß dem Konzentrationsbereich der Salpetersäure, in dem das Eisen passiv wird, d. h. zwischen ca. 40 Gew.% HNO_3 und 67 Gew.% HNO_3, ein Minimum der Löslichkeit des Ferri-Nitrates entspricht. Entlang der Löslichkeitskurve bis zu etwa 77 Gew.% HNO_3 hat der Bodenkörper die Zusammensetzung $Fe(NO_3)_3 \cdot 9H_2O$, während bei höheren HNO_3-Konzentrationen das Hexahydrat ausfällt. Es ist daher nicht ausgeschlossen, daß die Passivität des Eisens in Salpetersäure durch eine Ferrinitrat-$9H_2O$-Schicht verursacht wird. Diesbezügliche Ansichten wurden bereits früher vertreten [16]. Es soll in der Folge versucht werden, experimentelle Informationen über die Zusammensetzung und die Struktur der passivierenden Schicht, die sich auf Eisen in Salpetersäure bildet, zu erhalten.

Literatur

[1] GMELIN-Handbuch d. anorg. Chem., Eisen (A) 313ff. (1929/33)
[2] K. J. VETTER, Elektrochemische Kinetik, 602ff. (Springer 1961)
[3] M. KARŠULIN, Z. Elektrochem., **41**, 664 (1935)
[4] GMELIN-Handbuch, Eisen (B), Ltg. 4, 82 (1931)
[5] W. MANCHOT, zitiert nach: ABBEGS Handbuch d. anorg. Chem., Bd. IV, 98ff. (1930)
[6] E. ABEL, H. SCHMID und S. BABAD, Z. phys. Chem., **136**, 135 (1928)
[7] E. ABEL, H. SCHMID, Z. phys. Chem., **136**, 430 (1928)
[8] E. SCHRÖER, Z. anorg. allg. Chem., **202**, 382 (1931); Z. phys. Chem., A **176**, 20 (1936)
[9] K. J. VETTER, Z. Elektrochem., **55**, 279 (1951)
[10] A. R. KAZANIJAN, F. J. MINER, A. K. BROWN, Trans. Farad. Soc., **66**, 2192 (1970)
[11] E. E. GARCIA, Analyt. Chem., **39**, 1605 (1967)
[12] A. R. UBBELOHDE, Discuss. Farad. Soc. No 2, 215 (1947)
[13] N. URI, Discuss. Farad. Soc. No 2, 217 (1947)
[14] H. I. SCHLESINGER und H. B. VAN VALKENBURGH, J. Am. chem. Soc. **51**, 1323 (1929); zitiert nach: ABBEGS Handb. d. anorg. Chem., Bd. IV B, 103 (1929)
[15] ROBINSON und CAMERON, J. phys. Chem., **13**, 62 (1909); zitiert nach G. MALQUORI, Rend. Acad. Lincei, **9**, 324 (1929)
[16] H. BRACONNOT, Ann. Chim. Phys. [2] **52**, 286 (1833); Pogg. Ann., **29**, 174 (1833); E. BOUTMY, Chateau, Cosmos, **19**, 117 (1861); W. HELDT, J. phys. Chem., **90**, 257 (1863), zitiert nach GMELIN-Handbuch, Eisen, Teil A, 355 (1929/33)

M. Prazák and Trinh Xuan Sen

The G. V. Akimov State Research Institute for Material Protection, Prague

THE KINETICS OF REDOX REACTIONS ON SOME PASSIVE METALS AND ALLOYS

Abstract

The polarization resistance of the $[Fe(CN)_6]^{4-}-[Fe(CN)_6]^{3-}$ redox system on passive metals, Fe, Ni, Cr, FeCr16, FeCr18Ni9, FeCr18Ni10Mo2Ti and Pt was measured in dependence on the concentration of the redox system and of OH^- ions. The order of the catalytic efficiency of the studied metals was determined from the calculated exchange currents.

Electron transfer by tunneling is considered to be the controlling process in the reaction; its rate depends on the composition and thickness of the barrier film and on the surface concentration of the species from which the transfer takes place.

The importance of the dependences obtained for corrosion processes is pointed out and examples of the use of these measurements are given for the monitoring of extraction corrosion of alloys, for control of artificial passive films and for the determination of the surface roughness factor of steel.

1. Introduction

The presence of oxide films on passive metal electrodes exerts a strong influence on the kinetics of redox reactions occuring on them. This phenomenon is important not only when solid metals are used as electrode materials, but also during corrosion processes, especially when the process is controlled by the partial cathodic reaction.

Tomashov [1] dealt with the kinetics of the reduction of oxygen on Fe from this point of view. Reactions of other redox systems on Fe—Cr alloys were studied by Mackrides [2], who found a substantial decrease in the redox reaction exchange currents in the presence of passive oxide films. Under these conditions the contribution from the activation control of the reaction rate increases at the expense of the more common control by transport processes in the solution and the importance of the catalytic efficiency of the surface of passive metals for the redox reaction course thus increases.

The aim of the present paper is to compare the catalytic efficiencies of passive Fe, Cr, Ni and their alloys towards the redox reaction of a well-known system, ferrocyanide-ferricyanide. This system has already been employed with Fe by Moshtev [3], Lovreček [4], Agarwal [5] and Schultze [6]; its behaviour on Pt has been examined by Balej [7] and other authors.

The expression "catalytic efficiency" is used instead of electrode "activity", in order to avoid confusion with the concept of the active and passive states of metals.

2. Experimental

For the determination of the polarization resistance of the electrodes, the POLAROTRON instrument (VPZ Běchovice) was employed; the principle of the instrument operation is galvanostatic polarization by alternating symmetrical rectangular impulses with a frequency of $f = 0.25$, 0.5 or 1 Hz. The maximum deviation from the electrode stationary potential was ± 6 mV. A three-electrode circuit was used; a Pt gauze placed 3—5 mm away from the electrode surface served as the reference electrode. The maximum error of the instrument in measuring the ohmic resistance was 5%.

Potentiokinetic polarization curves were messured with a PLE 60 potentiostat (VPZ Běchovice) and recorded on an X-Y recorder ARITMA BAK 4 in both directions of the potential scan at a rate of 0.5 V/min.

The electrodes were made from the metals given in Table 1. Their working surface was planar, with a surface limited to 1 cm². They were placed vertically in the measuring cell, in a quiescent solution. The electrode surface was pretreated by grinding with fresh emery paper No. 400 and

Table 1: Electrode materials

No.	Metal	ČSN	Composition
2	Fe	—	99.99% Fe
3	Fe(C)	11320	construction steel, 0.1% C
4	Ni	—	99.95% Ni
5	Cr	—	99% Cr
6	FeCr16	—	16% Cr single-crystal, total impurity content below 0,05%*)
7	FeCr17	17041	cca 17% Cr, 0.12% C
8	FeCr18Ni9	17241	cca 18% Cr, 9% Ni, 0.1% C
9	FeCr18Ni10Mo2Ti	17347	cca 18% Cr, 10% Ni, 1.8% Mo, 0.4% Ti, 0,1% C

*) The single-crystal was produced in the Institute of Physical Chemistry of the Polish Academy of Sciences, Warsaw.

cleaned with ether. This procedure was preferred over polishing or chemical etching, which would lead to greater differences in the specific surface (roughness) — in view of the differences in the hardness and chemical composition of the studied metals. Still, it must be borne in mind that differences in the specific surfaces of various metals may be as much as 20% even when a sufficiently hard abrasive material is employed. Although these differences concern only the surface microgeometry, they can play a role with reactions whose rate is activation-controlled. These relationships are discussed in ref. [8], dealing with the determination of the surface roughness factor for steels by measuring their polarization resistance in a similar electrolyte.

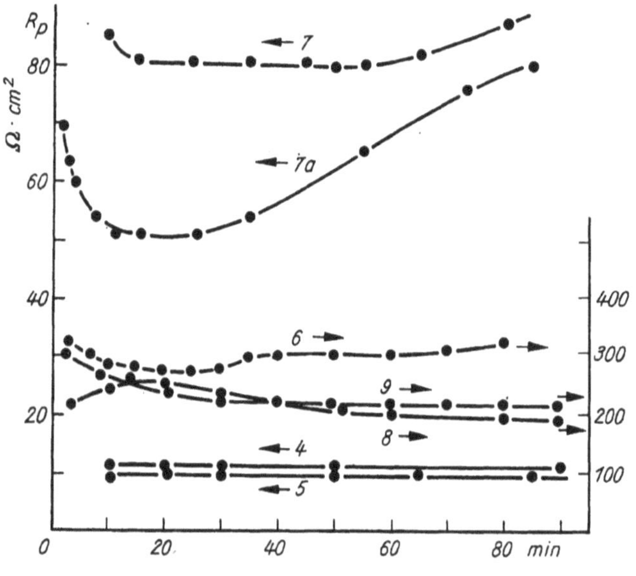

Fig. 1: The time-dependence of the polarization resistance in a solution 0.1 M in $K_4[Fe(CN)_6]$, $K_3[Fe(CN)_6]$ and NaOH. The numbers denote the materials according to Table 1. The arrows denote the R_p scale.

Solutions were prepared from p.a. compounds and replaced frequently because of their potential instability. The measurements were carried out in the air, at a temperature of 23 $\pm$ 2 °C.

The polarization resistance values were read after 40 minutes, when a quasistationary state was already established with most of the metals (see Fig. 1), depending chiefly on the stabilization of the properties of the passive film in the given solution. Regarding the differences among the metals compared, the thickness of the passive films was not determined.

Iron electrodes were immersed for 10 minutes in 1 M NaOH prior to the measurement and were thus pre-passivated; otherwise they sometimes underwent corrosion in the measuring solution.

3. Results and Discussion

The average values of the polarization resistance of the studied metals in a solution containing 0.1 M $K_4[Fe(CN)_6]$, 0.1 M $K_3[Fe(CN)_6]$ and 0.1 M NaOH are given in Table 2. For the sake of informative comparison, the value for non-activated platinum is also given, extrapolated to infinite frequency from the $R_p - \sqrt{f}$ dependence. This dependence is distinctly linear (see Table 3), in agreement with the assumed diffusion control of the kinetics of this process.

With the other passive metals R_p was not extrapolated as the (small) frequency dependence is no longer controlled by transport processes in the solution. The R_p values are generally higher than on Pt and thus indicate the controlling role of the electrode reaction in the kinetics of the process. The frequency dependence is then chiefly due to the electrode impedance in the R_p measuring procedure

Table 2: The polarisation resistence, R_p, and the exchange current, i_0, for 0.1 M $K_4[Fe(CN)_6]$ and $K_3[Fe(CN)_6]$ in 0.1 M NaOH, on passive metals
(The average values in the 40th minute)

No.	Metal	R_p [$\Omega \cdot cm^2$]	i_0 [A/cm^2]
1	Pt	1.2*)	2.2×10^{-2}
2	Fe	2,250	1.2×10^{-5}
3	Fe(C)	1,100	2.4×10^{-5}
4	Ni	11	2.4×10^{-3}
5	Cr	9.5	2.7×10^{-3}
6	FeCr 16	300	8.7×10^{-5}
7	FeCr 17	81	3.2×10^{-4}
8	FeCr 18 Ni 9	230	1.1×10^{-4}
9	FeCr 18 Ni 10 Mo 2 Ti	220	1.2×10^{-4}

*) Extrapolated to infinite frequency.

Table 3: The dependence of the polarization resistance, R_p, on the frequency in 0.1 M $K_4[Fe(CN)_6]$, $K_3[Fe(CN)_6]$ and 0.1 M NaOH

No.	Metal	Frequency Hz: 0.25	0.5	1	∞
		R_p [$\Omega \cdot cm^2$]			
1	Pt	3.7	2.9	2.4	1.2
2	Fe	2,250	2,150	2,000	—
4	Ni	10.9	10.35	9.95	—
5	Cr	9.5	9.1	8.6	—
7	FeCr 17	81	79.5	78	—
8	FeCr 18 Ni 9	182	180	178	—
9	FeCr 18 Ni 10 Mo 2 Ti	216	212	209	—

employed, which does not provide for compensation of the electrode capacity. For this reason, the R_p values measured at 0.25 Hz are further used for the passive metals, since the error is smallest at this value.

In the table are simultaneously given the values of the exchange current density, j_0, in the given solution, calculated from relationship [9]

$$j_0 = \frac{RT}{zFR_p}$$

assuming that R_p corresponds only to the charge-transfer reaction.

The studied metals can be classified roughly in three groups on the basis of comparison of the exchange current values, i.e. from the aspect of the catalytic efficiency of their surfaces for a single redox system. Relatively the highest exchange current for the redox reaction is exhibited by nickel and chromium, the lowest by pure iron. Alloys of these three components exhibit a medium value.

The high polarization resistance of iron is, however, decreased by other impurities in technical metals in addition to chromium and nickel, such as C and Mn; this also holds for commercial FeCr alloys.

The given order of the polarization resistance values for the studied metals can be compared with the corresponding order of the thicknesses of passive films on these metals. For the thickness of the oxide passive film in a borate buffer (pH = 8.42) and in the corresponding potential region SATO and KUDO [14] give about 3.3 nm (33 Å) for Fe and 0.8 nm for Ni on the basis of ellipsometric and other measurements. The passive film on Cr is assumed to be even thinner [9]. For pure metals (Fe or Pt) a relationship between the film thickness and the reaction rate was found by VETTER and SCHULTZE [10, 11], who explain this phenomenon (in the reaction of oxygen evolution) by electron-transfer through a tunnel effect between an adsorbed reaction intermediate and the metal atoms passing through electronically non-conductive oxide. The film thickness, as a barrier, has strong effect on this process. Simultaneously it is probable that the whole film thickness need not act as the barrier, but only that part, which exhibits poor conductivity due to a low content of defects, as explained in detail in ref. [12]. However, the barrier thickness alone does not determine the magnitude of the passing current, as this value also depends on the redox system concentration (see Table 4 and Fig. 2) and the concentration of OH^- ions (Table 5).

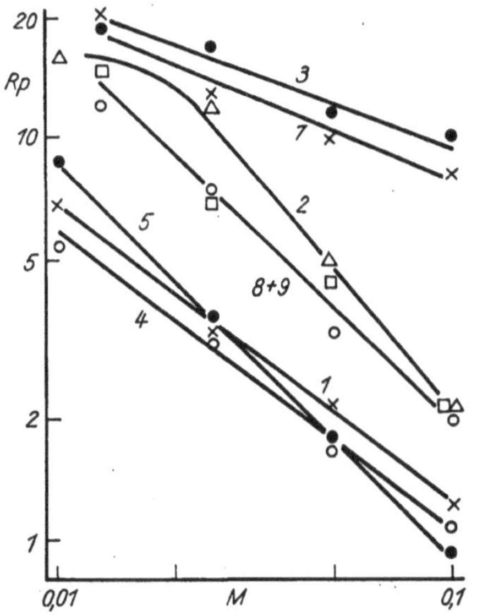

Fig. 2: The dependence of the polarization resistance on the redox system concentration in 0.1 M NaOH. The numbers denote the materials according to Table 1.

The R_p values are shifted by order of magnitude (except Pt).

With thicker synthetic Fe_2O_3 films LOVREČEK and BABIČ [4] found a reaction limiting current smaller than on Pt. They suppose that the cause of this phenomenon is depletion of mobile electric charge carriers in the films at higher potentials. On the contrary, this phenomenon did not occur with relatively thinner passive films examined here: the approximately-determined limiting currents (anodic and cathodic) on passive Fe (see Fig. 3) were virtually identical for other metals and even for platinum. Thus it can be assumed that the reaction rate in this case was controlled — after attaining a sufficient overpotential — by transport processes in solution and not in the oxide film.

Table 4: The dependence of the polarization resistance in 0.1 M NaOH on the equimolar concentrations of $K_4[Fe(CN)_6]$ and $K_3[Fe(CN)_6]$

No.	Metal	Redox *M*: 0.01	0.025	0.05	0.1
1	Pt	7[a])	3.6[a])	2.2[a])	1.2[a])
2	Fe	16,000	12,200	5,000	2,250
3	Fe(C)	1,860[b])	1,750	1,180	1,100
4	Ni	54	32	17	11
5	Cr	87	36	18	9.5
7	FeCr 17	207[b])	135	102	81
8	FeCr 18 Ni 19	1,200[b])	750	330	230
9	FeCr 18 Ni 10 Mo 2 Ti	1,500[b])	700	450	220

[a]) extrapolated to infinite frequency.
[b]) redox 0.0125 M.

The determination of the limiting current from the potentiodynamic curve (Fig. 3) is, of course, imprecise because of the interference from the maximum reaching to the region of oxygen evolution on the anodic branch. The current value in the straight part obtained with the reverse potential scan, however, gave sufficient information.

The dependence of the polarization resistance of the passive metals on the redox system concentration in the solution (Fig. 2) was found to be sufficiently linear, except for pure Fe, where, however, the highest measured resistance already lay in the region of decreased measuring precision. The slopes of these straight lines are not identical, but the cause for the differences cannot so far be unambiguously determined.

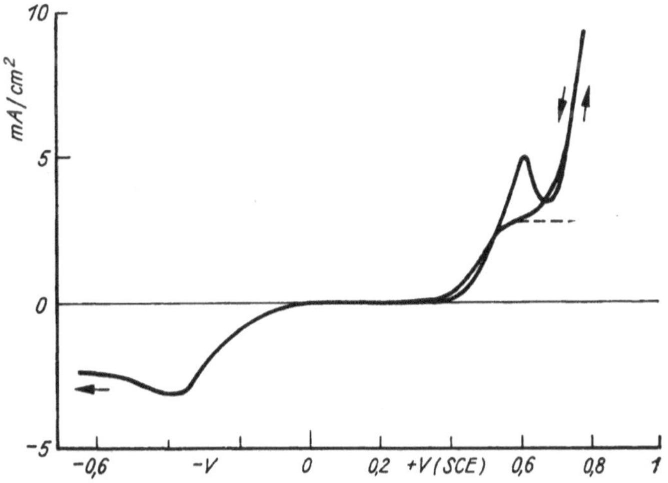

Fig. 3: A potentiodynamic polarization curve of passive Fe in a solution 0.1 M in $K_4[Fe(CN)_6]$, $K_3[Fe(CN)_6]$ and NaOH. The scan rate, ± 0.5 V/min.

4. The Effect of the OH^- Ion Concentration

The hydroxide concentration also affects the magnitude of the polarization resistance of metals with passive oxide films at a constant concentration of the redox system (see Table 5). The resistance generally decreased with increasing OH^- concentration. This effect appears to be substantially smaller for platinum, which indicates that it is not caused by changes in the behaviour of the redox system or in the ohmic resistance of the solution. It also cannot be assumed that the decrease in the resistance always corresponds to a simultaneous corrosion reaction course, since there is e.g. no corrosion of nickel under these conditions. Chromium forms an exception, since the potentiostatic measurements

Table 5: The dependence of R_p [$\Omega \cdot cm^2$] on the NaOH concentration with 0.1 M $K_3[Fe(CN)_6]$ and $K_4[Fe(CN)_6]$ at 0.25 Hz

Metal	NaOH M:		
	0.1	1	5
Pt	3.7	3.2	3.8
Fe	2,250	810	175
Fe(C)	1,090	367	160
Ni	11	4.1	3.9
Cr	9.5	5	0
FeCr16	300	68	55[a])
FeCr17	80	85	56[a])
FeCr18Ni9	230	47	6.2
FeCr18Ni10Mo2Ti	220	54	11

[a]) The values increase considerably in time.

carried out indicated that it can be corroded in the given potential range in 5 M NaOH: the parts of the polarization curve corresponding to the oxidation of ferrocyanide and to the Cr oxidation to hexavalency (corrosion) are connected. However, the resistance with chromium increases under these conditions, indicating the formation of a film of solid reaction products which were not identified in any detail.

In the other cases, when the resistance decreases with increasing OH^- concentration, an explanation should be sought in variations of the properties of the passive film surface. In accordance with VETTER's concepts [10] it can be assumed that an increase in the OH^- ion concentration at the passive film surface makes the electron-transfer by means of the described tunnel effect easier.

5. Extraction of Components of Passive Films on Alloys

With alloys of iron containing 16—17% Cr, the polarization resistance in 5 M NaOH increased so intensely and for such a long time (see Fig. 1) that the R_p values given in Table 1 (valid for the 40th minute) have only limited importance. In view of the instability of passive chromium in this strongly alkaline (and oxidizing) medium, this phenomenon can be explained by selective dissolution (extraction corrosion) of the chromium component from the passive film formed by mixed Cr and Fe oxide. The extraction of the chromium component must necessarily lead to an increase in the ratio of the oxides in the film in favour of Fe_2O_3. This change becomes manifest, in agreement with the behaviour of the other studied metals, in an increase in the polarization resistance.

This concept of the course of extraction corrosion and of its effect on the catalytic properties of the surface of passive alloys was also verified during experiments with extraction in strongly acidic media. In Table 6 is given a survey of changes in the polarization resistance of alloys, always measured in the same solution, but before and after extraction with 10% H_2SO_4 + 0.2 M $FeNH_4(SO_4)_2$

Table 6: The changes in the polarization resistance of alloys after 24 hour extraction in a solution containing 10% H_2SO_4 and 0.2 M $FeNH_4(SO_4)_2$

Metal	Initial R_p	R_p after extraction
FeCr16	300	107
FeCr17	81	50
FeCr18Ni9	230	96
FeCr18Ni10Mo2Ti	220	90

for 24 hours. The presence of the salt of tervalent iron ensured that the alloy remained in a passive state. In this acidic solution, iron is extracted from the surface of the passive alloy in contrast to the previous case, as was verified earlier by sensitive gamma-spectrometric analysis of the acid in which samples previously irradiated by neutrons in an atomic reactor were corroded [13]. Thus the Cr/Fe ratic in the passive oxide film increases, which is manifested — again in agreement with the behaviour of the pure metals — by a decrease in the polarization resistance of the sample after the extraction.

This procedure for the detection of extraction processes has an advantage that it can react to variations in the composition in a very thin zone (several atoms) on the passive film surface, which can hardly be achieved by other methods. On the other hand, this is, of course, an indirect method and so far its results cannot be quantitatively evaluated.

6. The Possibility of Evaluation of Artificial Passive Films

The catalytic efficiency of the surface of the studied passive metals is, of course, sensitive not only to changes in naturally formed passive films but also to the properties and their changes in deliberately — technically — formed films. For example, with electrodeposited chromium films on steels a suitable technological procedure permits formation of a thicker passivating layer, increasing the resistance of the system against corrosion. The detection of the existence of such a layer or information on its thickness cannot be obtained by common measuring methods because of its very small thickness. The differences in the polarization resistance of samples of chromium with natural passive films and of artificially passivated samples are shown in Table 7. The results indicate the possibility of relatively simple detection of the presence of such technologically formed passive films and even a possibility of certain quantitative evaluation of their properties.

Table 7: The polarization resistances of chromium with natural and artificial passive films in solutions:
1) 0.1 M in $[Fe(CN)_6]^{4-}$, $[Fe(CN)_6]^{3-}$ and NaOH (in the 5th minute),
2) 0.2 M in $[Fe(CN)_6]^{4-}$ and $[Fe(CN)_6]^{3-}$ in a phosphate buffer (pH = 10.5, in the 40th minute)

Sample	R_p [$\Omega \cdot cm^2$]	
	solution 1	*solution 2*
compact chromium, natural state	9.5	6.8
electrolytic chromium, natural state	9.7	13
electrolytic chromium, passivated	51	710

7. Conclusions

1. The catalytic efficiency of the surfaces of passive metals and alloys and of their natural passive films on the rate of the redox reaction of $[Fe(CN)_6]^{4-}$—$[Fe(CN)_6]^{3-}$ in the region of activation control depends on the chemical composition of the metal. It is substantially lower with all the studied metals than with platinum. Consequently, the potential region of the activation control

of the reaction rate is substantially broadened at the expense of control by transport process in solution for the redox processes on passive metals.
2. The catalytic efficiency is higher on metals with thinner passive films.
3. The exchange current of the redox reaction increases with increasing concentration of the redox system on all the studied electrodes.
4. The catalytic efficiency increases with increasing OH^- ion concentration.
5. With the studied electrodes, the limiting current of the redox reaction is controlled by transport processes in the solution and not in the passive film.
6. The measurement of the polarization resistance for a suitable redox system permits sensitive detection of changes in the surface properties of passive metals, caused by extraction corrosion of alloy components or by artificial formation of passive films.

References

[1] N. D. TOMASHOV, G. P. CHERNOVA, „*Passivnost i Zashchita Metallov ot Korrozii*", „Nauka" Publ., Moscow 1965
[2] A. C. MACKRIDES, J. Electrochem. Soc., **111**, 392, 400 (1964)
[3] R. V. MOSHTEV, Electrochim. Acta, **16**, 2039 (1971)
[4] B. LOVREČEK, R. BABIČ, "*Extended Abstracts of the 25th ISE Conference*", Brighton 1974, 86
[5] H. P. AGARWAL, S. QURESHI, ibid., 121
[6] J. W. SCHULTZE, P. KOHL, M. M. LOHRENGEL, U. STIMMING, ibid., 73
[7] A. REGNER, J. BALEJ, Coll. Czechoslov. Chem. Commun., **26**, 237 (1961)
[8] M. PRAZÁK, B. EREMIÁŠ, Corrosion Sci., **12**, 463 (1972)
[9] K. J. VETTER, „*Elektrochemische Kinetik*", (Springer) Berlin—Göttingen—Heidelberg 1961
[10] K. J. VETTER, J. W. SCHULTZE, Ber. Bunsenges. physik. Chem., **77**, 945 (1973); "*Extended Abstracts of the* 25th ISE *Conference*", Brighton 1974, 85
[11] J. W. SCHULTZE, K. J. VETTER, Electrochim. Acta, **18**, 889 (1973)
[12] M. PRAŽÁK, Collection Czechoslov. Chem. Commun., **29**, 1773 (1964)
[13] M. PRAŽÁK, D. KUCHYŇKA, M. LUKAŠOVSKÁ, "*Proceedings of the IInd International COMECON Conference on Corrosion and Anticorrosion Protection*", Prague, April 1975, 365 (in Russian)
[14] N. SATO, K. KUDO, Electrochim. Acta, **19**, 461 (1974)

W. Forker

Technische Universität Dresden, Sektion Chemie

ZUM AUFBAU UND VERHALTEN VON PASSIVIERENDEN OXIDSCHICHTEN

Die weitere Aufklärung des Mechanismus von Passivierungsvorgängen und des Stabilitätsverhaltens von Passivschichten erfordert u. a. eine Erweiterung unserer Kenntnisse über den stofflichen und strukturellen Aufbau sowie die physikalischen und elektrochemischen Eigenschaften der Passivschichten. Eine genauere Kenntnis der Korrelationen zwischen den genannten Vorgängen und Eigenschaften sowie dem Stabilitätsverhalten kann einerseits dazu beitragen, die jeweiligen Korrosionsvorgänge, die auf eine Behinderung der Ausbildung oder die Zerstörung von passivierenden Schichten zurückzuführen sind, besser zu verstehen, andererseits aber auch Maßnahmen zur Optimierung des Stabilitätsverhaltens solcher Schichten gezielter zu fördern.

Ich möchte heute in der Hauptsache über einige Fragen des Aufbaus und Verhaltens von Passivschichten auf Eisen sprechen und über einige diesbezügliche Ergebnisse berichten, die wir im Rahmen von Untersuchungen zur Stabilität von Eisen und niedriglegierten Stählen in neutralen bis schwach alkalischen Lösungen im Zusammenhang mit praktischen Fragestellungen durchgeführt haben.

Der passive Zustand des Eisens ist einer dünnen, porenfreien Oberflächenschicht von Eisenoxiden zuzuschreiben. Ihre inverse Spinell-Struktur kann dem Magnetit $Fe_{3-\Delta}O_4$ zugeordnet werden. Verschiedene experimentelle Befunde weisen darauf hin, daß die Schicht im Bereich der weitgehenden Potentialunabhängigkeit des Korrosionsstromes ein mit dem Abstand von der Metall/Oxid-Grenzfläche von etwa Fe_3O_4 bis $Fe_{2,67}O_4$ (bzw. Fe_2O_3) kontinuierlich zunehmendes Eisendefizit aufweist, wobei der Übergangsbereich verhältnismäßig schmal zu sein scheint (Abb. 1).

Eisen (1)	*Passivoxid (2)*			*Lösung (3)*
	(I)		*(II)*	
Fe	Fe_3O_4	$Fe_{3-\Delta}O_4$	$Fe_{2,67}O_4$ (γ-Fe_2O_3)	H_2O, $K^{z+}\cdot solv$, $A^{z-}\cdot solv$

Abb. 1: Phasenschema einer passiven Eisenelektrode.

Ein Strom durch die passive Eisenelektrode kann innerhalb der Oxidschicht sowohl ein Ionenstrom als auch ein Elektronenstrom sein. Der Ladungstransport durch die Phasengrenze Metall/Oxid kann durch Eisenionen- und gegebenenfalls Elektronendurchtritt erfolgen. Er ist offenbar nur wenig gehemmt, wenn das metallische Eisen mit seinem Gleichgewichtsoxid ($\approx Fe_3O_4$) im Kontakt steht. Für den Ladungstransport durch die Phasengrenze Oxid/Elektrolytlösung bestehen die drei Möglichkeiten:

a) Metallionendurchtritt (Korrosion),
b) Sauerstoffionendurchtritt (Schichtaufbau bzw. -abbau) und
c) Elektronendurchtritt (Redoxvorgang).

Zur Deutung der Ursachen für die Ortsabhängigkeit der stöchiometrischen Zusammensetzung der Passivschicht auf Eisen haben K. J. Vetter und etwas später C. Wagner ein detailliertes, hauptsächlich thermodynamisch orientiertes Modell entwickelt. Die wesentliche Basis dieses Modells ist die Vorstellung, daß bei Ausschluß eines Elektronendurchgangs durch die Phasengrenze Oxid/Lösung, d. h. bei Abwesenheit eines Redoxsystems in der Lösung, thermodynamisches Gleichgewicht der Elektronen in der Oxidschicht angenommen wird. Nach diesem Modell ist für den stationären Zustand des passiven Eisens eine vom Elektrodenpotential unabhängige Ortsabhängigkeit der Schichtzusammensetzung zu erwarten. Die mit dem Elektrodenpotential veränderliche Gesamtdicke der Passivschicht sollte im wesentlichen nur unterschiedlichen Schichtanteilen an Fe_2O_3 zugeschrieben werden können, während der Anteil an Fe_3O_4 konstant bleiben sollte. Beim instationären Schichtaufbau sollte man andererseits eine Verminderung des Fe_3O_4-Anteils wegen der gegenüber dem stationären Zustand erhöhten Stromdichte und damit Feldstärke erwarten.

Im Zusammenhang mit diesem Problem haben wir im Rahmen von Untersuchungen in neutralen und schwach alkalischen Medien coulometrische und ellipsometrische Bestimmungen der Passivschichtdicke und -zusammensetzung durchgeführt.

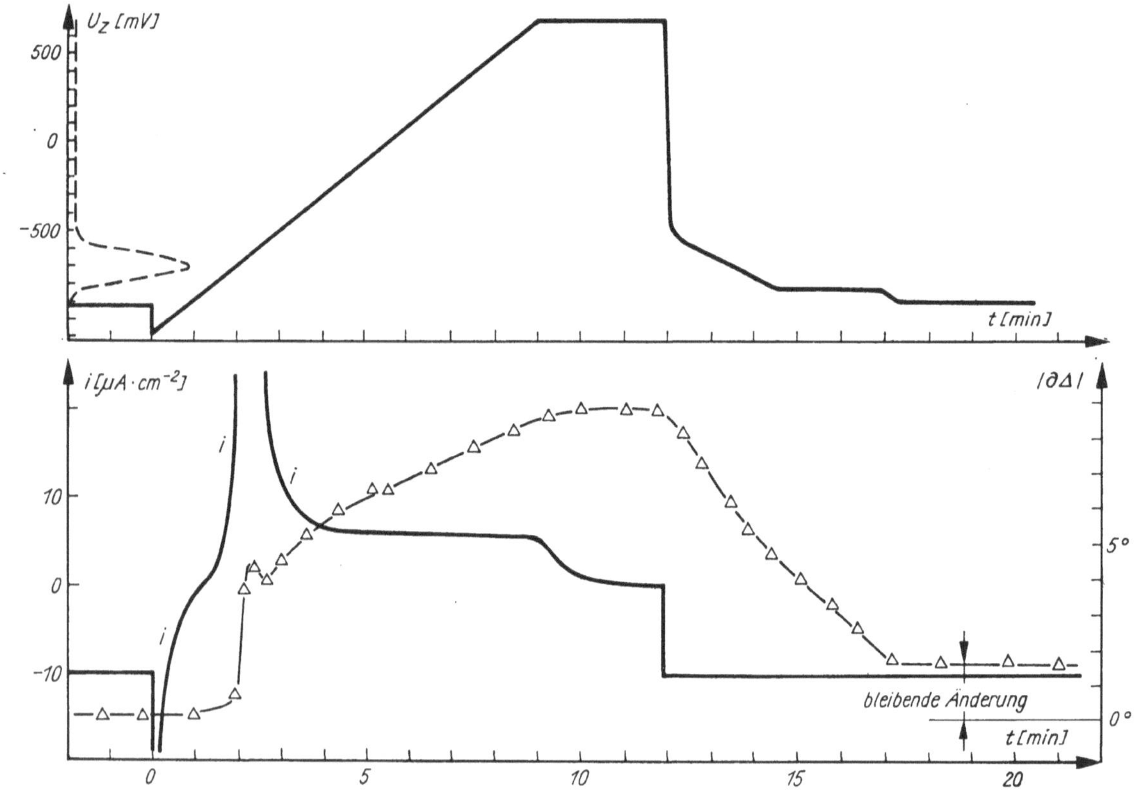

Abb. 2: Zeitabhängigkeit des Elektrodenpotentials U_Z (SCE), der Stromdichte i und der Schichtdicke $\delta\Delta$ bei potentiodynamischer Oxydation und anschließender galvanostatischer Reduktion.

Reinsteisen, elektropoliert, Boratpufferlösung pH 8,4, gestrichelte Linie — Strom-Potential-Kurve.

Als Untersuchungslösungen dienten vorrangig Borsäure/Borat-Pufferlösungen vom pH-Wert 6,8 und 8,4. In diesem pH-Wertbereich ist gegenüber sauren Medien in stärkerem Maße zu beachten, daß die anodische Auflösung des Eisens zur Ausfällung von Hydroxiden oder Oxidhydraten führen kann, die das einwandfreie Aufwachsen des Passivoxids erheblich stören können.

In Abb. 2 sind zur Verdeutlichung dieses Sachverhaltes Ergebnisse der potentiodynamischen Oxydation und anschließenden galvanostatischen Reduktion einer elektropolierten Reinsteisenprobe bei pH 8,4 dargestellt. Die ellipsometrischen Meßwerte zeigen im Bereich des aktiven Strompeaks ein Schichtdickenmaximum. Weiterhin fällt auf, daß der optische Bezugszustand (die oxidfreie Metalloberfläche) durch Reduktion nicht zurückerhalten werden kann. Diese Erscheinung ist meist dann zu beobachten, wenn die Elektroden einige Zeit im Aktivbereich verweilten. Bei potentiostatischem Einschaltvorgang wird im Passivbereich der Ausgangszustand durch kathodische Reduktion hingegen in der Regel vollständig zurückerhalten. Aus Voruntersuchungen dieser Art ergab sich, daß zur weitgehenden Herabminderung von Störungen infolge Hydroxidbildung durch die in Lösung gehenden Eisenionen drei Maßnahmen geeignet sind:

1. Verwendung rotierender Elektroden und strömender Elektrolytlösungen,
2. Zugabe von Komplexbildnern,
3. Erzeugung der Passivschicht durch potentiostatischen Einschaltvorgang.

Es bereitet gegenwärtig noch erhebliche Schwierigkeiten, die kontinuierliche örtliche Änderung der stöchiometrischen Zusammensetzung einer Passivschicht im Kontakt mit einer Elektrolytlösung in situ exakt zu verfolgen. Sowohl die ellipsometrische als auch die coulometrische Methode liefern neben Aussagen über die Schichtdicke vorrangig quantitative Informationen über zwei diskrete Schichtanteile, die den beiden Grenzzusammensetzungen Fe_3O_4 und Fe_2O_3 zugeordnet werden können.

Abb. 3 zeigt den Potential- und Schichtdickenverlauf als Funktion der kathodisch geflossenen Ladungsmenge Q für Passivschichten verschiedener Materialien in Boratpufferlösungen vom pH-

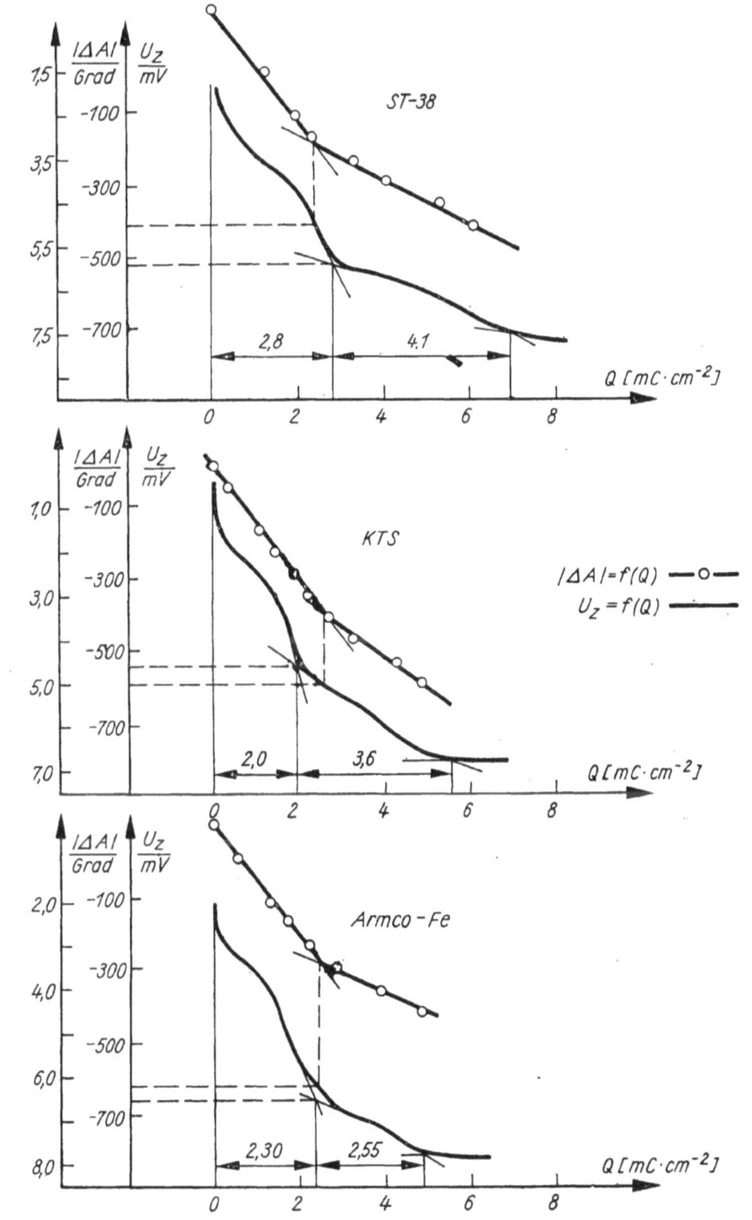

Abb. 3: Abhängigkeit des Elektrodenpotentials U_Z (SCE) und der Oxidschichtdicke (proportional [ΔA]-Analysatoränderung) von der kathodischen Ladungsmenge Q. Stromdichte: 100 µA/cm². Boratpufferlösung pH 6,8.

Wert 6,8. Sowohl die coulometrischen als auch die ellipsometrischen Kurven lassen übereinstimmend und sehr deutlich die beiden Schichtanteile erkennen. Der Abbau der Fe_3O_4-Schicht erfolgt mit geringerer Geschwindigkeit, was auf eine kleinere Stromausbeute ihrer Reduktion zurückgeführt werden kann.

Den folgenden Betrachtungen liegen hauptsächlich Schichtanalysen zugrunde, die auf coulometrischem Wege durchgeführt wurden. Als Maß für die beiden Schichtanteile bzw. ihrer relativen

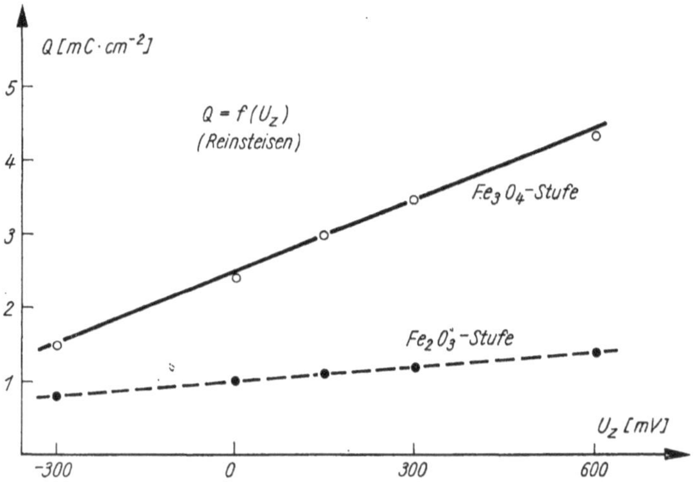

Abb. 4: Ladungsmengen Q der Fe_2O_3- und Fe_3O_4-Reduktion als Funktion des Elektrodenpotentials U_Z (SCE) im Passivbereich.
Reduktionsstromdichte: 50 μA/cm²; Boratpufferlösung pH 8,5.

Änderungen dienen die zur Reduktion erforderlichen Ladungsmengen. Für die Angabe absoluter Schichtdicken sind noch genauere quantitative Kenntnisse der Dichten, der Rauhigkeiten sowie der Stromausbeute und des Mechanismus der Reduktion erforderlich.

Abb. 4 zeigt Ergebnisse der coulometrischen Analyse der Schichtzusammensetzung als Funktion des Passivpotentials. Die Ergebnisse beziehen sich auf quasistationäre Passivschichten, die ausgehend vom Ruhepotential durch sprunghafte Einstellung des Passivpotentials erzeugt wurden.

Abb. 5 zeigt Ergebnisse bei potentiodynamisch durchgeführter Schichtbildung, ausgehend von einer quasistationären Schicht im Anfangsteil des Passivbereichs.

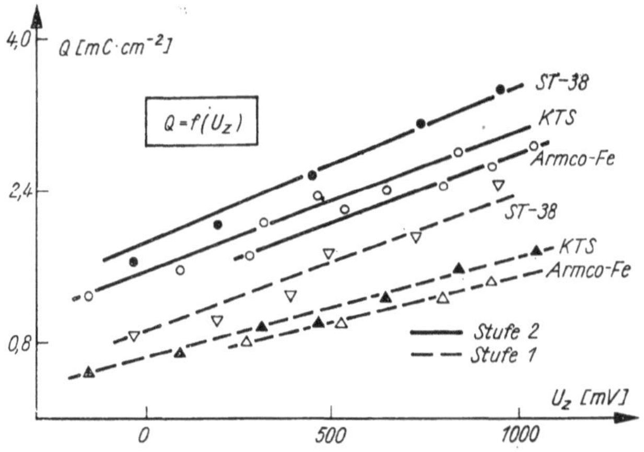

Abb. 5: Ladungsmengen Q der Fe_2O_3 (Stufe 1) und Fe_3O_4 (Stufe 2) Reduktion als Funktion des Elektrodenpotentials U_Z (SCE) bei potentiodynamischer Schichtbildung.
dU_Z/dt = 1,67 mV/s; Reduktionsstromdichte: 100 μA/cm²; Boratpufferlösung pH 6,8.

Als Maß für die beiden Oxidanteile ist jeweils die für ihre kathodische Reduktion erforderliche Ladungsmenge angegeben. Es wird deutlich, daß in allen Fällen sowohl die Menge an Fe_2O_3 als auch die Menge an Fe_3O_4 linear mit dem Passivpotential zunimmt.

Dieses zunächst unerwartete Ergebnis wirft die Frage auf, inwieweit der coulometrisch bestimmte Zuwachs an Fe_3O_4 tatsächlich beim anodischen Schichtaufbau und nicht etwa erst durch die Reduktion des darüberliegenden Fe_2O_3 entsteht. In früheren Arbeiten anderer Autoren, z. B. COHEN u. Mitarb., wurde mehrfach gefunden, daß die Reduktion des Fe_2O_3 mit nahezu hundertprozentiger Stromausbeute zu einem Durchtritt von Fe^{2+}-Ionen in die Lösung führt. Das Auftreten von Fe^{2+}-Ionen in der Lösung wurde von uns auch mit der Ring-Scheiben-Elektrode bestimmt. Insbesondere aber treten keine signifikanten Unterschiede im Fe_3O_4-Schichtanteil auf, wenn der Abbau des darüberliegenden Fe_2O_3 einerseits reduktiv, andererseits außenstromlos erfolgt. Weiterhin tritt häufig der Fall auf, daß annähernd gleich langen Fe_2O_3-Stufen ganz verschiedene Fe_3O_4-Stufen entsprechen. Alle diese Befunde berechtigen zu dem Schluß, daß beim anodischen Aufbau der Passivschicht neben

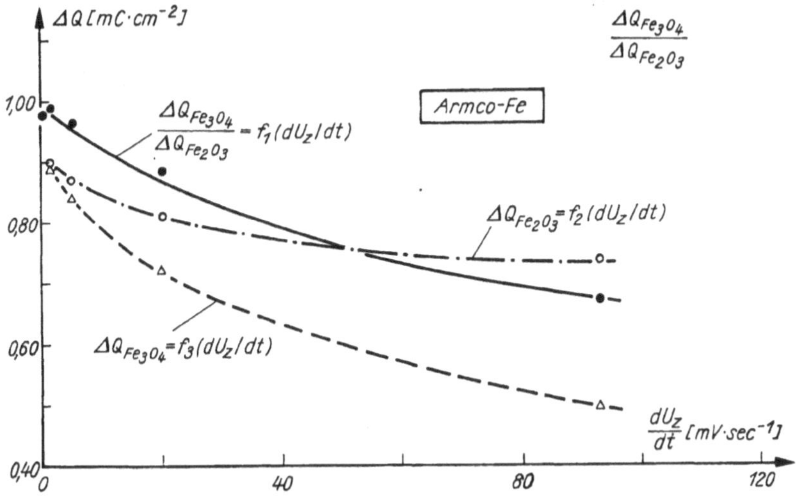

Abb. 6: Ladungsmengen ΔQ der Fe_2O_3- und Fe_3O_4-Reduktion als Funktion der Potentialänderungsgeschwindigkeit bei potentiodynamischer Passivschichtbildung.

dem Fe_2O_3-Anteil auch der Schichtanteil mit niedrigerem Oxydationsgrad zunimmt. Dieser Sachverhalt läßt die Frage aufkommen, inwieweit der stöchiometrische Aufbau der Passivschicht des Eisens rein thermodynamisch im Sinne eines Gleichgewichtes der Elektronen mit dem Ionengitter determiniert sein kann oder inwieweit möglicherweise kinetische Einflüsse, Einflüsse der Art des Ladungstransportes in der Passivschicht eine Rolle spielen könnten.

Die Annahme eines Elektronengleichgewichtes $\Delta\tilde{\mu}_e = \Delta\mu_e - F \cdot \Delta\varphi = 0$ im Schichtanteil niederen Oxydationsgrades ($Fe_{3-\Delta}O_4$ mit $\Delta < 0{,}33$) — kurz Fe_3O_4-Anteil genannt — führt unter der Voraussetzung eines eindeutigen Zusammenhangs zwischen Oxydationsgrad und chemischem Potential der Elektronen im stationären Zustand zu einer vom Elektrodenpotential unabhängigen Fe_3O_4-Schichtdicke, da wegen der Konstanz des Passivstromes die Feldstärke elektrodenpotentialunabhängig ist. Der Strom in der Passivschicht müßte dann ausschließlich ein Fe^{3+}-Ionenstrom sein. Ein mit dem Elektrodenpotential zunehmender Schichtanteil an Fe_3O_4 ist offenbar nur denkbar, wenn $\Delta\tilde{\mu}_e = \Delta\mu_e - F \cdot \Delta\varphi \neq 0$, d. h. kein Elektronengleichgewicht vorliegt. Ein Gradient des elektrochemischen Potentials der Elektronen aber hätte zur Folge, daß Elektronen am Ladungstransport durch die innere Schicht beteiligt sein müssen, und zwar auch dann, wenn bei Abwesenheit eines Redox-Systems in der Lösung ein Elektronenübergang an der Grenzfläche Oxid (Fe_2O_3)/Lösung nicht stattfindet. Eine solche Möglichkeit erscheint gegeben, wenn im inneren Teil der Passivschicht die Migration von Fe^{2+}-Ionen in Betracht gezogen wird, so daß für die Gesamtstromdichte $i = i_{Fe^{3+}} + i_{Fe^{2+}} + i_{e^-}$ gilt. Auf die prinzipielle Möglichkeit einer Migration von Fe^{2+}-Ionen im Fe_2O_4-Anteil sollte auch daraus geschlossen werden können, daß die anodische Oxydation sowohl im Vorpassiv — als auch im Transpassivgebiet zu Fe^{2+}-Ionen führt, wofür die Beteiligung von $Fe2^{+}$-Ionen beim Ladungstransport durch die Oxidschicht eine notwendige Voraussetzung ist.

Die Mitberücksichtigung eines Fe^{2+}-Ionenstromes im Modell der Passivschicht macht verständlich, daß ein Schichtbildungsstrom sowohl zum Weiterwachsen des Fe_2O_3- als auch des Fe_3O_4-Anteils führen kann. Die Anteile beider Teilschichten am Gesamtwachstum sollte dann wesentlich von den Überführungszahlen aller am Ladungstransport durch die Passivschicht beteiligten Ladungsträger bestimmt werden. Auf der Basis eines solchen Modelles sehen wir auch eine Möglichkeit, Ergebnisse von Untersuchungen zum potentiodynamischen Passivschichtwachstum zu deuten.

Aus Abb. 6 ist zu erkennen, daß der relative Anteil des Zuwachses an Fe_2O_3 mit steigender Vorschubgeschwindigkeit, d. h. zunehmender Feldstärke, zunimmt. Wir sehen die mögliche Ursache für dieses Verhalten in der Feldstärkeabhängigkeit der Überführungszahlen der Fe^{2+}- und Fe^{3+}-Ionen in dem Sinne, daß die Migrationsgeschwindigkeit der Fe^{3+}-Ionen mit der Feldstärke exponentiell stärker zunimmt als die der Fe^{2+}-Ionen.

Diese Ergebnisse weisen erneut deutlich auf die Notwendigkeit hin, zu einer qualitativ wie quantitativ besseren Kenntnis des Ionentransports und des Einflusses des elektrischen Feldes in passivierenden Schichten und damit zu einem besseren Verständnis der Rolle auch dieser Faktoren im Hinblick auf das Stabilitätsverhalten zu gelangen.

CH. WEISSMANTEL, P. MORZECK und G. REISSE

Technische Hochschule Karl-Marx-Stadt

PASSIVIERUNGS- UND KORROSIONSSCHUTZ MITTELS PLASMA- UND IONENINDUZIERTER PROZESSE

Zusammenfassung

Zur Erzeugung von Schutzschichten auf Werkstoffoberflächen werden zunehmend Verfahren entwickelt und eingesetzt, bei denen durch Ionen ausgelöste Anregungs-, Stoß- und Zerstäubungsprozesse von wesentlicher Bedeutung sind. Dies gilt insbesondere für die Verfahren der Ionenzerstäubung (Ion-sputtering), der Ionenplattierung (Ion-assisted film formation), der Ionenimplantation sowie der plasmainduzierten Schichtabscheidung aus der Gasphase (Plasma-induced-deposition). Ausgehend von einer vergleichenden Darstellung dieser Verfahren werden Anwendungsmöglichkeiten für die Passivierung und den Korrosionsschutz erörtert. Die Untersuchung wichtiger Eigenschaften von Al_2O_3-, Si_3N_4- und Polytetrafluoräthylen-Schichten lieferte Aufschlüsse über die Struktur und den Wachstumsmechanismus sowie für einen praktischen Einsatz bei der Veredlung hochwertiger Bauelemente und Werkstücke.

1. Einführung

Dünne Schichten und Oberflächenüberzüge können durch physikalische Abscheidung, chemische Reaktion oder durch kombinierte Anwendung beider Wirkprinzipien hergestellt werden. Für die letzten beiden Fälle ist allgemein die Bezeichnung „Reaktive Schichtbildung" gebräuchlich, deren Varianten in der Tab. 1 dargestellt sind. In der industriellen Anwendung werden klassische CVD-Verfahren, elektrochemische Beschichtungsmethoden und die Vakuumbedampfung am häufigsten angesetzt, wie z. B. in der Halbleiterindustrie und für Zwecke des Korrosionsschutzes.

Tabelle 1: Reaktive Beschichtungsverfahren

Methode	Variante
Chemische Abscheidung aus der Gasphase (chemical vapour deposition)	Gas- und Transport-Reaktionen plasmainduzierte Reaktionen einschließlich Glimmpolymerisation
Gas-Festkörper und Flüssig-Festkörper-Reaktion	Deckschichtwachstum anodische und kathodische elektrochemische Beschichtungsprozesse
Reaktive Zerstäubung (reactive sputtering) und ionengestützte Beschichtung (ion-assisted film formation)	Reaktive Gleichstrom- und Hochfrequenzzerstäubung Reaktives Ionenplattieren Reaktive Ionenstrahl-Beschichtung

Neuerdings werden jedoch verstärkte Anstrengungen unternommen, um die reaktive Schichtbildung durch Ionenstöße oder in Plasmaentladungen anzuregen. In diesem Zusammenhang spricht man von einer plasma- bzw. ioneninduzierten Abscheidung [1, 2]. Wenn Ionen ausreichender Energie erzeugt und auf Festkörper geschossen werden, so findet infolge der Impulsübertragung ein Abstäuben von Material statt. Diese Abtragung wird seit langem für die Reinigung von Oberflächen genutzt. Von den neueren Anwendungen der Ionenätzung sind das Abdünnen von Proben für elektronenmikroskopische Untersuchungen sowie die als leistungsfähige Methode der Analytik von Oberflächenschichten entwickelte Sekundärionenmikroanalyse (SIMS) anzuführen [8].

Weiterhin eignen sich ionentechnologische Verfahren in besonderem Maße zum Auf- oder Einbringen von Stoffen in oberflächennahe Gebiete der Festkörper. Die Wirkung der Ionen, die unter Umständen nur einen kleinen Bruchteil der am Prozeß beteiligten Teilchen ausmachen, läßt sich im wesentlichen durch die folgenden Teilprozesse beschreiben.

1.1. Stoßprozesse auf einem Target (Ion-sputtering)

Dieser Effekt wird beim Beschichten mittels Ionenzerstäubung angewendet. Im Ergebnis des Auftreffens energiereicher Ionen auf eine Targetelektrode werden sekundäre Neutralteilchen, Ionen und Cluster freigesetzt, deren Anzahl von der Massenzahl der Ionen und des Targets, von der Ionenenergie, vom Inzidenzwinkel und von der chemischen Natur der Partner sowie in geringerem Maße auch von der Temperatur abhängt. Als Zerstäubungsrate S bezeichnet man die Gesamtzahl der abgestäubten Atome und Ionen je einfallendes Ion; für Argon-Ionen mit Energien von 10 keV liegt die Zerstäubungsrate von Metallen beispielsweise zwischen 2 und 10 Atome/Ion in Abhängigkeit vom Targetmaterial.

Theoretisch läßt sich die Zerstäubung durch Stoßfolgen primärer und sekundärer Teilchen erklären [14]. Im Gegensatz zur thermischen Verdampfung wird bei Legierungen und Verbindungen ein weitgehend stöchiometrischer Materialabtrag beobachtet, jedoch können, besonders beim Abstäuben der ersten Atomlagen, erhebliche Abweichungen von dieser Regel auftreten. Die Energie der abgestäubten Teilchen weist eine breite Verteilung auf; im Mittel beträgt sie 10 eV, jedoch tritt ein kleiner Anteil wesentlich energiereicherer Teilchen auf. Auf Grund dieser hohen mittleren Energie zeichnen sich aufgestäubte Schichten durch eine gute Haftfestigkeit aus, außerdem können sekundär durch die abgestäubten Teilchen chemische Prozesse aktiviert werden, was bei der chemisch-reaktiven Zerstäubung genutzt wird [3].

1.2. Stoßprozesse im Plasma (Plasma-induced-deposition)

Bei der Beschichtung mittels chemischer Abscheidung durch Reaktionen in der Gasphase (CVD-Verfahren) werden häufig hohe Temperaturen benötigt, was zu unerwünschten Nebenwirkungen im Substrat führen kann. Läßt man die Reaktion hingegen in einer elektrischen Gleichspannungs- oder Hochfrequenzentladung ablaufen, so bildet sich ein Plasma aus, in dem infolge von Elektronen- und Ionenstößen eine zusätzliche Aktivierung der Reaktionspartner erfolgt. Dadurch kann die Abscheidung bei erheblich niedrigeren Temperaturen des Reaktionsgefäßes und speziell der Substrate durchgeführt werden. Derartige plasmainduzierte CVD-Verfahren wurden erfolgreich für die Abscheidung von Polymerschichten durch Glimmpolymerisation und Si_3N_4-Schichten entwickelt [16].

1.3. Stoßprozesse im Substrat und in der wachsenden Schicht (Ion assisted film formation)

Ionen genügender Energie sind imstande, in die Oberflächenschicht von Festkörpern einzudringen, wobei bei kristallinen Stoffen eine ausgeprägte Bevorzugung bestimmter Richtungen (Kanäle) mit geringer Packungsdichte der Gitterbausteine festzustellen ist. In der Halbleitertechnik wird diese Erscheinung zur Dotierung mittels Ionenimplantation bereits technisch genutzt. Bei der Abbremsung der Ionen werden sekundär zahlreiche Atome von ihren normalen Gitterpositionen verschoben, wodurch bei hoher Teilchendichte eine Gefügeänderung sowie eine erhöhte Festkörperreaktivität hervorgerufen wird. Analoge Prozesse finden in einer wachsenden Aufdampfschicht statt, wenn den Neutralteilchen ein Anteil energiereicher Ionen beigemengt wird. Man bezeichnet diese Verfahren allerdings nicht sehr glücklich als „Ion-plating"; besser erscheint uns die Bezeichnung „Ion-assisted film formation", d. h. ionengestützte Beschichtung.

Die gezielte Nutzung sekundärer Gefügeänderungen, Zerstäubungsprozesse und der erhöhten Festkörperreaktivität für das Ein- und Aufbringen von passivierenden oder härtenden Schichten sehr guter Haftfestigkeit steht noch am Anfang, doch es läßt sich einschätzen, daß diesen Verfahren in der künftigen Entwicklung der Oberflächentechnik eine große Bedeutung zukommen wird.

2. Anlagen und Verfahren

Die bisher verfügbaren Anlagen und Verfahren der Ionentechnologie sind vorwiegend für Anwendungen im Bereich der Elektronik und des wissenschaftlichen Gerätebaues entwickelt worden. Bei einer vergleichenden Diskussion der Anwendungsmöglichkeiten für die Passivierung und den Korrosionsschutz auch in anderen Gebieten der Technik sind daher vor allem die Besonderheiten herauszuarbeiten, die sich als Forderungen an die künftige apparative Gestaltung ergeben. Dies betrifft in erster Linie die Problematik einer Steigerung der Leistungsfähigkeit hinsichtlich Ionenstrom und Behandlungsfläche sowie die häufig auftretende Notwendigkeit einer gleichmäßigen Behandlung der Oberfläche geometrisch unterschiedlich geformter Bauteile. Am weitesten fortgeschritten ist die technische Entwicklung gegenwärtig bei den Verfahren der Ionenzerstäubung („sputtering"), die in Plasma- und Ionenstrahlmethoden untergliedert werden können.

2.1. Plasma-Zerstäubung

Bei ihr werden die Ionen im Beschichtungsraum durch eine selbständige elektrische Entladung zwischen dem Target und der Substrathalterung erzeugt. Die wesentliche Verbesserung gegenüber der früheren Kathodenzerstäubung besteht in der Einführung eines zusätzlichen Magnetfeldes, das mit Hilfe äußerer Spulen senkrecht oder parallel zu den Elektroden aufrechterhalten wird. Da die Elektronen in diesem Feld auf helizentrische Bahnen abgelenkt werden, läßt sich auch bei Drücken von $5 \cdot 10^{-4}$ bis $1 \cdot 10^{-2}$ Torr Argon noch eine stabile Entladung mit einer für technische Beschichtungsprozesse ausreichenden Ionenstromdichte betreiben. Das ist für die Qualität der erzeugten Schichten wichtig, denn die relativ große Energie der abgestäubten Teilchen wird am Substrat nur wirksam, wenn die mittlere freie Weglänge größer oder gleich dem Abstand Target—Substrat ist. Für das Aufbringen elektrisch leitender, insbesondere metallischer Schichten bevorzugt man wegen der einfachen Ausführung den Betrieb mit Gleichspannung von einigen kV. Kommerziell sind Anlagen bis zu mehreren m² Beschichtungsfläche mit teilautomatisiertem Betrieb entwickelt worden.

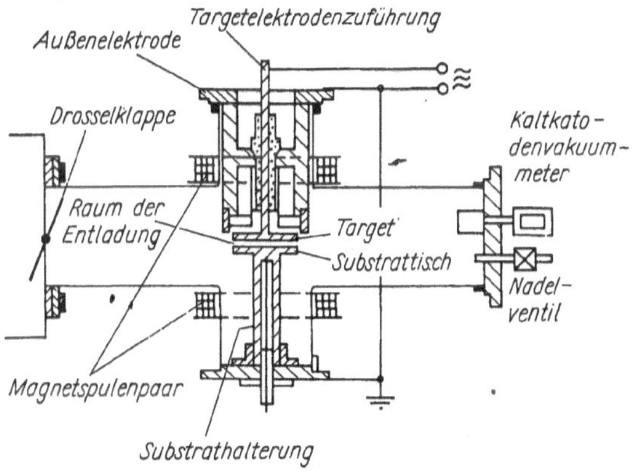

Abb. 1: Schema der Hochfrequenz-Plasmazerstäubung („plasma-sputtering"; *PS*).
Teilschnitt durch die in Abb. 2 dargestellte Apparatur.

Bei der Abscheidung isolierender Passivierungsschichten führt der Gleichstrombetrieb zu Schwierigkeiten, da sich Raumladungen aufbauen, die eine merkliche Abnahme der Beschichtungsrate und eine mangelhafte Schichtqualität zur Folge haben. Man benutzt dann die Hochfrequenz-Plasma-Zerstäubung, die in Abb. 1 schematisch und in Abb. 2 in einer für Laborzwecke entwickelten Ausführungsform dargestellt ist. In einer früheren Arbeit konnte gezeigt werden, daß die Kompensation der Targetaufladung bei Hochfrequenzbetrieb nicht, wie vielfach angenommen, durch eine echte Umpolung in jeder zweiten Halbwelle, sondern durch einen Diffusionseffekt bei geringer negativer Targetspannung zustande kommt [4]. Die abgebildete Anlage, die in mehreren Exemplaren gebaut und unter anderem für die chemisch-reaktive Abscheidung der im folgenden beschriebenen Al_2O_3- und Si_3N_4-Schichten eingesetzt wurde, weist folgende Daten auf:

Frequenz: 27,3 MHz, *HF-Spannung:* 1000…1700 V, *HF-Leistung:* maximal 2 kW; *Magnetfeld:* 10…100 Gauß, *Target-Durchmesser:* 150 mm

Bei einem Argon-Druck von $7 \cdot 10^{-3}$ Torr wurden für SiO_2 Beschichtungsraten von 20…40 nm/min gemessen. Für chemisch-reaktive Zerstäubungen kann mit einem Dosiersystem eine konstante Beimengung von Sauerstoff bzw. Stickstoff zum Zerstäubungsgas Argon eingestellt werden. Die Verwendung eines Glasrezipienten, der mit viton-gedichteten Glas-Metallflanschen angeschlossen ist, bietet nach unserer Auffassung den Vorteil, daß Störungen durch ioneninduzierte Reaktionen an Metallwänden angeschlossen werden. Eine analoge Anlage für eine Beschichtungsfläche von 0,75 m² wurde gemeinsam mit dem VEB Elektromat, Dresden, aufgebaut. Eine andere Ausführung ist für

Abb. 2: Hochfrequenz-Plasma-Zerstäubungsapparatur mit kreuzförmigen Glasrezipienten.

die Beschichtung von Rasierklingen mit dem Ziel einer Stabilisierung der Schneiden gemeinsam mit dem VEB Feintechnik Eisfeld entwickelt worden.

Es kann eingeschätzt werden, daß passivierende Schichten mit der HF-Plasmazerstäubung für viele Anwendungen aufgebracht werden können. Einschränkungen für den praktischen Einsatz ergeben sich jedoch bei Werkstücken, die nicht eben sind, da eine gleichmäßige Beschichtung in diesem Falle nicht ohne weiteres möglich ist. Außerdem ist natürlich stets der technologisch-ökonomische Vergleich mit anderen Schutzverfahren zu beachten. Einen anderen Lösungsweg zur Durchführung von Zerstäubungsprozessen bietet die

2.2. *Ionenstrahlzerstäubung*

Bei dieser Variante werden die Ionen in einer externen Ionenquelle erzeugt und über Druckstufen in den auf Hochvakuum gehaltenen Rezipienten eingespeist. Für die Durchführung von Grundlagenuntersuchungen zum Zerstäubungsprozeß bei kleinen Strahlleistungen wird diese Methode wegen der definierten Bedingungen vorzugsweise eingesetzt. Auch bei der Sekundärionenmikroanalyse wird das Strahlprinzip verwendet, wobei die Proben mit einem gut ausgeblendeten Strahlbündel rasterförmig abgetastet werden.

Für Zerstäubungsprozesse höherer Leistung bei einem Vakuum unterhalb 10^{-5} Torr wurden von uns in den letzten Jahren eine Reihe von Anlagen entwickelt, deren Prinzip in Abb. 3 dargestellt ist [15]. Die Ionen werden in einer modifizierten Quelle vom Finkelstein-Typ bei einem Edelgasdruck von $5 \cdot 10^{-4}$ Torr erzeugt. Durch Absaugen und Beschleunigen der Ionen über ein Elektrodensystem mit siebförmigen Einsätzen gelingt es, ein breites Ionenstrahlbündel zu extrahieren, wobei zwischen dem Entladungsraum in der Quelle und dem Zerstäubungsraum ein Druckgefälle von 1···2 Größenord-

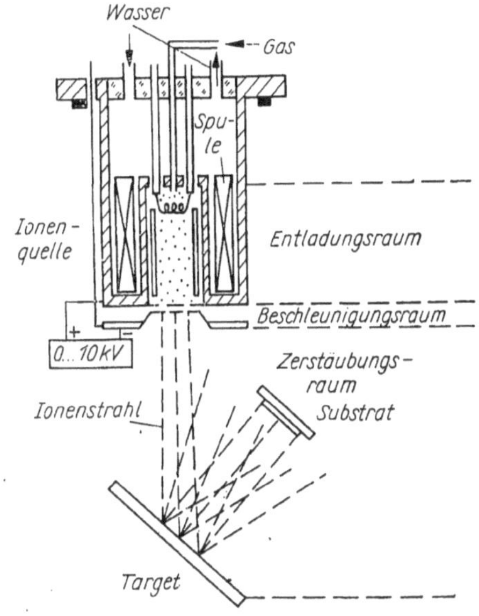

Abb. 3: Schema der Ionenstrahl-Zerstäubung („ion-beam-sputtering"; *IBS*).

nungen je nach Versuchsbedingungen gehalten werden kann. Aus einer Ionenquelle vom Standardtyp läßt sich ein Ionenstrom von 15 mA bei 10 keV Ionenenergie und einem Durchmesser des Strahlbündels auf dem Target von ca. 50 mm extrahieren.

Mit einem verbesserten Versuchssystem konnte bei gleichen Abmessungen eine Steigerung auf 40 mA Ionenstrom erreicht werden.

Um größere Flächen gleichmäßig zu beschichten, wurden 4 Ionenquellen in einem System mit gemeinsamer magnetischer Stabilisierung zusammengefaßt. Abb. 4 zeigt diese Paketionenquelle während des Betriebes. Mit dieser Anordnung können auch Verbindungsschichten durch reaktive Mehrstrahlzerstäubung hergestellt werden, wenn man die Targets aus den Feststoffkomponenten zusammensetzt. Eine andere Variante der Mehrstrahltechnik wurde durch Einbau von zwei getrennten Ionenquellen in einen Rezipienten realisiert [15]. Die eine der beiden Quellen wird dabei zum Abstäuben

Abb. 4: Ionenstrahl-Zerstäubungssystem, während des Betriebes: Paketionenquelle (oben Mitte), Substrathalter (unten rechts) und Target (vorn).

eines Feststofftargets, beispielsweise Silizium, genutzt, während die andere Quelle mit Stickstoff betrieben wird und einen direkt auf das Substrat gerichteten Strom von Stickstoffionen liefert. Am Substrat kreuzen sich beide Strahlbündel, und es lassen sich Nitridschichten definierter Stöchiometrie erzeugen. Analog ist natürlich der Aufbau von anderen für Passivierungs- und Härtungszwecke geeigneten Verbindungsschichten, etwa von Boriden und Karbiden, möglich und teilweise bereits realisiert worden.

Die Anwendung der Ionenstrahltechnik für die Passivierung und den Korrosionsschutz dürfte in absehbarer Zeit vor allem im Bereich der Forschung, speziell der Herstellung und Untersuchung komplizierter Oberflächenschichten unter definierten Bedingungen liegen, jedoch sind technische Anwendungen bei hochveredelten Erzeugnissen bereits jetzt denkbar, wofür im folgenden an Hand der Polytetrafluoräthylen-Schichten ein Beispiel gebracht wird.

Für das Einbringen von Stoffen in die oberen Atomlagen der Werkstücke, beispielsweise zur Nitrierung oder Karburierung mit dem Ziel der Erhöhung der Korrosions- und Verschleißbeständigkeit bei hochbeanspruchten und wertvollen Maschinenbauteilen, ist die Strahltechnik bei hohen Leistungen und nicht zu kleinen Bearbeitungsflächen zweifellos eine besonders geeignete Methode. Mit Hilfe einer neuerdings entwickelten Ringstrahlionenquelle, bei der im Unterschied zur Paketionenquelle die einzelnen Einheiten nicht mehr getrennt sind, lassen sich Ionenströme von einigen hundert mA extrahieren, was auch für technische Anwendungen, beispielsweise für die Ionennitrierung, ausreichen dürfte [5].

2.3. Ionengestützte Beschichtung (IAFF)

Im Bemühen um die Entwicklung produktiver Verfahren für die Abscheidung festhaftender korrosions- und verschleißmindernder Überzüge hat neuerdings die ionengestützte Beschichtung besonderes Interesse erlangt. Das vor allem von MATTOX begründete Verfahren beruht im Prinzip auf einer Vereinigung der hohen Produktivität des Aufdampfens im Vakuum mit der dichteren Struktur und der besseren Haftfestigkeit von implantierten oder gestäubten Schichten. Im einfachsten Falle, der in Abb. 5 dargestellt ist, werden die verdampften Teilchen in einer intensiven Gasentladung zu einem Bruchteil ionisiert und mit einer Spannung von etwa 1 kV in Richtung auf die Substrate beschleunigt [9, 10]. Bei anderen Varianten erfolgt die anteilige Nachionisation im Impulsbetrieb, durch Elektronenbeschuß oder sogar nach der Zusammenballung von Clustern in einer Entspannungsdüse.

In jedem Falle kommt die Schichtbildung durch ein Gemisch von überwiegend Neutralteilchen mit nahezu thermischer Energie und einem relativ kleinen Prozentsatz von energiereichen Ionen zu-

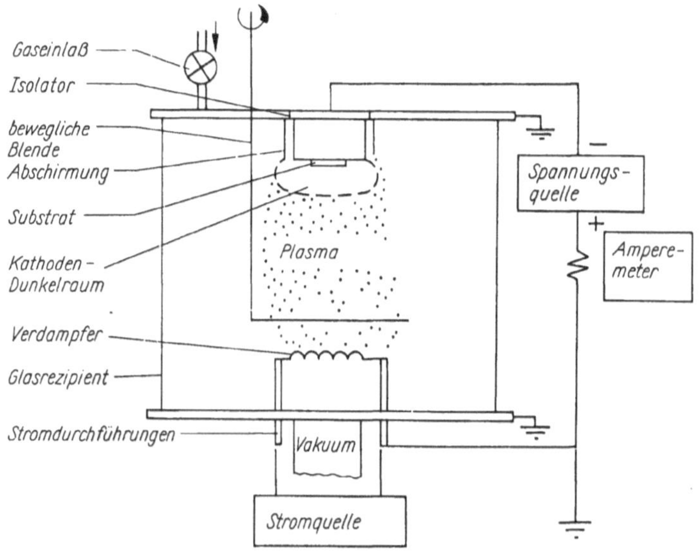

Abb. 5: Schema der ionengestützten Beschichtung („ion-assisted-film-formation"; *IAFF* oder auch „ion plating") nach MATTOX [9].

stande. Die von den Ionen mitgebrachte beträchtliche Aktivierungsenergie dürfte, wie von MATTOX angenommen wird, eine erhöhte Oberflächenmigration und teilweise ein sekundäres Abstäuben in der wachsenden Schicht bewirken. Darüber hinaus halten wir jedoch einen Einfluß der Implantation. eine Art „Verhaken" der Schicht auf dem Substrat sowie bei Anwesenheit reaktionsfreudiger Gase eine erhöhte Festkörperreaktivität für wahrscheinlich.

Obwohl der Mechanismus im einzelnen noch wenig geklärt ist, wurde an zahlreichen Anwendungsbeispielen nachgewiesen, daß die erhaltenen Schichten bezüglich Haftfestigkeit, Permeation und anderer Parameter sehr gute Eigenschaften aufweisen. Auch bei diesem Verfahren dürfte der praktische Einsatz zunächst bei der Beschichtung von wertvollen Maschinenbauteilen zum Zweck gleichzeitiger chemischer Passivierung und Senkung der Reibung und des Verschleißes gegeben sein. Besonders verlockend erscheint die Aussicht, Werkzeuge durch Ionenplattieren bei gleichzeitiger Nutzung der erhöhten Festkörperreaktivität mit standfesten Hartstoffschichten zu überziehen.

2.4. Plasmainduzierte Schichtabscheidung aus der Gasphase (PID)

Dieses Verfahren stellt ein weiteres Beispiel ionentechnologischer Beschichtungsverfahren dar [6]. Die Verhältnisse liegen hier besonders kompliziert, da die Teilprozesse kaum aufgeklärt sind, und die Einflüsse der Anregung im Plasma dürften je nach der speziell herausgegriffenen Reaktion sehr unterschiedlich sein. Im Anschluß an Arbeiten anderer Gruppen wurde von uns besonders die Bildung von Siliziumnitrid aus Silan und Ammoniak bei Anregung durch eine HF-Entladung untersucht [11]. Ein Schema der benutzten Anordnung ist in Abb. 6 dargestellt. Aus den im linken Teil angedeuteten

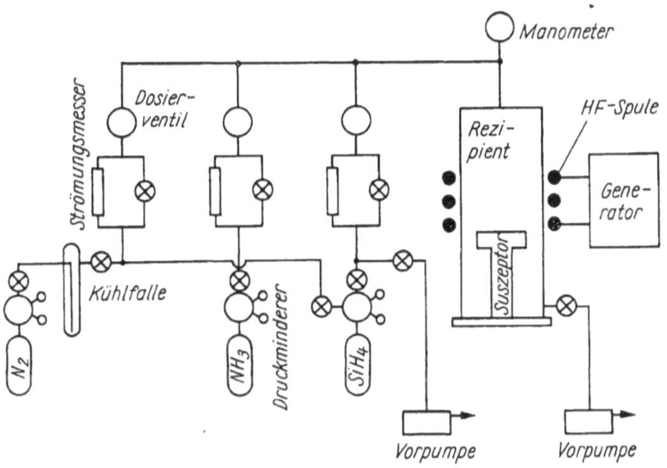

Abb. 6: Schema zur plasmainduzierten Abscheidung aus der Gasphase („plasma-induced-deposition"; *PID*).

Vorratsgefäßen werden mittels Dosierventilen und genauer Kontrolle durch Strömungsmesser die Reaktanten in einem definierten Verhältnis eingelassen. Die Abscheidung erfolgt in einem Quarzgefäß, das von der HF-Spule umgeben ist, auf scheibenförmigen Substraten. Es zeigte sich, daß die Reaktion $3\,SiH_4 + 4\,NH_3 \rightarrow Si_3N_4 + 12\,H_2$, die normalerweise erst bei Substrattemperaturen über 600 °C abläuft, unter Plasmaanregung bereits bei 170 °C auf Siliziumsubstraten reproduzierbar vor sich geht.

3. Herstellung und Eigenschaften spezieller Schichten

Im Rahmen des laufenden Versuchsprogramms wurden Herstellungsbedingungen und Schichteigenschaften folgender Materialien untersucht:

- Al_2O_3 mittels reaktiver HF-Plasmazerstäubung;
- Si_3N_4 mittels reaktiver HF-Plasmazerstäubung, durch ioneninduzierte CVD-Abscheidung sowie durch Ionenstrahlzerstäubung im Zweistrahlverfahren;
- PTFE (Polytetrafluoräthylen) mittels HF-Plasmazerstäubung und durch Ionenstrahlzerstäubung.

3.1. *Aluminiumoxid*

Es wurde durch reaktive Zerstäubung eines Reinstaluminiumtargets in einer Gasmischung von 80% Sauerstoff und 20% Argon bei einem Gesamtdruck von $5 \cdot 10^{-3}$ Torr hergestellt. Als Substrate wurden überwiegend einkristalline Siliziumscheiben verwendet. Wie die Angaben in Tab. 2 belegen, bilden sich geschlossene Schichten aus, die im Isolationsvermögen und anderen Parametern den Eigenschaften des reinen kompakten Materials nahekommen. Die Untersuchungen belegen ferner, daß es

Tabelle 2: Eigenschaften von Al_2O_3-Schichten, die durch reaktive Hochfrequenz-Plasmazerstäubung aufgebracht wurden

Dichte	4,68 g/cm³
Brechungsindex	1,85
spezifischer Widerstand	10^{16} Ohm · cm
relative Dielektrizitätskonstante	8
Durchbruchfeldstärke	$4,5 \cdot 10^6$ V/cm

möglich ist, Verbindungsschichten mittels HF-Zerstäubung auf reaktivem Wege aus den Elementen herzustellen, vorausgesetzt, daß die Abscheidung nicht zu schnell erfolgt, denn nur bei langsamen Schichtwachstum (3 nm/min) kann der Oxydationsprozeß dem Schichtwachstum folgen. Parallel dazu durchgeführte Abscheidungen mit einem Al_2O_3-Verbindungstarget (gepreßt und gesintert aus reinstem Al_2O_3-Pulver ohne Bindemittel) lieferten auch bei Abscheidungsraten bis zu 30 nm/min Schichten guter Qualität.

Beim Zerstäuben vom Verbindungstarget wurde eine Gasmischung von etwa 90% Argon und 10% Sauerstoff verwendet. Die Zerstäubung erfolgt dabei im wesentlichen durch Argonionen, und der Sauerstoffgehalt bewirkt lediglich einen Ausgleich des beim Abstäuben auftretenden Sauerstoffdefizits. Generell ist also die Herstellung von passivierenden Schichten durch Zerstäubung vom Verbindungstarget der reaktiven Zerstäubung vom Elementtarget vorzuziehen. Diese Regel wurde auch bei Untersuchungen an anderen Verbindungen bestätigt.

3.2. *Siliziumnitrid*

Diese Verbindung wurde in vergleichenden Untersuchungen mit 3 verschiedenen Methoden abgeschieden. Die reaktive HF-Plasma-Zerstäubung lieferte nur bei der Zerstäubung von Verbindungstargets Schichten guter Qualität. Auch aus der Literatur ist bekannt, daß die chemisch-reaktive Zerstäubung, ausgehend von den Elementen, infolge der Reaktionsträgheit des Stickstoffes nicht zu brauchbaren Schichten führt [7]. Die besten Ergebnisse wurden bei einem Gesamtdruck von $9 \cdot 10^{-3}$ Torr und einer Zusammensetzung des Zerstäubungsgases von 90 bis 95% Stickstoff und 10 bis 5% Argon erzielt. Bei einer HF-Leistungsdichte von 5 Watt/cm² wurden Aufwachsraten von 28 nm/min erhalten. Wie die Zusammenstellung der Schichtparameter in Tab. 3 erkennen läßt, sind die Schichten

Tabelle 3: Eigenschaften von Si_3N_4-Schichten, die durch Hochfrequenz-Plasmazerstäubung eines Verbindungs-Targets aufgebracht wurden

Brechungsindex	1,8···1,9
relative Dielektrizitätskonstante	6,0···7,5
Verlustfaktor tan	$7,0 \cdots 2,0 \cdot 10^{-3}$
spezifischer Widerstand	$4,0 \cdots 10 \cdot 10^{14}$ Ohm · cm
Durchbruchfeldstärke „intrinsic“	$7,0 \cdots 11 \cdot 10^6$ V/cm
Durchbruchfeldstärke „Dotwerte“	$1,0 \cdots 8 \cdot 10^6$ V/cm
Ätzrate in gepufferter Flußsäure (75 g NH_4F, 250 ml H_2O, 35 g HF 40%)	0,6···1,0 nm/s
Absorptionsbande (Si—N)	11,5 μm
mechanische Spannungen	$1,5 \cdot 10^{10}$ dyn/cm²

sehr beständig, sie weisen selbst in Flußsäuren nur geringe Ätzraten auf, jedoch wurden mit Hilfe des Blasentestes noch Schichtdefekte einer Dichte in der Größenordnung von $10^2/cm^2$ bei Schichtdicken von 300 bis 400 nm nachgewiesen.

Ähnliche Ergebnisse ergab die plasmainduzierte Schichtabscheidung aus der Gasphase. Als Reaktanten werden 1 Vol.-% Silan in Argon und reinstes Ammoniak eingesetzt. Bei einem Totaldruck von 2 Torr, einem Volumendurchsatz von 1 l/min, einer HF-Leistung von 200 W und einer Substrat-

Tabelle 4: Eigenschaften von Si_3N_4-Schichten, die durch plasmainduzierte Abscheidung aus der Gasphase aufgebracht wurden

Brechungsindex	1,9
spezifischer Widerstand	10^{14} Ω cm
Dielektrische Durchschlagsfestigkeit	$5 \cdot 10^7$ V cm^{-1}
Bandabstand	5,25–5,30 eV
Ätzrate	0,1 ··· 0,6 nm s^{-1}
Infrarotabsorptionsband	11,5 μm
mechanische Spannungen	$1{,}5-3{,}0 \cdot 10^{10}$ dyn cm^{-2}
Struktur	amorph

temperatur von 170 °C wurde das Konzentrationsverhältnis Silan/Ammoniak in Stufen 0,01, 0,02 und 0,10 verändert. Für die Abscheidungsrate W ergaben sich, entsprechend dem steigenden Konzentrationsverhältnis $W_{0,01} = 30$ nm/min, $W_{0,02} = 60$ nm/min und 110 nm/min. Die Untersuchung der Schichten hinsichtlich ihrer Stöchiometrie und ihres Sauerstoffgehalts erfolgte mittels Elektronenmikrostrahlanalyse und Infrarotsproskopie.

Innerhalb der Nachweisgrenzen beider Verfahren sind die Schichten als sauerstofffrei zu bezeichnen und die Stöchiometrie entsprach nahezu dem formelmäßigen Wert. Durch röntgenografische Untersuchungen konnten keine polykristallinen Anteile in den Schichten nachgewiesen werden. Ob dies, wie Kusnezow begründet, mit dem Einbau von Wasserstoff zusammenhängt, konnte von uns nicht entschieden werden. Es zeigte sich, daß die reproduzierbare Beherrschung dieser Methode eine sehr genaue Konstanz aller Versuchsparameter voraussetzt. Wenn diese Forderung erfüllt ist, bilden sich Schichten sehr guter Homogenität und Beständigkeit.

Die Bestimmung der Defektdichte wurde mit dem von uns weiterentwickelten Flüssigkristalltest durchgeführt [13].

Unter Ausnutzung des dynamischen Streueffektes nematischer Flüssigkristalle gestattet dieses Verfahren den Nachweis von Defekten in isolierenden Schichten. Es ist möglich, verschiedene Defektarten, wie z. B. topographische und stöchiometrische, der Größe $d \approx 1$ μm mit Sicherheit nachzuweisen. Das Verfahren ist für die verschiedensten Isolator-, Passivierungs- und Korrosionsschichten anwendbar und gestattet integrale Aussagen über die Defektdichte sowie deren lokale Verteilung. Präparation und Messung sind einfach und schnell durchführbar. Voraussetzung für die Untersuchung ist, daß sich die Isolator- oder Passivierungsschicht auf einem leitenden Substrat befindet.

Zur Defektanalyse wird eine Meßanordnung, wie sie schematisch in Abb. 7 dargestellt ist, verwendet. Dabei befindet sich zwischen der Isolatorschicht und einer transparenten Deckelektrode eine etwa 20 μm dicke nematische Flüssigkristallschicht. Bei stufenweisem Anlegen einer elektrischen Spannung werden Defekte in der Schicht als Inhomogenitäten im reflektierten Licht unter dem

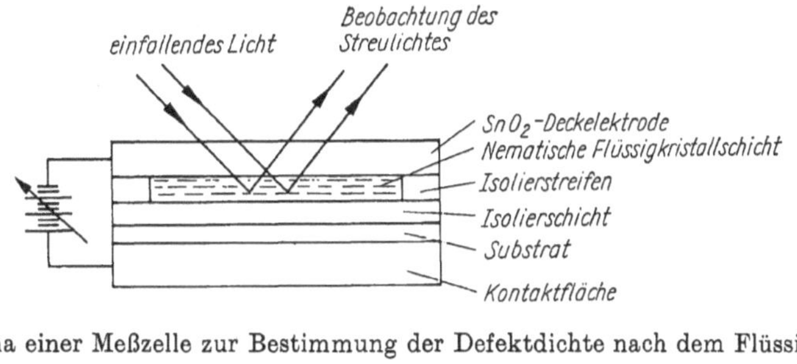

Abb. 7.: Schema einer Meßzelle zur Bestimmung der Defektdichte nach dem Flüssigkristall-Test.

Mikroskop sichtbar, da in den Gebieten mit Stromfluß die von Anzeigenelementen bekannte Umorientierung der Moleküle nematischer Flüssigkristall-Moleküle erfolgt. Das Verfahren bietet den Vorteil, daß es rasch und zerstörungsfrei arbeitet, und daß durch Einregeln der Spannung die Defektdichte bis zu der im Anwendungsfall interessierenden kritischen Feldstärke sichtbar gemacht wird. Abb. 8 zeigt den Verlauf der mittleren Defektdichte in Abhängigkeit von der Schichtdicke. Danach wird eine steile Abnahme der Zahl der Defekte mit zunehmender Schichtdicke beobachtet, so daß

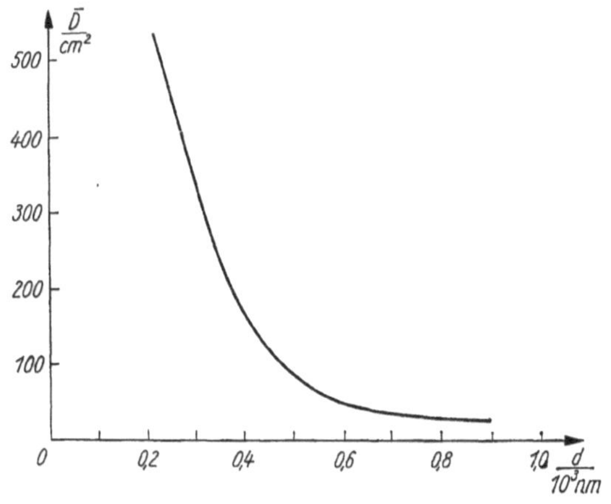

Abb. 8: Mittlere Defektdichte als Funktion der Schichtdicke für Siliziumnitridschichten.

für einen wirksamen Oberflächenschutz mindestens Schichten einer Dicke von 500 nm erforderlich sind. Bei dieser Schichtdicke erhielten wir Werte von etwa 100 Defekten pro cm^2, das ist sehr günstig, wenn man bedenkt, daß die bisherigen Untersuchungen nicht in staubfreien Räumen durchgeführt werden konnten.

Ein wesentliches Kriterium für die passivierende Wirkung einer Schicht bildet die bisher leider nur wenig untersuchte Permeation bei der Einwirkung von Ionenlösungen. Zur Bestimmung der Eindiffusion von Na-Ionen wurde die Radioisotopenmethode herangezogen. Die Markierung erfolgte

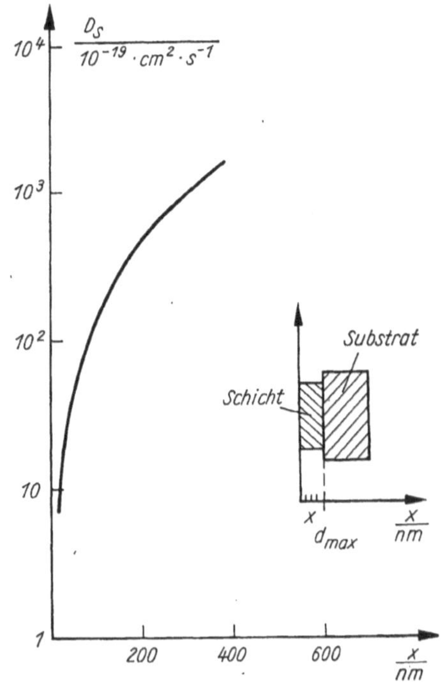

Abb. 9: Mittlerer Diffusionskoeffizient als Funktion der Schichtdicke für die Diffusion von Natrium in Siliziumnitrid (bestimmt bei $\vartheta = 350\,°C$).

durch Aufbringen von Natrium-Azetat, als Trägersubstanz für Natrium-22, auf die Schichtoberfläche. Nach 14 Tagen Temperung bei 350 °C wurde durch definiertes Abätzen die Verteilung des Natriums in der Schicht bestimmt. Dabei ergab sich die in Abb. 9 dargestellte Abhängigkeit des Diffusionskoeffizienten von der Eindringtiefe in der Schicht. Danach setzen die obersten Lagen der Diffusion von Na-Ionen den höchsten Widerstand entgegen, während im Schichtinneren dieser wesentlich geringer ist. Diese Untersuchungen müssen weiter ausgebaut und durch Einbeziehung von Anionen ergänzt werden.

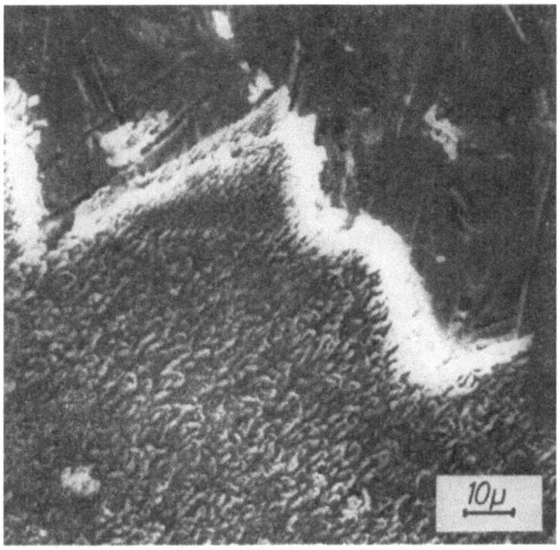

Abb. 10: Faserstruktur einer geätzten teilkristallinen Fluor-Kohlenstoffschicht.

Eine direkte Bildung von Siliziumnitrid-Schichten mit annähernd idealer Stöchiometrie konnte durch die Ionenstrahlzerstäubung im Zweistrahlverfahren aus den Elementen verwirklicht werden. Mit diesem Verfahren, bei dem sich ein strahlgestäubtes Silizium-Teilchenbündel und ein direkt auf die Substrate gerichtetes Stickstoffionenbündel treffen, lassen sich auch aus reaktionsträgen Stoffen Verbindungsschichten aufbauen. Neben Siliziumnitrid wurden auch Chrom-, Indiumnitrid sowie Carbide abgeschieden.

3.3. *Polytetrafluoräthylen*

Als letztes Beispiel soll die Ionenstrahlzerstäubung von Polytetrafluoräthylen betrachtet werden. Überraschenderweise beobachteten wir an diesem System ungewöhnlich hohe Zerstäubungsraten von mehr als 10^2 CF_2-Gruppen pro einfallendes Argonion, und auf den Targets bildete sich eine charakteristische Faserstruktur aus (Abb. 10 und Tab. 5), die nach den Ergebnissen der Struktur-

Tabelle 5: Eigenschaften von aufgestäubten Fluor-Kohlenstoff-Schichten

	Kompaktes PTFE	Fluor-Kohlenstoff-Schicht	
Verfahren		HF-Zerstäubung	Strahlzerstäubung
Struktur	kristallin	amorph	kristallin
Farbe	weiß	gelb	milchig-weiß
Ionenätzung	Nadelbildung	keine Strukturierung	Nadelbildung
Durchschlagsfestigkeit (V/cm)	$2-3 \cdot 10^5$ (bei 1 mm Dicke)	$7 \cdot 10^5$	$2 \cdot 10^6$
spezifischer Widerstand (Ohm · cm bei Raumtemperatur)	10^{16}	10^{16}	10^{16}
Dielektrizitätskonstante	2,1 (bei 800 Hz)		2,0···2,1
Verlustfaktor	$3 \cdot 10^{-4}$ (bei 800 Hz)		1%
Reibungskoeffizient	0,39 (bei $F_n = 148$)	0,61	0,50

untersuchungen den orientierten Quasikristallen des Polymers entspricht. Die Deutung dieser Erscheinung ist wahrscheinlich darin zu suchen, daß die Ionen nur zur Sprengung der festen innermolekularen Bindungen benötigt werden, während die Herauslösung ganzer Molekülfragmente unter Überwindung der schwachen zwischenmolekularen Kräfte im wesentlichen durch thermische Schwingungen erfolgt [12].

Im Gegensatz dazu lieferte die HF-Plasmazerstäubung nur amorphe gelbliche Schichten ohne erkennbare Strukturierung. Zweifellos können die mit Hilfe der Strahlmethode abgeschiedenen PTFE-Schichten für spezielle Zwecke der Passivierung und der Reibungsminderung interessante Anwendungen finden.

4. Schlußfolgerungen

Die bisher durchgeführten Untersuchungen zeigen, daß die Methoden der Ionentechnologie geeignet sind, auf Metallen, Halbleitern und anderen Materialien fest haftende passivierende Verbindungsschichten aufzubringen. Die bisher erhaltenen Ergebnisse wurden allerdings vorwiegend im Ergebnis von Untersuchungen verschiedener Passivierungsschichten für elektronische Bauelemente gewonnen. In künftigen Arbeiten sollte das Schwergewicht bei der Ausarbeitung von Verfahren von vornherein stärker auf den Korrosionsschutz metallischer Werkstoffe und Bauteile ausgerichtet werden. Dabei sind auch die neueren Methoden der Ionenplattierung in das Programm mit einzubeziehen. Im ganzen gesehen halten wir es für aussichtsreich, die Methoden der Ionentechnologie für den Schutz nicht zu großer, wertvoller Metallbauteile auch technisch einzusetzen. Im Vergleich zu den herkömmlichen Verfahren werden sich die Anwendungen dabei wohl zunächst auf Sonderfälle beschränken, bei denen durch Anforderungen an eine kombinierte Verringerung der Korrosion und des Verschleißes der relativ hohe Aufwand von Vakuumbeschichtungsverfahren gerechtfertigt erscheint. Darüber hinaus dürften die Möglichkeiten für den gezielten Aufbau auch komplizierterer Oberflächenschichten, beispielsweise mit Hilfe der Mehrstrahltechnik, interessante Aspekte für die weitere Grundlagenforschung zum Verhalten passivierender Oberflächenschichten eröffnen.

Abschließend ist es uns ein tiefempfundenes Bedürfnis, unserem verehrten Lehrer, Herrn Professor Dr.-Ing. habil. Dr. rer. nat. h.c. mult. Kurt SCHWABE für die stete Förderung und Unterstützung unserer Arbeit herzlich zu danken.

Literatur

[1] „Berichtsband des VI. Internationalen Vakuumkongresses Kyoto 1974", herausg. als Supplement Nr. 2 (1974) der Zeitschr. J. appl. Phys.
[2] Berichtsband Symposium: „Ionenzerstäubung", TH Karl-Marx-Stadt, 25. bis 27. 9. 1973, herausg. 1974
[3] CARTER G. und J. S. COLLIGON, "*Ion Bombardment of Solids*", Elsevier Publishing Co., Inc., New York 1968
[4] O. FIEDLER, G. KREISELMEIER und Chr. WEISSMANTEL, Experim. Techn. d. Physik XX, Heft 6, 467 (1972)
[5] O. FIEDLER und G. REISSE, DDR-Patent WPC 23c/172711
[6] L. V. GREGOR, *In "Thin Film Dielectrics.*" The Electrochem. Soc., New York 1969
[7] A. R. JANUS und G. A. SHIRN, J. Vac. Sci. Techn., V. **4**, Nr. 1 (1967)
[8] KIENITZ, H. (Hrsg.): „*Massenspektrometrie*", Verlag Chemie, Weinheim/Bergstraße 1968, S. 233
[9] MATTOX, D. M., J. Appl. Phys. V. **34**, 2493 (1963)
[10] MATTOX D. M., „Berichtsband des VI. Intern. Vakuumkongresses Kyoto 1974", herausg. als Supplement Nr. 2 (1974) der Zeitschr. Japan. J. app. Phys.
[11] P. MORZECK, J. TUPHORN, S. ERNST, D. KNOLL, D. ROTH und Chr. WEISSMANTEL, Wiss. Z. d. Techn. Hochsch., Karl-Marx-Stadt, **17**, 3, 327 (1975)
[12] M. ROST, H.-J. ERLER, H. GIEGENGACK, O. FIEDLER und Chr. WEISSMANTEL, Thin Solid Films **20**, 15—19 (1974)
[13] ROTH, D., Promotion A (1977) Techn. Hochsch. Karl-Marx-Stadt
[14] P. SIEGMUND, "*Theory of Sputtering I.*", Phys. Rev. **184**, 383 (1969)
[15] Chr. WEISSMANTEL, O. FIEDLER, G. REISSE, H.-J. ERLER, U. SCHEIT, M. ROST und J. HERBERGER, „Berichtsband des VI. Internationalen Vakuumkongresses Kyoto 1974", herausg. als Supplement Nr. 2 (1974) der Zeitschr. Japan. J. appl. Phys.
[16] WEISSMANTEL Chr., Wiss. Z. der TH Karl-Marx-Stadt, XV (1973) H. 1, Sonderausgabe, S. 65

M. Śmiałowski

Institut für physikalische Chemie der Polnischen Akademie der Wissenschaften, Warschau

EINFLUSS VON KATALYSATORGIFTEN AUF DIE WASSERSTOFF-DURCHLÄSSIGKEIT VON KATHODISCH POLARISIERTEN STAHLMEMBRANEN

Zusammenfassung

Mit einer rotierenden Scheibenelektrode wurde der Einfluß der den schwach-sauren wäßrigen Elektrolyten in kleinen Mengen beigefügten Verbindungen von P, As, Sb, Bi, S, Se und Te auf die Wasserstoffdurchlässigkeit von kathodisch polarisierten Stahlmembranen untersucht.

Es wurde festgestellt, daß die beschleunigende Wirkung der Zusätze bei der Wasserstoffdurchlässigkeit am stärksten auftritt, wenn Hydride obiger Elemente sich an der Kathode bilden und eine ausreichende Beständigkeit aufweisen. So ist z. B. die Wirksamkeit von S-, Se- und Te-Verbindungen auf saure Lösungen beschränkt, während P-, As- und Sb-Verbindungen im gesamten pH-Bereich aktiv sein können. Für jedes Element gibt es ein bestimmtes Elektrodenpotential sowie eine bestimmte optimale Konzentration, bei der die größten Beschleunigungseffekte beobachtet werden. Höhere Konzentrationen veranlassen die Entstehung von Deckschichten, die die Wasserstoffdurchlässigkeit hemmen. Es besteht eine direkte Proportionalität zwischen dem Logarithmus der optimalen Konzentration des gegebenen Elementes im Elektrolyten und dem chemischen Potential des betreffenden Hydrids. Je höher die Beständigkeit des Hydrids, um so größer ist die optimale Konzentration. Im Falle optimaler Konzentration im Elektrolyt beobachtet man Wasserstoffdurchlässigkeiten, die in direkter Proportionalität zu den Bindungskräften der betreffenden Hydride stehen. Die stärksten Beschleunigungseffekte werden durch H_2S, die schwächsten durch BiH_3 bewirkt.

Höchstwahrscheinlich werden die Moleküle der genannten Hydride so stark an der Kathodenoberfläche adsorbiert, daß sie Wasserstoffatome abspalten. Dadurch wird die Aktivationsenergie des Wasserstoffeindringens in die Metallphase erniedrigt.

Obwohl der beschleunigende Einfluß von gewissen Katalysatorgiften auf das Eindringen des Wasserstoffs in Metallen schon seit mindestens 50 Jahren bekannt ist, ist der Mechanismus dieser Reaktion noch nicht völlig geklärt.

Auf Grund von Dilatationsmessungen an Eisenproben, die in verdünnter Schwefelsäure in Anwesenheit von verschiedenen Giften kathodisch polarisiert wurden, hat man eine gerade Proportionalität zwischen der Bindungskraft mancher Hydride der Elemente der VA und VIA periodischen Gruppe (PH_3, AsH_3, SbH_3, H_2S und H_2Se) und der Größe des durch die Wasserstoffbeladung hervorgerufenen Dilatationseffektes gefunden [1].

In einer späteren Arbeit von Newman und Shreir [2] wurden diese Ergebnisse bestätigt. McCright und Staehle sind [3] der Meinung, daß die Wirksamkeit von Arsen nicht in der Hydridform, sondern in der Elementarform zum Ausdruck kommt.

Um diesen Widerspruch endgültig zu klären, wurden systematische Versuche unter Benutzung einer rotierenden Scheibenelektrode durchgeführt [4, 5]. Diese Methode hat den Vorteil, daß man den an der polarisierten Elektrode herrschenden pH-Wert in weiten Grenzen ändern kann, ohne die Zusammensetzung des Gesamtelektrolyten wechseln zu brauchen. Auf die Benutzung von Pufferlösungen muß man verzichten, da verschiedene Anionen die Wasserstoffbeladung verschiedentlich beeinflussen können.

Zur Messung der Wasserstoffdurchlässigkeit der Membranen wurde die äußerst empfindliche elektrochemische Methode benutzt*). Abb. 1 zeigt die im Längsschnitt benutzte Scheibenelektrode.

Die untere Seite der vor jedem Versuch sorgfältig gereinigten Stahlmembran wurde mit kathodisch entwickeltem Wasserstoff beladen, während man die obere, mit einer dünnen Palladiumschicht bedeckte Seite anodisch in 0,1 n Natronlauge polarisierte. Zur Messung der Wasserstoffdurchlässigkeit

*) Die Versuche wurden durch Herrn Dipl.-Chem. T. Zakroczymski durchgeführt.

hat man sich der anodischen Stromdichte bedient, die zur Oxidation des durch die Membrane diffundierenden Wasserstoffs verbraucht wurde. Die Membranen wurden aus einem handelsüblichen weichgeglühten Stahl mit 0,305% C hergestellt. Wesentlich ist es, für die Herstellung der Membranen kein Blech zu benutzen, wie es vorher üblich war, sondern dünne, aus Rundstahl senkrecht zur Stabachse geschnittene Scheiben. Im Blechmaterial sind nämlich die nichtmetallischen Einschlüsse in Schichten verteilt, die parallel zur Oberfläche, also senkrecht zu dem hineindiffundierenden Wasserstoffstrom liegen. Molekularer Wasserstoff scheidet sich an diesen Schichten aus und führt zur Blasenbildung. Darum ist die Durchlässigkeit solcher Membranen klein und unregelmäßig. Bei dem anderen Material sind die Einschlüsse in senkrecht zur Stabachse geschnittenen Scheiben parallel zu dem Wasserstoff-

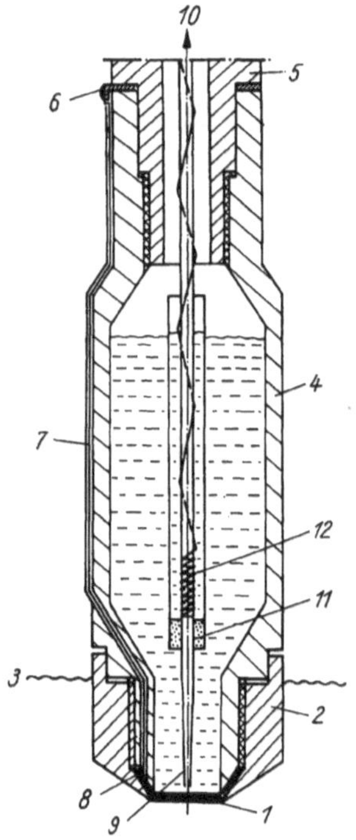

Abb. 1: Längsschnitt durch die rotierende Scheibenelektrode

1 – Stahlmembrane; 2 – Haube aus Polytetrafluoräthylen; 3 – Niveau der Kathodenflüssigkeit; 4 – Hülse aus Plexiglas; 5 – Stahlwelle; 6 – Messingring; 7 – Cu-Draht; 8 – Lötmetall; 9 – Luggins-Kapillare; 10 – Salzbrücke zur Bezugselektrode; 11 – Glasfritte; 12 – Pt-Draht.

strom verteilt und hemmen ihn praktisch nicht. Eine derartige Membrane kann oftmals mit Wasserstoff beladen werden, ohne dabei ihre Durchlässigkeit wesentlich zu ändern. In den meisten Versuchen wurden Membranen von 11 mm Durchmesser und 0,6 mm Dicke verwendet.

Als Grundelektrolyt für die Kathodenflüssigkeit hat man gewöhnlich 0,1 normale, mit Schwefelsäure auf pH 2,6 angesäuerte, wäßrige Natriumsulfatlösung benutzt.

Die Katalysatorgifte wurden in kleinen Mengen dem Grundelektrolyten zugesetzt. Arsen, Antimon, Selen und Tellur wurden als Oxide benutzt. Phosphor wurde als Natriumhypophosphit zugesetzt oder von außen her als Phosphorwasserstoff dosiert. Schwefel wurde als Natriumsulfidlösung zugegeben. Der Elektrolyt war vom gelösten Luftsauerstoff durch Durchleiten von reinem Argon befreit.

Die Drehzahl der Scheibenelektrode war konstant gehalten und betrug 540 Touren pro Minute. Die Temperatur war auch konstant, 25° ± 0,1 °C.

Bei einer konstanten Tourenzahl hängt der an der Oberfläche der Scheibenelektrode herrschende pH-Wert (pH_s) von der kathodischen Stromdichte ab. In einem schwach sauren Elektrolyt, z. B. mit pH 2,6, kann man leicht den Grenzstrom der Wasserstoffionenentladungsreaktion erreichen. Damit steigt der pH_s-Wert auf 7 und höher. Die Wasserstoffionenentladung hört praktisch auf und wird durch die Entladung von Wassermolekeln ersetzt. Durch Regelung der kathodischen Stromdichte kann man also die pH_s-Werte und die sich an der Kathodenoberfläche abspielende Prozesse in großem Umfang ändern und deren Einfluß auf die Wasserstoffdurchlässigkeit der Membranen systematisch verfolgen.

Als Beispiel werden hier einige Versuchsergebnisse vorgelegt.

In Abb. 2 *a* ist der Einfluß des negativen Potentials auf die kathodische Stromdichte dargestellt. Bei ca. 4,6 mA/cm² wird der Grenzstrom der Wasserstoffionen-Entladungsreaktion erreicht. Im mit SeO_2 versetzten Elektrolyt steigt die Strom-Potentialkurve viel steiler als in der mit Sb_2O_3 versetzten

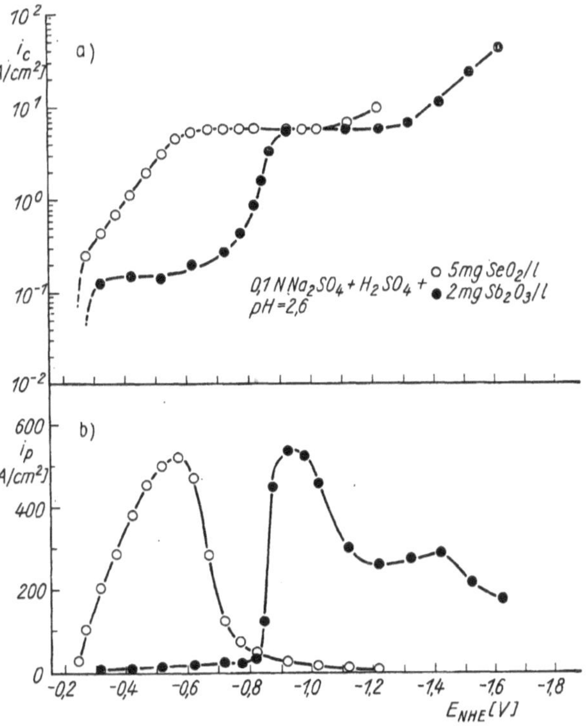

Abb. 2: Einfluß des Kathodenpotentials E auf
(*a*) kathodische Stromdichte i_c, (*b*) Wasserstoffdurchlässigkeit i_p.

Lösung. In Abb. 2 *b* ist die in μA/cm² ausgedrückte Wasserstoffdurchlässigkeit i_p als Funktion des Kathodenpotentials angegeben. In der Anwesenheit von Selenium steigt die Kurve wiederum steiler als in der Gegenwart von Antimon. Maximale i_p-Werte werden in beiden Fällen bei jenem Potential erreicht, das dem Anfang des Grenzstromgebietes entspricht. In der mit Antimonoxid versetzten Lösung ist bei niedrigen Polarisationsgraden der Kathode die Wasserstoffdurchlässigkeit sehr gering. Plötzlich steigt sie bei ca. −0,82 V an. Bei diesem Potential beginnt die elektrochemische Ausscheidung des Antimonwasserstoffs. Bei Selen scheint die Bildung von H_2Se bei niedrigeren Polarisationsgraden, nämlich schon beim Korrosionspotential des Stahls, zu beginnen.

Bei höheren Polarisationsgraden als jenen, die dem Anfang des Grenzstromgebietes entsprechen, nimmt die Wasserstoffdurchlässigkeit bedeutend ab. In der Anwesenheit von Antimon ist der Wert i_p noch bei hohen Polarisationsgraden beträchtlich, während der von Selen fast Null beträgt. Die Maxima der Durchlässigkeits-Potentialkurven liegen bei den gleichen Potentialwerten, die dem Anfang des Grenzstromgebietes bei der Anwesenheit des betreffenden Katalysatorgiftes entsprechen.

Die beobachteten Unterschiede zwischen dem Verhalten des Antimons und dem des Selens im alkalischen Gebiet können auf folgende Weise gedeutet werden: Die Hydride der Elemente der VA periodischen Gruppe, nämlich des Phosphors, Arsens und Antimons, sind sowohl in sauren wie in alkalischen Lösungen verhältnismäßig stabil, während die Beständigkeit der Hydride der Elemente der VIA-Gruppe sich nur auf saure Lösungen beschränkt. Bei einem bestimmten kritischen pH-Wert findet die Dissoziation statt:

$$H_2X = HX^- + H^+;$$

X = S, Se oder Te.

In Tab. 1 sind die kritischen pH-Werte (nach POURBAIX [6]) mit den experimentellen pH_s-Werten zusammengestellt, bei denen die beschleunigende Wirkung des betreffenden Katalysatorgiftes abzunehmen begann.

In Abb. 3 wird der Einfluß von Natriumsulfid mit dem von Telluroxid verglichen. Das Natriumsulfid reduziert schon beim Korrosionspotential des Eisens zum Schwefelwasserstoff, die i_p—E-Kurve

Tabelle 1: Vergleich der pH/XH^--Werte mit den experimentell gefundenen kritischen pH_s-Werten

Element	pH/XH^-	krit. pH_s
S	7,00	7,0
Se	3,74	3,3
Te	2,64	2,9

steigt gleichmäßig, während die Durchlässigkeit von TeO_2 am Anfang sehr gering ist und die Kurve erst bei etwa —0,38 V steiler anzusteigen beginnt. Die Maxima beider Kurven entsprechen wiederum dem Beginn des Grenzstromgebietes, das von Na_2S bei etwa 6 mA/cm² auftritt. Ein Teil des Stromes wird für die Reduktion beträchtlicher Mengen von Na_2S verbraucht.

Bei höheren Stromdichten als beim Grenzstrom nimmt die Wasserstoffdurchlässigkeit bedeutend ab. Im Falle von Tellur wird das durch die Ausscheidung des Elementes an der Kathodenoberfläche bewirkt.

Abb. 4 *a* zeigt den Einfluß der Konzentration verschiedener Promotoren im Elektrolyt auf die Wasserstoffdurchlässigkeit bei einer Stromdichte von 2,25 mA/cm². Sie ist also niedriger als die des

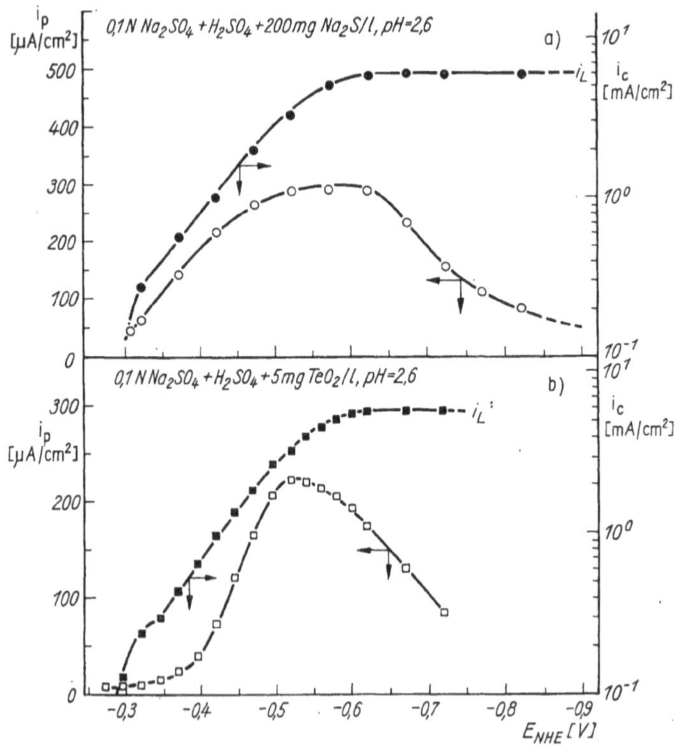

Abb. 3: Einfluß des Kathodenpotentials auf die kathodische Stromdichte und die Wasserstoffdurchlässigkeit in Anwesenheit von Na_2S (*a*) und TeO_2 (*b*) im Elektrolyt.

Grenzstromes. In Lösungen, die mit Arsen-, Antimon-, Selen- und Telluroxid versetzt wurden, beobachtet man Maximalwerte der Durchlässigkeit, die bei weiteren Zugaben dieser Oxide bedeutend abnimmt. Verursacht wird dies durch die Bildung von Deckschichten, die das Eindringen des Wasser-

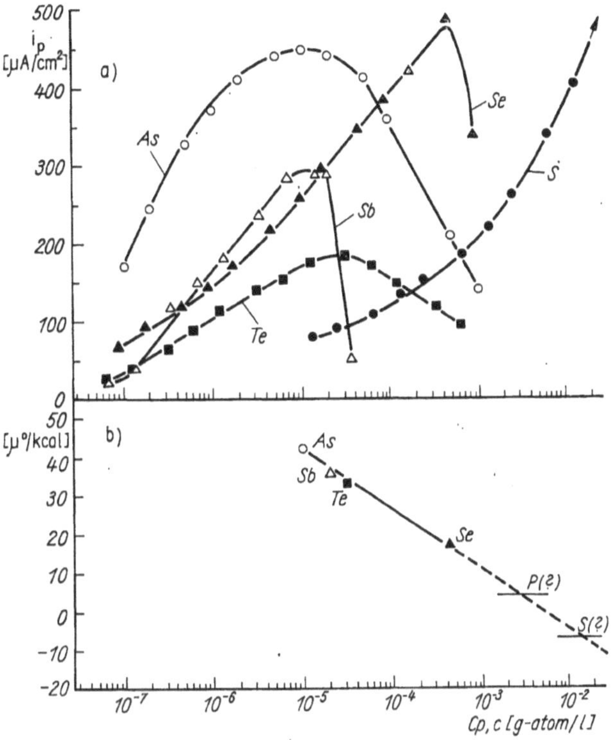

Abb. 4: Zusammenhang zwischen der Konzentration verschiedener Promotoren im Elektrolyt ($c_{p,c}$) und der Wasserstoffdurchlässigkeit (*a*) sowie dem Standardpotential μ^0 verschiedener Promotorhydride (*b*).

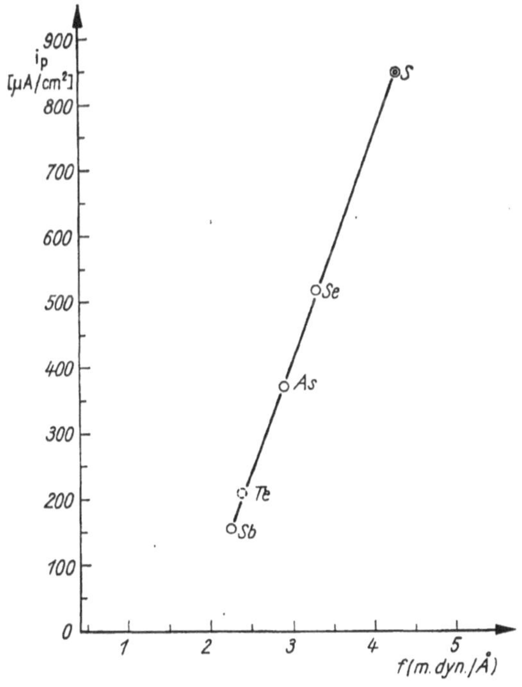

Abb. 5: Zusammenhang zwischen der maximalen Wasserstoffdurchlässigkeit i_p und der Bindungskraft verschiedener Promotorhydride *f*.

stoffs in die Kathode stark hemmen. Bei Antimon, das einen dichten metallischen Überzug bildet, ist das Sinken der Kurve besonders steil. Die für Na_2S erhaltene Kurve zeigt kein Maximum auf, da in diesem Fall der Schwefel nicht als kompakter Niederschlag, sondern als feine Suspension ausscheidet.

Das Ausfallen von As, Sb, S, Se und Te aus Lösungen, die katalytische Gifte in höheren Konzentrationen als $c_{p,c}$ enthalten, wird wahrscheinlich durch die Zersetzung der primär gebildeten Hydride bewirkt. Die in Abb. 4 *b* dargestellte lineare Abhängigkeit der kritischen Konzentration vom chemischen Standardpotential der Hydride stimmt mit dieser Annahme überein. Mit Phosphor und Schwefel konnten die Maximalwerte von i_p nicht erreicht werden.

In Abb. 5 sind die maximalen Wasserstoffdurchlässigkeiten gezeigt, die in der Gegenwart verschiedener Promotoren bei 2,25 mA/cm², also unterhalb des Grenzstromes, beobachtet wurden, als Funktion der Bindungskräfte der betreffenden Hydride. Die Daten von *f* wurden der Arbeit von Siebert [7] entnommen. Für Tellurwasserstoff gibt es im Schrifttum keine Angaben. Der in Abb. 5 aufgetragene Wert wurde durch Extrapolation der Abhängigkeit der Bindungskraft von der Ordnungszahl der Elemente der VIA-Gruppe gebildet. Für Schwefel ist der Maximalwert von i_p unsicher. Wegen experimenteller Schwierigkeiten konnten mit Phosphorverbindungen keine systematischen Messungen durchgeführt werden. Das Wismutoxid weist nur einen ganz schwachen Einfluß auf die Wasserstoffdurchlässigkeit auf, da sich der Wismutwasserstoff erst bei hohen Polarisationsgraden, also weit oberhalb des in schwachsauren Elektrolyten vorkommenden Grenzstromes, bildet.

Diskussion

Alle vorgetragenen Resultate scheinen eindeutig zu beweisen, daß die beschleunigende Wirkung der geprüften Katalysatorgifte hauptsächlich, wenn nicht ausschließlich, in Hydridform vorkommen. In der Elementarform wird keine Beschleunigung, sondern in den meisten Fällen eine starke Hemmung der Wasserstoffdurchlässigkeit beobachtet. Die Oxide üben einen stark beschleunigenden Einfluß nur bei jenen Potentialwerten aus, bei denen sie zu Hydriden kathodisch reduziert werden.

Somit erhebt sich die Frage, wie diese Einwirkungsart der Hydride gedeutet werden kann?

Höchstwahrscheinlich ist das mit der dissoziativen Chemisorption der Hydridmoleküle verbunden. Seit langem ist bereits bekannt, daß die an reiner Metalloberfläche wirksamen Adsorptionskräfte so stark auftreten, daß sie die sich adsorbierenden Molekeln spalten. So haben z. B. Suhrmann u. Mitarb. [8] festgestellt, daß die an dünnen, im Hochvakuum aufgedampften Eisenfilmen adsorbierten H_2O-Molekeln schon bei 77 °K teilweise in H-Atome und OH-Radikalen zerfallen. Die nicht zerfallenen Wassermolekeln werden so an der Eisenoberfläche adsorbiert, daß der positive Pol der molekularen Dipole vom Metall fortweist. Wird der mit mehreren Wassermolekelschichten bedeckte Film auf Zimmertemperatur erwärmt, so bildet sich eine Oberflächenverbindung, und der Wasserstoff wird abgegeben. Ein Teil des abgeschiedenen Wasserstoffs wird am Eisen unter Abnahme des elektrischen Widerstandes des Films adsorbiert.

Inai und Kemball [9] nehmen an, daß sie sich an reiner Wolframoberfläche bei Zimmertemperatur in Gegenwart von Wasserdampf abspielenden Prozesse in den folgenden zwei Stufen verlaufen können:

1. $H_2O + 2W = WH + WOH$,

2. $H_2O + WH = WOH + H_2$.

Nach Waclawski und Hughey [10] beginnt ein meßbarer Verlauf der Stufe 2 erst nach dem Erreichen eines genügenden Bedeckungsgrades der Metalloberfläche mit OH-Radikalen, was während der ersten Stufenreaktion auftritt. Durch Auffahlen weiterer H_2O-Molekeln wird die Desorption des Wasserstoffs stark gefördert.

Wie bekannt, ist im Wolfram die Wasserstofflöslichkeit sehr gering, so daß in diesem Fall praktisch kein merkliches Eindringen der H-Atome in die Metallphase berücksichtigt werden muß. Beim Eisen sind Wei und Simmons [11] der Ansicht, daß die dissoziative Chemisorption des Wassers, die die

Wasserstoffbeladung hervorruft, für die beobachtete Versprödung der hochfesten Stähle verantwortlich ist. Newman und Shreir [2] haben dem Wasser auch eine gewisse Fähigkeit zugeschrieben, das Eindringen des Wasserstoffs ins Eisen zu fördern.

Ebenso wie das Wasser werden auch andere, stärker wirkende Promotoren des Wasserstoffeindringens, wie Schwefelwasserstoff und Arsenwasserstoff, an reinen Metalloberflächen chemisorbiert und zerlegt.

Nach Roberts und Ross [12] verläuft die Chemisorption des gasförmigen Schwefelwasserstoffs am Eisen schon bei 193 °K mit unmeßbar großer Geschwindigkeit. Bei höheren Bedeckungsgraden als ca. 0,7 wird bei dieser Temperatur die Desorption von Wasserstoff beobachtet. Die zitierten Verfasser nehmen an, daß die adsorbierten, während der Dissoziation gebildeten Wasserstoffatome sich in einem metastabilen Zustand gegenüber dem gasförmigen Wasserstoff befinden.

Wie Perdereau und Oudar [13] mit der Diffraktion langsamer Elektronen bewiesen haben, ruft die Einwirkung des Schwefelwasserstoffs auf verschiedene Kristallflächen des Nickels den Zerfall der H_2S-Moleküle und die Adsorption von Schwefelionen hervor. Diese Ionen besetzen allmählich jedes zweite Nickelatom, bis sich schließlich eine zweidimensionale Sulfidschicht bildet. Das Verhalten des dabei freiwerdenden Wasserstoffs ist unbekannt.

Nach Hayward [14] kann die dissoziative Chemisorption solcher Hydride wie NH_3, PH_3 und AsH_3 in folgenden zwei Stufen verlaufen:

1. $$Y\mathrm{H}_3 + 2M = MY\mathrm{H}_2 + M\mathrm{H},$$

2. $$M\mathrm{H} + Y\mathrm{H}_3 = MY\mathrm{H}_2 + \mathrm{H}_2.$$

Hierin bedeuten: M Metallatom an der Oberfläche, Y — Element der VA-Gruppe.

Alle bisherigen Untersuchungen über die dissoziative Chemisorption verschiedener Moleküle wurden im Hochvakuum unter Benutzung physikalischer Meßmethoden durchgeführt, die es nicht erlauben, das etwaige Eindringen des Wasserstoffs in die Metallphase zu verfolgen. Die meisten Verfasser sind der Auffassung, daß sich die abgespalteten H-Atome entweder an der Metalloberfläche befinden oder als H_2-Moleküle in die Gasphase entweichen. Im Gegensatz nimmt Hayward [14] an, daß ein Teil des abgeschiedenen Wasserstoffs ins Innere des Metalls gelangt.

Die Situation ist jedoch äußerst kompliziert. Wie z. B. die Messungen der Desorptionskinetik bewiesen haben, kommt der an der Metalloberfläche adsorbierte Wasserstoff in mehreren Formen vor, und über den Zustand des Wasserstoffs an der Metall-Elektrolyt-Grenzfläche ist noch weniger als über die Metall-Gas-Phasengrenze bekannt.

Gewöhnlich wird angenommen, daß die sich entladenden Wasserstoffatome primär an der Kathodenoberfläche adsorbiert werden müssen, bevor sie in das Metall eindringen können. Die Eindringungsgeschwindigkeit sollte also vom Bedeckungsgrad der wirksamen Metalloberfläche mit adsorbierten H-Atomen abhängig sein. Bei Eisen gibt es jedoch keine zuverlässigen Angaben über diesen Wert. Meßergebnisse, die man im Schrifttum finden kann, stimmen mit der obigen Annahme nicht.

Manche Verfasser, wie z. B. Frumkin [15], nehmen an, daß die elektrolytische Entladung der Wasserstoffionen in einem gemeinsamen Schritt mit dem Eindringen der H-Atome in das Metall verlaufen kann.

Auf Grund gegenwärtiger Erkenntnisse kann man leider nicht entscheiden, welche Anschauung richtig ist. Über den Mechanismus der Einwirkung von Promotoren auf die Kinetik des Wasserstoffeindringens in Metalle können nur Vermutungen angestellt werden.

Über die in dieser Arbeit besprochenen Katalysatorgifte kann man vermuten, daß die dissoziative Chemisorption der Hydride von Arsen, Schwefel usw. eine Erniedrigung der Aktivierungsenergie und damit eine Beschleunigung des Wasserstoffeindringens hervorrufen. Nach dieser Auffassung würde das Promotorhydrid als Transportmittel des Wasserstoffs in die Metallphase von außen her wirken.

Es kann auch vermutet werden, daß die Chemisorption der Promotorhydride eine Schwächung der Anziehungskräfte zwischen den in der Oberflächenschicht vorhandenen Metallatomen hervorruft; das könnte das Eindringen des Wasserstoffs erleichtern.

Man kann nicht ausschließen, daß die Hydridradikalen fähig sind, die Austrittspunkte der Versetzungen zu dekorieren, die in der Oberflächenschicht bestehenden Strukturdefekte zu vertiefen und vielleicht sogar neue Störstellen zu bilden. Dadurch würde der Eintritt des Wasserstoffs in die Metallphase auch wesentlich erleichtert werden.

Die Frage, ob diese Auffassungen richtig sind, kann nur durch weitere Untersuchungen beantwortet werden.

Literatur

[1] M. ŚMIAŁOWSKI und Z. SZKLARSKA-ŚMIAŁOWSKA, Roczniki Chem., **29**, 85 (1955)

[2] J. F. NEWMAN und L. L. SHREIR, Corr. Sci., **9**, 631 (1969)

[3] R. D. McCRIGHT und R. W. STAEHLE, J. Electrochem. Soc., **121**, 609 (1973)

[4] T. ZAKROCZYMSKI, Z. SZKLARSKA-ŚMIAŁOWSKA und M. ŚMIAŁOWSKI, Werkstoffe u. Korrosion, **26**, 617 (1975)

[5] E. LUNARSKA, Z. SZKLARSKA-ŚMIAŁOWSKA und M. ŚMIAŁOWSKI, Werkstoffe u. Korrosion, **26**, 624 (1975)

[6] M. POURBAIX, "*Atlas of Electrochemical Equilibria in Aqueous Solutions*", Pergamon Press and Cebelcor, 1966

[7] H. SIEBERT, Z. anorg. allgem. Chem., **274**, 24 (1953)

[8] R. SUHRMANN, J. M. HERAS, L. V. HERAS und G. WEDLER, Ber. Bunsenges. Phys. Chem., **72**, 855 (1968)

[9] H. INAI und C. KEMBALL, Proc. Roy. Soc. (London), A **302**, 399 (1968)

[10] J. WACLAWSKI und L. R. HUGHEY, Surface Sci., **19**, 464 (1970)

[11] R. P. WEI und G. W. SIMMONS, Proc. Int. Conf. on Stress Corrosion Cracking and Hydrogen Embrittlemen, Unieux-Firminy (1973)

[12] M. W. ROBERTS und J. R. H. ROSS, Trans. Farad. Soc., **62**, 2301 (1966)

[13] M. PERDEREAU und J. OUDARD, Surface Sci., **20**, 801 (1970)

[14] D. O. HAYWARD, "*Chemisorption and Reactions on Metallic Films*", Academic Press (1971), Vol. 1

[15] A. N. FRUMKIN, "*Advances in Electrochemistry and Electrochemical Engineering*", Vol. 3, Interscience, New York 1963, p. 93

H.-J. ENGELL, K. BOHNENKAMP, E. RIECKE, E.-G. DAFFT und F. KUSTER*)

Max-Planck-Institut für Eisenforschung GmbH, Düsseldorf

WASSERSTOFFAUFNAHME UND WASSERSTOFF-VERSPRÖDUNG BEI DER KORROSION VON STAHL

Bei vielen Korrosionsvorgängen wirkt als kathodischer Teilvorgang die Reduktion der Wasserstoffionen des angreifenden Elektrolyten mit. Intermediär entstehender atomarer Wasserstoff oder Protonen können in das Metall eindringen und zu verformungsarmer Rißbildung führen. Versuche an Stählen mit gleicher Festigkeit und sehr unterschiedlichem Gefüge zeigen, daß die Gefügeausbildung starken Einfluß auf diesen Vorgang hat. Während der vergütete Werkstoff die übliche, mit steigender Festigkeit zunehmende Empfindlichkeit gegen Versprödung durch Wasserstoff zeigt, erwies sich der gleiche Stahl im patentiert-gezogenen Zustand bei gleicher Festigkeit als weitgehend unempfindlich. Dies trifft allerdings nur bei Belastung in Richtung der Zugverformung zu, nicht z. B. bei Torsionsbeanspruchung. Eine Glühbehandlung des gezogenen Drahtes von nur 5 Sekunden bei 500 °C führt bereits zu deutlicher Steigerung der Empfindlichkeit.

Untersuchungen der Rißausbreitung in höherfesten Stählen bei Korrosionsangriff werden durchgeführt, um den Beitrag der Wasserstoff-Versprödung von dem der Spannungsrißkorrosion durch anodische Metallauflösung am Rißgrund zu unterscheiden. Offenbar kann die Rißausbreitung grundsätzlich durch beide Mechanismen erfolgen. Bei kathodischem Schutz höherfester Stähle, z. B. in Seewasser, sind daher enge Potentialgrenzen einzuhalten.

Weiter wird über Untersuchungen zum Einfluß von Inhibitoren und Promotoren auf den Mechanismus der Reduktion von Wasserstoffionen und auf die stationäre Aktivität der Wasserstoffatome an Metalloberflächen berichtet. Hierzu wird eine Meßzelle verwendet, die durch eine Membran des untersuchten Metalls in zwei Teile geteilt ist. Auf beiden Seiten der Membran werden Strom- und Potentialmessungen vorgenommen.

Versuche an Pd ergaben, daß die besonders wirksamen Promotoren wie z. B. As_2O_3 sowohl auf die Reduktion der H^+-Ionen zu H-Atomen (VOLMER-Reaktion) wie auf die nachfolgende chemische oder elektrochemische Rekombination zu H_2 (TAFEL- bzw. HEYROWSKI-Reaktion) einwirken.

*) Der Vortrag war ein Auszug aus einer größeren Untersuchung. Deshalb ist er hier nur in der Zusammenfassung wiedergegeben.

V. I. Spitzin, M. M. Kurtepow und R. L. Baru

Institut für physikalische Chemie der Akademie der Wissenschaften der UdSSR, Moskau

ELEKTROCHEMISCHE EIGENSCHAFTEN UND KORROSIONS-CHARAKTERISTIKEN VON METALLISCHEM TECHNETIUM UND MIT TECHNETIUM LEGIERTEM STAHL

Zusammenfassung

1. Die anodischen Auflösungsgeschwindigkeiten des Technetiums haben in 10 n Schwefelsäure und 2 n Salzsäure bei 25 °C sehr niedrige Werte von 10^{-3} bzw. 10^{-4} g/m²h.
2. Technetium besitzt eine niedrigere Wasserstoffüberspannung im Vergleich zu Gebrauchsmetallen und Legierungen. Es kommt in seinem Verhalten im kathodischen Bereich den Platinmetallen sehr nahe.
3. Das Legieren eines 18.8. kohlenstoffarmen Stahls mit Technetium steigert seine Anodenpassivierung und Kathodeneffektivität und, als Folge, die Neigung zur Selbstpassivierung in Schwefel- und Salzsäurelösungen.
4. Das Legieren des 18.8. kohlenstoffarmen Stahls mit Technetium erhöht seine Beständigkeit gegen Pittingkorrosion in Chloridlösungen. Technetium besitzt eine ähnliche Wirkung auf die Lochfraßneigung von rostfreiem Stahl wie Molybdän.

In letzter Zeit steht das in der Atomindustrie anfallende Abfallprodukt, das langlebige Isotop ^{99}Tc ($T_{1/2} = 2{,}12 \cdot 10^5$ Jahre) (Technetium) in ziemlich großen Mengen zur Verfügung [1].

Seine physikalischen Eigenschaften wurden schon mehrmals untersucht [2]—[4].

Angaben über elektrochemische Eigenschaften und die Korrosionscharakteristik des metallischen Technetiums wurden bisher von Cartledge [5] gemacht. Wie aus dieser Arbeit hervorgeht, hat Technetium eine sehr hohe Korrosionsbeständigkeit im breiten Potentialbereich in 1 n Schwefelsäurelösung bei 25 °C. Unter diesen Bedingungen ist die anodische Oxydationsgeschwindigkeit des Technetiums bis zum Potential +650 mV der Stromdichte von 10^{-6} A/cm² äquivalent. Bei positiveren Potentialen wird eine starke anodische Stromdichtesteigerung festgestellt, die dem Anfang der Oxydation des Technetiums zu Technetiumsäure entspricht. Die Neigung der Tafel-Geraden, die der kathodischen Entwicklung von Wasserstoff auf Technetium in dem genannten Medium entspricht, beträgt 42 mV. In der zitierten Arbeit wurde jedoch die Oxydationsgeschwindigkeitsbewertung von Technetium im Bereich der Passivierungspotentiale nur nach Strommessung der Anodencharakteristik des Technetiums ohne Rücksicht auf den Selbstauflösungsprozeß (mit Wasserstoff-Depolarisation) durchgeführt. Es fehlte außerdem ein sicheres Kriterium für die völlige Entfernung des im Laufe der elektrochemischen Bearbeitung auf der Elektrode adsorbierten Wasserstoffs, dessen anodische Oxydation beim Technetium genügend schnell verläuft. Das begrenzte auch wesentlich die Genauigkeit dieser Messung.

In den vorliegenden Untersuchungen wurden die tatsächlichen Anodencharakteristiken des Technetiums durch unmittelbare Geschwindigkeitsmessung seines Übergangs in die Lösung erhalten. Die Lösungsgeschwindigkeiten des Technetiums wurden nach Analysenresultaten der Arbeits- und Eichlösungen durch die Methode der Elektroabscheidung auf einem Substrat errechnet. Ihre relative Radioaktivität wurde mit Hilfe des Stirnzählers gemessen.

Für die Vorbereitung der Eichlösungen diente Kaliumpertechnetat. Die Substrate bestanden aus einem titanlegierten rostbeständigen Stahl. Die Kathodenstromdichte betrug 0,9 A/cm². Der Elektrolyseprozeß verlief bei einer konstanten Temperatur von 25 °C. Bei der weiteren Untersuchung wurde festgestellt, daß die Ausbeute der Technetiumabscheidung in 10 n Schwefelsäure- und 2 n Salzsäurelösung 90% betrug. Die ausgearbeitete Methode erlaubt die Bestimmung eines Technetiumgehaltes in Lösungen mit einer Empfindlichkeit von $0{,}55 \cdot 10^{-6}$ g/*l*. Die in dieser Untersuchung gegebenen Prüfbedingungen (50 m*l* der Arbeitslösung, Elektrodenfläche — 5 cm², Prüfungszeit — 5 Stunden) entsprechen einer Korrosionsgeschwindigkeit des Technetiums von $1{,}1 \cdot 10^{-5}$ g/m²h.

Für die elektrochemischen Versuche wurde eine aus Technetium-Folie (Reinheitsgrad 99,92%) erzeugte spatenförmige Elektrode mit einer Arbeitsfläche von 16 × 16 mm angewandt. Der Musterfuß wurde mit Hilfe des Halters aus Teflon isoliert. Die Prüfungen wurden bei 25 °C in potentiostatischer Arbeitsweise in den belüfteten 10 N Schwefelsäure- und 2 N Salzsäurelösungen durchgeführt. Vor den Untersuchungen wurde die Elektrode sorgfältig in folgender Reihenfolge präpariert: zuerst mit feinkornigem Schmirgel bearbeitet, dann mit Wasser gewaschen, mit Azeton entfettet, einige Sekunden in konzentrierter Salzsäure geätzt und endlich wieder mit destilliertem Wasser gewaschen. Die im Text angegebenen Potentiale beziehen sich auf die Normalwasserstoffelektrode.

Vorläufige Untersuchungen zeigten, daß die anodische Oxydation des Technetiums nach 30 bis 60 Minuten stationär verlief, die Geschwindigkeit des Oxydationsprozesses verminderte sich in dieser Zeit nur unwesentlich. Um die Genauigkeit der Messung der stationären Lösungsgeschwindigkeit des Technetiums zu verbessern, wurde die Elektrode bei jedem gegebenen Potential 5 Stunden gehalten.

Abb. 1 zeigt ein „Potential-Korrosions-Diagramm" des Technetiums in 10 N Schwefelsäure bei 25 °C, das aus Untersuchungen der radiometrischen Messung der Oxydationsgeschwindigkeit entworfen wurde.

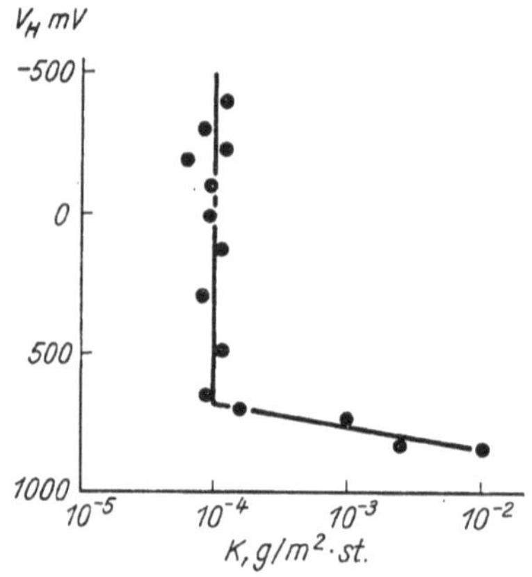

Abb. 1: Potential-Korrosions-Schaubild von Technetium in 10 N Schwefelsäure bei 25 °C.

Die anodische Oxydationsgeschwindigkeit des Technetiums bleibt praktisch im breiten Potentialbereich von −500 bis +700 mV ohne Veränderung und hat einen Wert von 10^{-4} g/m²h.

Bei Potentialen, die positiver als +700 mV sind, ist eine sehr intensive anodische Oxydation des Technetiums zu beobachten. Dieser Potentialbereich entspricht laut Diagramm [6] dem Übergang des Technetiums in die Lösung in den siebenwertigen Zustand, d. h., einer Oxydation bis zum Pertechnetat-Ion. In Salzsäure ist die Anodencharakteristik des Technetiums dieselbe.

Bei Potentialen, die negativer als das Transpassivpotential des Technetiums sind, war seine Lösungsgeschwindigkeit in Salzsäurelösung bedeutend höher als die in Schwefelsäurelösung und betrug 10^{-3} g/m²h. Unter diesen Bedingungen ist der Valenzzustand des Technetiumübergangs in die Lösung kleiner als 7+, sein genauer Wert konnte jedoch noch nicht festgestellt werden. Bei der Berechnung der Ströme der anodischen Oxydationsgeschwindigkeit des Technetiums in diesem Potentialbereich war nach radiometrischen Lösungsanalysen die Valenz der Elektrodenreaktion als 7+ angenommen worden. Unter Berücksichtigung dieser Voraussetzung sind die errechneten Werte des Stromverlaufes höher als die wirklichen.

In Schwefelsäure- und Salzsäurelösungen hatte der die anodische Oxydationsgeschwindigkeit repräsentierende Strom des Technetiums im angegebenen Potentialbereich eine Höhe von 10^{-8} und 10^{-7} A/cm². Demnach sind im Potentialbereich unterhalb der Transpassivpotentials der nach Angaben der radiometrischen Lösungsanalyse in 10 N Schwefelsäure berechneten Werte der Oxydationsgeschwindigkeit des Technetiums (in Stromdichteeinheiten) zwei Zehnerpotenzen niedriger im Vergleich zu den in der Arbeit [5] angegebenen, die durch eine Messung des Anodenstromes des Metalls in 1 N Schwefelsäure ermittelt worden sind. Es ist dabei zu berücksichtigen, daß bei einer Schwefelsäurekonzentration von 10 N fast die maximale Korrosionsaktivität dieses Systems vorliegt. Das stationäre Potential des Technetiums in belüfteten Säurelösungen wurde nach etwa 15 Stunden erreicht und hatte einen Wert von +680 mV.

In Abb. 2 ist eine in potentiostatischer Arbeitsweise gemessene Kathodencharakteristik des metallischen Technetiums in 10 N Schwefelsäurelösung dargestellt.

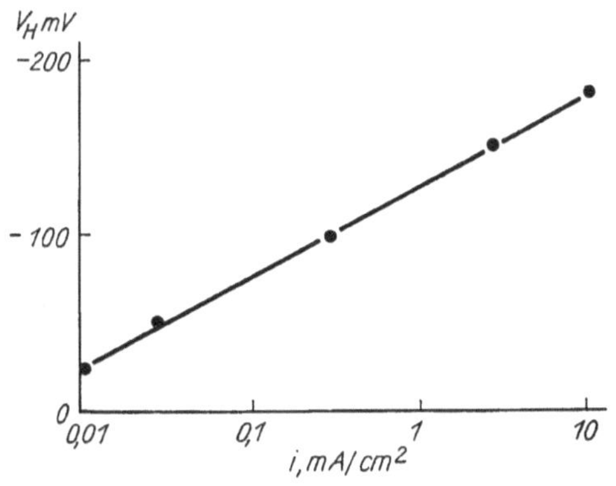

Abb. 2: Kathodische Charakteristik des Technetiums in 10 N Schwefelsäure bei 25 °C.

Der Koeffizient „b" der TAFEL-Gleichung für die kathodische Wasserstoffentwicklung und der Überspannungswert bei einer Stromdichte von 10 mA/cm² sind für Technetium 0,052 bzw. 0,22 V (bezogen auf den Wert des Gleichgewichtspotentials der Wasserstoffelektrode in 10 N Schwefelsäure). Laut Literaturangaben [7] sind diese Parameterwerte für die meisten Metalle und Legierungen bedeutend höher (z. B. für Eisen 0,12 und 0,56 V).

Andererseits kommt das Technetium entsprechend der Charakteristik der kathodischen Wasserstoffentwicklung den Platinmetallen nahe, deren Koeffizient „b" = 0,04 V beträgt und die eine Überspannung von 0,1 bis 0,3 V bei einer Stromdichte von 10 mA/cm² besitzen [7]. Somit ist für Technetium eine niedrigere kathodische Überspannung der Wasserstoffentwicklung im Vergleich zu breit anwendbaren Metallen und Legierungen kennzeichnend.

Anoden- und Kathodencharakteristiken des Technetiums bieten die prinzipielle Möglichkeit seiner Anwendung als Legierungselement, das die Passivierbarkeit und die Kathodeneffektivität der Legierungen erhöhen kann. Damit vergrößern sich auch die Neigung zur Selbstpassivierung und die Korrosionsbeständigkeit. In diesem Zusammenhang verdient die Untersuchung der Korrosionsprozesse von Technetiumlegierungen besonderes Interesse. Als Untersuchungsmethode kann auch hier die sehr empfindliche radiometrische Analyse angewandt werden, da sie wichtige Informationen über den Wirkungsmechanismus der Legierungszusätze zu geben imstande ist.

In der vorliegenden Arbeit werden Untersuchungsergebnisse über die Wirksamkeit des Technetiums als Legierungselement in einem kohlenstoffarmen rostfreien Stahl mitgeteilt. Der Stahl war mit 1% Technetium legiert. Die Elektroden für die Untersuchung waren zylinderförmig (Durchmesser — 5 mm, Höhe — 15 mm). Sie hatten Innengewinde, um in einem aus Teflon hergestellten Spezialhalter befestigt werden zu können. Die mit Technetium legierten als auch unlegierten Stahlmuster wurden zuerst einer mechanischen Bearbeitung, dann einer Wärmebehandlung unterzogen, Glühen bei 1050 °C, 15 min und nachfolgendes Abschrecken in Wasser. Vor den Untersuchungen wurden die Elektroden sorgfältig vorbehandelt: zuerst mit Schmirgelpapier geschliffen, dann mit Azeton entfettet und endlich mit destilliertem Wasser gewaschen. Die vergleichenden Untersuchungen zur Korrosionsneigung der mit Technetium legierten als auch unlegierten Stähle werden in Schwefelsäure- und Salzsäurelösungen bei 25 °C durchgeführt.

Die Elektroden wurden nach Einsetzen in die Lösung mit einem Kathodenstrom von 10 mA/cm² 5 Minuten aktiviert. Nach Abschalten des Stromes blieben die Elektroden in der Lösung bis zur Einstellung eines stationären Potentials. Die Potentiale wurden mit Hilfe eines Röhrenvoltmeters (mit

Schreiberausgang) gemessen. Aus der Höhe des stationären Potentials wurde auf den Zustand (aktiv — passiv) der Legierung in der jeweiligen Lösung geschlossen.

Diese Messungen wurden durch die Bestimmung der Auflösungsgeschwindigkeit des Stahls ergänzt. Die Anoden- und Kathodencharakteristiken in 10 N Schwefelsäurelösung bei 25 °C der mit Technetium legierten als auch unlegierten Stähle wurden gegenübergestellt.

Es wurden „Potential-Korrosionsdiagramme“ (Anodencharakteristiken) aufgestellt auf Grund von Messungen der stationären Auflösungsgeschwindigkeit des Stahls (Gewichtsverlustmessungen) mit Hilfe der Gewichtsmethode unter porentiostatischen Bedingungen.

Bei der Untersuchung der Anodencharakteristik des mit Technetium legierten Stahls wurden die Technetiumanreicherung auf der Elektrodenoberfläche und sein Gehalt in der Lösung mit speziell ausgearbeiteten Methoden gleichzeitig bestimmt. Für die Beurteilung der Technetiumanreicherung auf der Stahloberfläche wurden die Elektroden (nach den Versuchen) in 6 N Salpetersäure getaucht und dann die absolute Aktivität dieser Lösung mit Hilfe der Einrichtung „Protoka“ („Durchfluß“) bestimmt.

Die Partialauflösungsgeschwindigkeit von Technetium wurde über eine Lösungsanalyse mit Hilfe der oben genannten Methode der elektrolytischen Abscheidung des Metalls aus dem Substrat und nachfolgender Messungen seiner relativen Radioaktivität mit einem Stirnzähler errechnet.

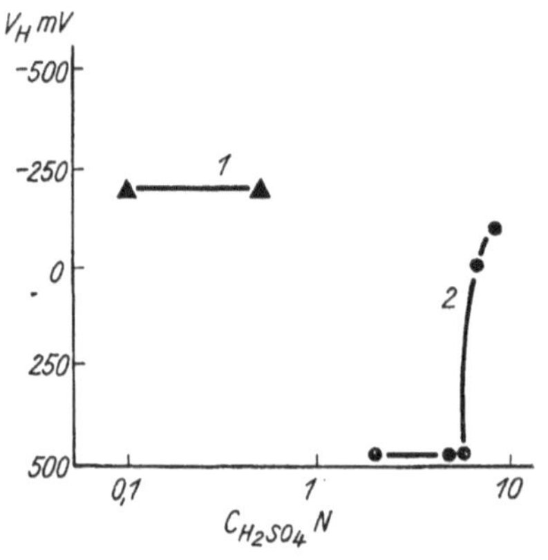

Abb. 3: Einflußwirkung der Konzentration von Schwefelsäure auf die stationären Potentiale des kohlenstoffarmen Stahls 18-8 (*1*) und desselben mit 1% Tc (*2*) bei 25 °C.

Bei Konzentrationen unterhalb von 6 N in Schwefelsäure und 0,5 N in Salzsäure tritt, wie Abb. 3 zeigt, bei dem technetiumlegierten Stahl Selbstpassivierung auf. Unter diesen Bedingungen stellte sich das Stahlpotential in Schwefelsäurelösungen bei + 500 mV ein. Die Korrosionsgeschwindigkeit dieses Stahls liegt z. B. in 5 N Schwefelsäure auf einem Niveau von 0,02 mm/Jahr. In Salzsäurelösungen geht jedoch der Stahl während der Selbstpassivierung von aktivem Zustand in einen solchen über, in dem Lochfraßkorrosion möglich ist. Andererseits befindet sich der nicht mit Technetium legierte Stahl sogar in 0,05 N Schwefelsäure- (Abb. 3) und Salzsäurelösung im aktiven Zustand und zeigt keine Neigung zur Selbstpassivierung. In 5 N Schwefelsäure löste sich dieser Stahl mit einer Geschwindigkeit von 1 mm/Jahr.

Das Legieren des Stahls mit Technetium erhöhte seine anodische Passivierbarkeit (Abb. 4) und Kathodeneffektivität (Abb. 5). Diese elektrochemischen Parameter, die die Neigung des Stahls zur Selbstpassivierung charakterisieren, verändern sich mit der Zeit und nehmen ihre stationären Maximalwerte in 40—60 Minuten an. Die Legierung des Stahls mit Technetium besitzt eine verminderte Auflösungsgeschwindigkeit am Passivierungspotential (Abb. 4). Die Überspannung des Kathodenprozesses der Wasserstoffentwicklung (Abb. 5) in 10 N Schwefelsäure nimmt ebenfalls unter stationären Bedingungen ab (um 100 mV bei einer Stromdichte von 10 mA/cm²).

Laut der im Potentialbereich des aktiven Stahlzustandes durchgeführten radiometrischen Untersuchungen gehen nur 3% von der gesamten Menge des während des Korrosionsprozesses frei werdenden Technetiums in 10 N Schwefelsäurelösung über. Der Hauptteil des Technetiums, das beim Lösungsvorgang freigesetzt wird, häuft sich auf der Oberfläche des Stahls. Dabei ist zu erwähnen, daß die Abscheidung des Technetiums aus der Lösung auf dem Stahl bei Potentialen, die positiver als —250 mV (Abb. 4) liegen, entsprechend unserer Untersuchungen nicht mit merklichen Geschwindigkeiten vor sich geht.

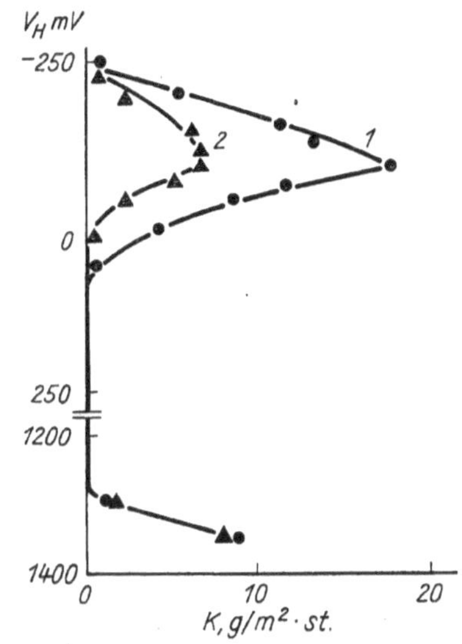

Abb. 4: Potential-Korrosions-Schaubild „Potential—Korrosion" für einen kohlenstoffarmen Stahl 18-8 (*1*) und desselben mit 1% Tc (*2*) in 10 N Schwefelsäure bei 25 °C.

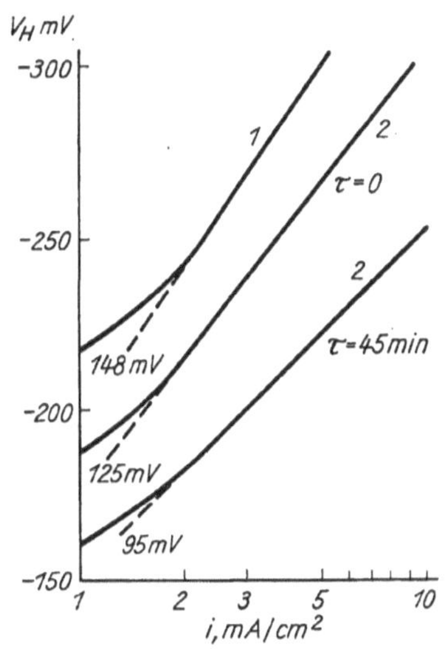

Abb. 5: Kathodische Charakteristiken des kohlenstoffarmen Stahls 18-8 (*1*) und desselben mit 1% Tc (*2*) in 10 N Schwefelsäure bei 25 °C.

τ = 0, Messung sofort nach Einsetzen der Elektrode in die Lösung: τ = 45, Messung nach einer Wartezeit von 45 min.

Eine Abschätzung zeigte, daß das Verhältnis zwischen den Mengen des Technetiums, die unmittelbar in der Oberflächenschicht des Stahls bzw. im Korrosionsfilm gebunden sind, 1 : 9 beträgt. Für die Durchführung dieser Abschätzung wurde das nur lose am Metall anhaftende poröse Korrosionsprodukt zuerst mit Hilfe einer azetongetränkten Watte und dann mit destilliertem Wasser behandelt. Danach erfolgte die Bearbeitung der Elektrode in Salpetersäure. Es wurde z. B. 2,1 mg/cm² Technetium nach einer Dauer von 9 Stunden unmittelbar in der Oberflächenschicht des Metalls am Passivierungspotential des Stahls in 10 N Schwefelsäure (—125 mV) nachgewiesen. Berücksichtigt man die niedrige Adhäsion des Korrosionsfilms mit dem Metall, dann ist die Steigerung der Anodenpassi-

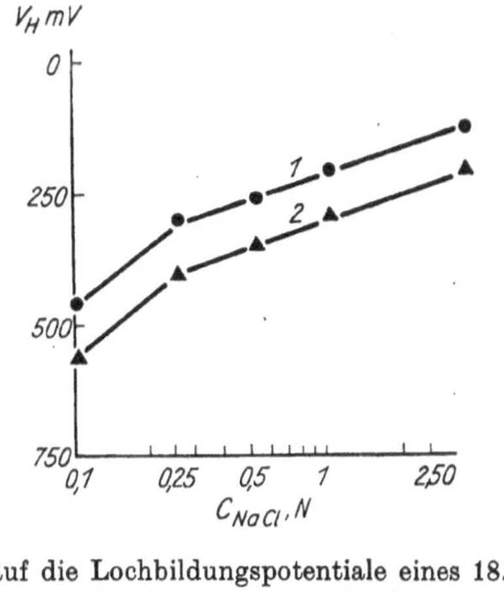

Abb. 6: Einfluß von Technetium (1%) auf die Lochbildungspotentiale eines 18.8. kohlenstoffarmen Stahls in Natriumchloridlösungen bei 25 °C.

1 — ohne Tc; *2* — mit Tc.

vierung und Kathodeneffektivität des mit Technetium legierten Stahls in großen Maße auf die Anreicherung des Technetiums unmittelbar in der Oberflächenschicht zurückzuführen.

Es wurde eine bestimmte Verteilung des während des Korrosionsprozesses im Passivbereich (bis 200 mV, Transpassivpotential von Te) freigesetztem Technetium zwischen Lösung und Probenoberfläche festgestellt. Eine ausreichende Reproduzierbarkeit für eine richtige qualitative Bewertung dieser Verteilung lag jedoch nicht vor. Das Technetium geht bei Potentialen positiver als +700 mV in Lösung entsprechend seinem Gehalt im Stahl und wird nicht auf der Elektrodenoberfläche angereichert.

Das Legieren des Stahls mit Technetium erhöht auch seine Beständigkeit gegen Pittingkorrosion in Chloridlösungen.

Entsprechend (Abb. 6) veredelt sich das amperostatisch bei einer Stromdichte von 0,05 mA/cm² bestimmte Lochbildungspotential des Stahls in 0,1—2,5 N Natriumchloridlösungen bei 25 °C um 75 mV. Ein Vergleich unserer Ergebnisse mit Literaturangaben (siehe Abb. 6) zeigt, daß das Technetium auf die Lochfraßneigung von rostbeständigen Stählen eine ähnliche Wirkung wie Molybdän besitzt.

Literatur

[1] V. I. SPITZIN, A. F. KUSINA, A. F. ZARENKO u. a., Radiochemie, **12**, 617 (1970)
[2] D. E. BARKER, J. Less-Comm. Met., **8**, 435 (1965)
[3] V. I. SPITZIN, Ju. N. GOLOWANOW und O. A. BALACHOWSKIJ, Berichte der Akademie der Wissenschaften der UdSSR, **205**, 1421 (1972)
[4] V. I. SPITZIN, W. E. SINOWIEW, P. W. GELD und O. A. BALACHOWSKIJ, Berichte der Akademie der Wissenschaften der UdSSR **221**, 145 (1975)
[5] G. H. CARTLEDGE, J. Electrochem. Soc., **118**, 1752 (1971)
[6] M. POURBAIX, "*Atlas of Electrochemical Equilibria in Aquaeus Solution*", Pergamon Press (1966)
[7] K. J. VETTER, „*Elektrochemische Kinetik*", Kap. 4., Springer-Verlag (1961)

V. I. SPITZIN, A. F. KUSINA, A. A. OBLOWA, M. I. GLINKINA und L. I. STEPOWAJA*)

Institut für physikalische Chemie der Akademie der Wissenschaften der UdSSR, Moskau

ÜBER DEN MECHANISMUS DER ELEKTROCHEMISCHEN REDUKTION DES TcO_4^--IONS

Die Kathodenreduktion von Anionen ist ein höchst komplizierter Prozeß und noch nicht ausreichend untersucht worden. Das betrifft besonders die elektrochemische Reduktion von sauerstoffhaltigen Anionen der Elemente der VII. Gruppe des Periodischen Systems, darunter des Technetiums in Form des TcO_4^--Ions. In diesem Zusammenhang ist die vorliegende Arbeit von besonderem wissenschaftlichen Interesse.

Bei Technetium läßt sich eine Anzahl von Oxydationsstufen nachweisen. Es gehört zu einer Gruppe des Periodensystems, die durch eine ausgeprägte Neigung zur Komplexbildung gekennzeichnet ist. Entsprechend ihren chemischen und elektrochemischen Eigenschaften sind Technetium und Rhenium analog und weisen im Vergleich zu Mangan wesentliche Unterschiede auf. Für Technetium und Rhenium ist der siebenwertige Zustand mit einer schwach ausgesprochenen Tendenz zur Reduktion charakteristisch.

In dieser Veröffentlichung werden Ergebnisse unserer Untersuchung von Produkten der elektrolytischen Reduktion von Technetium (VII) in Form des TcO_4^--Ions an einer Platinkatode vorgelegt, wobei die Methoden der präparativen und analytischen Chemie, der Röntgenphasenanalyse, der elektronischen und Schwingungsspektroskopie, der Elektrophorese und der Dialyse angewandt wurden. Literaturangaben über analoge Untersuchungen liegen bisher nicht vor.

Die Elektrolyse wurde bei einer Stromdichte von 1 A/cm^2 unter galvanostatischen Bedingungen durchgeführt. Als Elektrolyt wurde eine Lösung mit folgender Zusammensetzung gewählt: 0,07 M $KTcO_4$, 0,71 M H_2SO_4. Die Dauer der Elektrolyse betrug 30 Minuten.

Nach Abschluß der Versuche war es gelungen, die Produkte der Elektrolyse zu isolieren und zu identifizieren. Dabei wurden einige Besonderheiten des Prozesses festgestellt, die mit der Bildung von metallischem Technetium und von Zwischenkomplexen zusammenhängen, die die Reaktion zur Bildung des elementaren Technetiums erschweren und die Stromausbeute herabsetzen.

Tabelle 1: Röntgenographische Untersuchung des metallischen Technetium mittels der Pulvermethode (Cu-Strahlung)

Reflexions-intensität	Netzebenen-abstände [Å]	Reflexions-indexe
15	2,368	100
50	2,201	002
100	2,094	101
20	1,61	102
40	1,375	110
40	1,252	103
40	1,166	112
20	1,145	201
30	0,928	203
30	0,883	211
30	0,861	114

*) Als Ergänzung wird an dieser Stelle der Vortrag gedruckt, den V. I. SPITZIN vor dem Plenum der Sächsischen Akademie der Wissenschaften zu Leipzig in der Sitzung am 13. Juni 1975 gehalten hat.

Durch die Röntgenphasenanalyse wurde die Existenz von metallischem Technetium an der Kathode bestätigt (Tab. 1, Abb. 1).

Wie aus der röntgenographischen Untersuchung hervorgeht, liegt das elektrolytische abgeschiedene metallische Technetium feinkristallin vor, wodurch das Röntgenogramm die Netzebenenabstände weniger deutlich wiedergibt.

Man kann leicht erkennen, daß die ermittelten Kennzahlen der Reflexionsintensitäten mit den Intensitäten für Rhenium und für weitere Metalle mit hexagonal dichter Packung übereinstimmen.

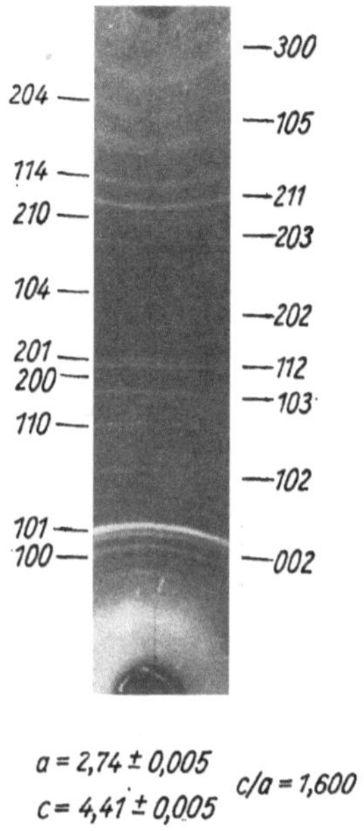

Abb. 1: Debyegramm der Probe des metallischen Technetiums (s. auch Tab. 1).

Die Werte der Gitterparameter des metallischen Technetiums entsprechen den aus der Literatur bekannten Gitterparametern und betragen $a = 2{,}737$, $c = 4{,}402$, $c/a = 1{,}610$.

Im Laufe der Elektrolyse läßt sich eine Verfärbung der Lösung nach Graubraun infolge der Anreicherung eines entstehenden Hydrolyseproduktes von Technetium(IV) beobachten.

Unsere Aufgabe war die Feststellung seiner Zusammensetzung, da gerade diese Verbindung für die Stromausbeute des metallischen Technetiums von Bedeutung ist.

Es wurden die Absorptionsspektren der Lösungen vor und nach der Elektrolyse aufgenommen (Abb. 2).

Es sind charakteristische Absorptionsbanden mit Maxima bei 250, 320 und 500 nm vorhanden. Bei Alkalisierung der Lösung verschwinden sie bei einem pH-Wert von 4 (Abb. 3), da eine weitere Hydrolyse der Verbindung erfolgt und das dunkelbraune Dioxid $TcO_2 \cdot 2H_2O$ ausfällt, dessen Zu-

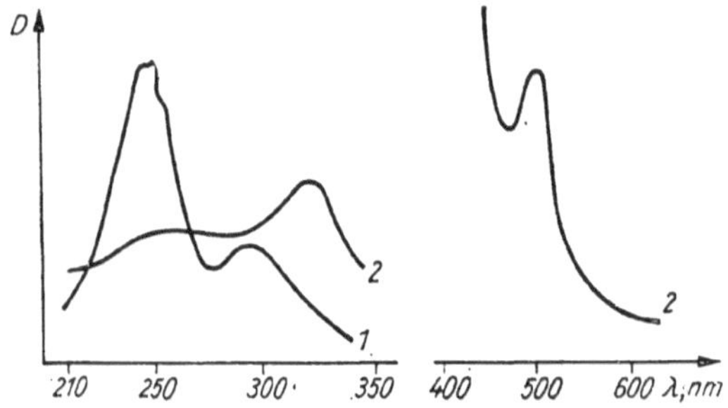

Abb. 2: Absorptionsspektren.

1 — Spektrum der Ausgangslösung von $KTcO_4$; *2* — Spektrum der Lösung nach der Elektrolyse sammensetzung durch eine chemische Analyse festgestellt wurde. Das IR-Spektrum dieser Verbindung ist durch die Bande bei $\bar{\nu} = 900\ cm^{-1}$ (Abb. 4) der Tc—O-Bindung gekennzeichnet.

Es wurde auch der Oxydationsgrad des Technetiums in der graubraunen gelösten Substanz durch cerimetrische Titration festgestellt. Dabei wurde bei einem Überschuß von Cer(IV)sulfat bis zum Pertechnetat oxydiert und mit einer geeichten Lösung von Eisen(II)sulfat rücktitriert.

Eine Rückauflösung des ausgefällten Dioxids, TcO_2-2H_2O, Schwefelsäure (0,71 M H_2SO_4) hat die Bildung des graubraunen Ausgangsstoffes zur Folge, dessen Absorptionsspektrum dem auf der Abb. 2, Kurve *2*, dargestellten Absorptionsspektrum entspricht. Diese Eigenschaft der bisher unbekannten Verbindung haben wir genutzt und ihre chemische Zusammensetzung durch Messung des pH-Wertes festgestellt. Auf Grund entsprechender Berechnungen wurde der Verbindung die Formel $[Tc(OH)_2(SO_4)_2]$ zugeschrieben.

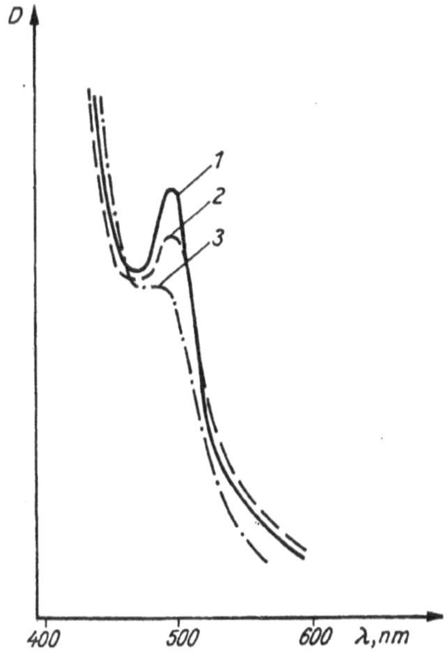

Abb. 3: Absorptionsspektren der graubraunen Lösung bei verschiedenen Aziditätswerten.
1 – pH = 0,8; *2* – pH = 1,5; *3* – pH = 3.

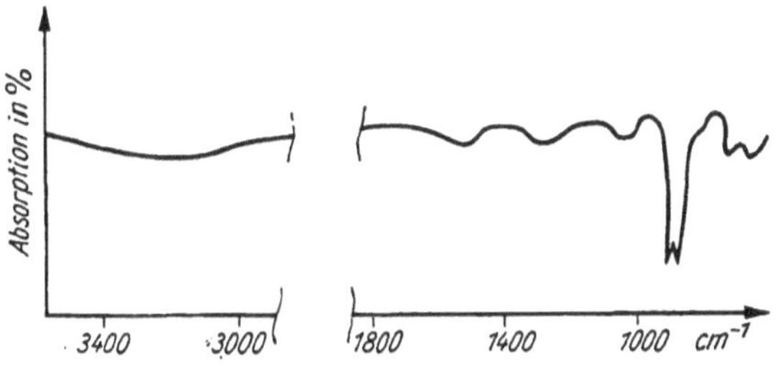

Abb. 4: IR-Spektrum von $TcO_2 \cdot 2H_2O$, gefällt aus der Elektrolytlösung.

Bei der weiteren Untersuchung der physikalischen und chemischen Eigenschaften des Komplexes wurde durch Analyse der Absorptionsspektren im sichtbaren Spektralbereich die Schlußfolgerung gezogen, daß der Träger der chromophoren Eigenschaften der Verbindung der hydrolisierte Teil des Komplexes, die $Tc(OH)_2$-Gruppe ist, da der Ersatz des SO_4^{2-}-Liganden durch Cl^- zu einem analogen elektronischen Spektrum führt.

Durch Zonenelektrophorese an Chromatografiepapier wurde die Ionenladung festgestellt. Da der Stoff radioaktiv ist, war es möglich, die Verlagerung der aktiven Zone mit Hilfe des GEIGER-MÜLLER-Zählrohrs zu beobachten, durch welches die β-Strahlung des Isotops Technetium-99 registriert wird. Auf Grund der ermittelten Angaben konnte das Verhalten der radioaktiven Ionen im elektrischen Feld nach einem zweistündigen Experiment ausgewertet werden (die Zeit wurde aus einer Versuchsreihe als eine optimale gewählt) (Abb. 5).

Wie aus Abb. 5 folgt, migrieren die beiden Ionen zur Kathode mit einer unterschiedlichen Geschwindigkeit, so daß eine Trennung möglich war. Die Komplexverbindung der graubraunen Farbe mit einer Zusammensetzung von $[Tc(OH)_2(SO_4)_2]^{2-}$ weist im Vergleich zum TcO_4^--Ion eine geringe Beweglichkeit auf, ist jedoch kein Kolloid, da sie bei der Dialyse einen dichten Zelluloidfilm passiert und sich nicht zentrifugieren läßt, wodurch ihr Ionenzustand bestätigt wird. Durch Ionenaustausch

$[Tc(OH)_2(SO_4)_2]^{2-}$ an den stark basischen Harzen AB-17 und AH-25 mit einer hohen Basizität wird die Aussage der Elektrophorese hinsichtlich der negativen Ionenladung bestätigt.

Zusammenfassend läßt sich aussagen, daß im Laufe der Elektrolyse neben der Abscheidung des metallischen Technetiums eine hydrolysierte Komplexverbindung angereichert wird, die bei einem Kathodenpotential von 0,82 V (N. W. E.), welches für die Abscheidung des Technetiums in Form eines

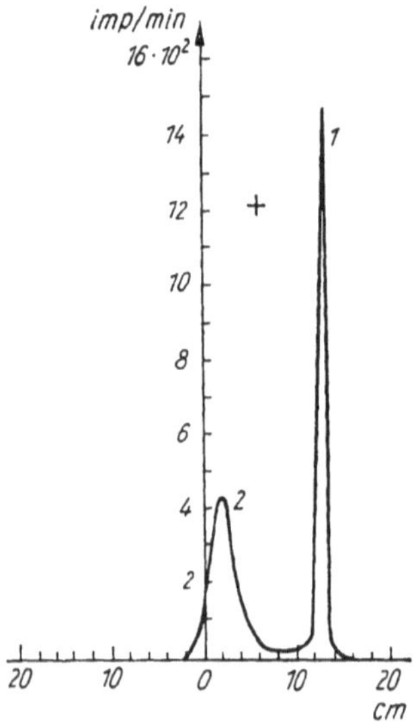

Abb. 5: Trennung der Technetium-Ionen durch Elektrophorese an chromatographischem Papier.
1 – TcO_4^-; *2* – $[Tc(OH)_2(SO_4)_2]^{2-}$.

Metalls notwendig ist, nicht entladen wird; daher wird die Bildungsreaktion des elementaren Technetiums gehemmt.

Somit gehen an der Kathode gleichzeitig drei Prozesse vor sich:

1. eine Entladung der Wasserstoffionen $2H^- \rightarrow H_2$;
2. eine Entladung der Ionen $Tc(VII) \rightarrow Tc^0$;
3. eine Entladung der Ionen $Tc(VII) \rightarrow Tc(IV)$.

I. ROSENFELD, L. FROLOWA und B. LARKIN

Institut für physikalische Chemie der Akademie der Wissenschaften der UdSSR, Moskau

CHEMISCHE PASSIVIERUNG VON STAHL DURCH INHIBITOREN

In dem vorliegenden Vortrag möchte ich über die von uns entwickelte neue Methode zur Untersuchung der Passivierungseigenschaften der Inhibitoren und der sich für das Verständnis ihres Wirkungsmechanismus ergebenden Folgen berichten.

Das Wesen der Methode besteht in der Konstruktion besonderer Diagramme der Abhängigkeit der Geschwindigkeit der Anodenreaktion (Korrosion) von dem Potential, welches der Elektrode nicht durch äußere Anodenpolarisation, sondern durch innere Anodenpolarisation, hervorgerufen direkt oder indirekt durch Korrosionsinhibitoren (chemische Passivierung), erteilt wird. Nach diesen Diagrammen kann über die wahren Auflösungsgeschwindigkeiten des Metalls bei vorgegebenen Konzentrationen der Inhibitoren oder über die Potentialwerte und den von dem Inhibitor auf die Kinetik der elektrochemischen Reaktionen ausgeübten Einfluß geurteilt werden.

Diese Methode zur Untersuchung der Inhibitoren macht es möglich, wertvolle Aussagen über den Wirkungsmechanismus der Inhibitoren in der Nähe des stationären Potentials zu erhalten, was die Methode der äußeren Anodenpolarisation nicht gestattet, die für die Konstruktion vollständiger Polarisationskurven eine merkliche Verschiebung des Potentials nach der positiven Seite hin erfordert. Es ist außerdem bei der äußeren Anodenpolyrisation schwierig, die durch die Abscheidung von Sauerstoff entstehenden Passivierungseffekte von den durch die Adsorption der Inhibitoren zustandekommenden Passivierungseffekten zu trennen. Dies ist darauf zurückzuführen, daß diese Effekte nicht additiv sind; die äußere Anodenpolarisation erleichtert meistens die Adsorption der inhibierenden Anionen und verfestigt deren chemische Bindung mit dem Metall, weshalb ihre Resultate nicht immer eindeutig sind.

Die vorgeschlagene Methode macht es auch möglich, Anodenpolarisationskurven in verdünnten Elektrolyten zu erhalten, auf welche die äußere potentiostatische Anodenpolarisation auf Grund des hohen OHMschen Potentialabfalls nicht anwendbar ist.

In Abb. 1, 2 und 3 sind Polarisationsdiagramme für Vertreter von drei Inhibitorenklassen dargestellt:

1. Bei Korrosionspotentialen leicht reduzierbare Inhibitoren.
2. Schwerlösliche Verbindungen bildende Inhibitoren.
3. Ein bei Korrosionspotentialen nichtreduzierbares gemeinsames Anion vom Typ MeO_4^{n-} aufweisende Inhibitoren.

Die Methodik der Konstruktion der Korrosionsdiagramme besteht in der Untersuchung der Abhängigkeit der Korrosionsgeschwindigkeit, des Potentials und des aktiven Teils der Elektrode von der Konzentration des Inhibitors und in der Konstruktion nach diesen Angaben von Polarisationsdiagrammen, die die Abhängigkeit der wahren Geschwindigkeit der Anodenauflösung von dem Potential widerspiegeln [1, 2].

Wie zu ersehen ist, zeigen die Polarisationsdiagramme qualitativ die gleichen Gesetzmäßigkeiten der Auflösung wie die potentiostatischen Kurven, erhalten nach der Methode der äußeren Anodenpolarisation; sie weisen alle drei Abschnitte auf, die für ein Metall kennzeichnend sind, das geneigt ist, in den passiven Zustand überzugehen, nämlich den Abschnitt der aktiven Auflösung, den des aktiv-passiven Zustandes und den des passiven Zustandes. Was die Kinetik des Prozesses anbelangt, so werden hier große Unterschiede festgestellt. Die Passivierung durch Inhibitoren führt zu viel größeren kinetischen Effekten als die anodische Passivierung; bei ein und demselben Potential kann die Auf-

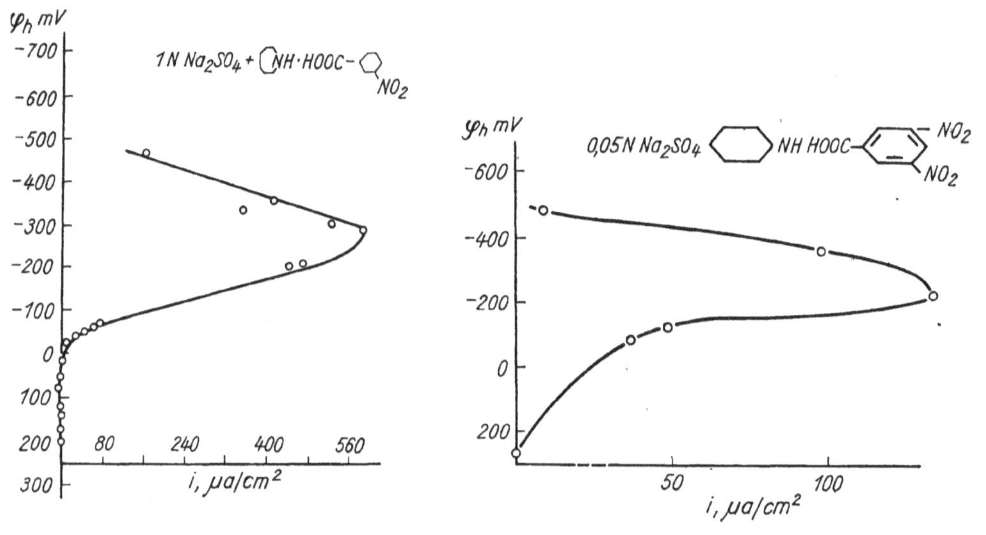

Abb. 1: Abhängigkeit der Korrosionsgeschwindigkeit von Stahl von dem Potential, das der Elektrode mit Hilfe von Hexamethyleniminnitrobenzoat und Piperidindinitrobenzoat vorgegeben wird.

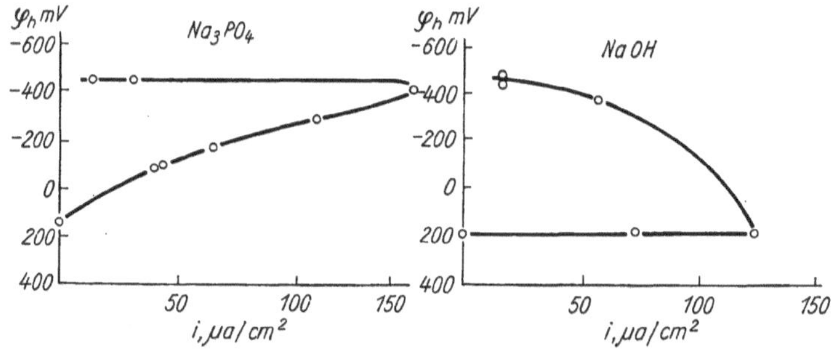

Abb. 2: Abhängigkeit der Korrosionsgeschwindigkeit von Stahl von dem Potential, das der Elektrode mit Hilfe von Alkali und Natriumorthophosphat vorgegeben wird.

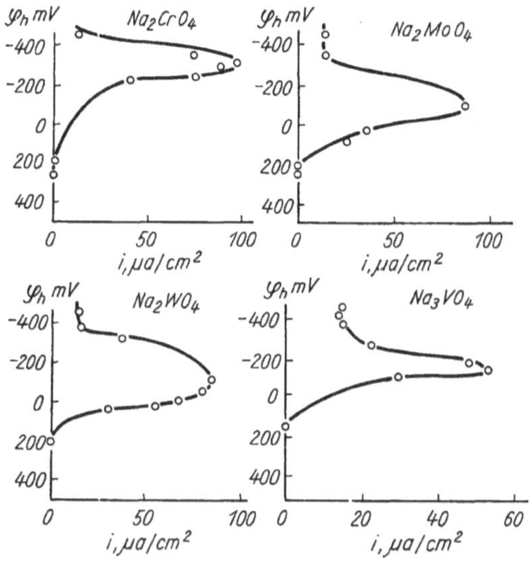

Abb. 3: Abhängigkeit der Korrosionsgeschwindigkeit von Stahl von dem Potential, das der Elektrode mit Hilfe von Natriumchromat, Natriummolybdat, Natriumwolframat, Natriumorthovanadat vorgegeben wird.

lösungsgeschwindigkeit in Gegenwart von Inhibitor um einige Größenordnungen niedriger sein als bei der anodischen Passivierung.

Mit der Verschiebung des Potentials nach der positiven Seite hin nimmt die Auflösungsgeschwindigkeit zu, beginnt jedoch nach dem Erreichen eines bestimmten Potentials, welches wir als Potential der teilweisen Passivierung bezeichnen, stark zu fallen. Die vollständige Passivierung der Elektrode tritt ein für die untersuchten Inhibitoren bei ungefähr gleichen Potentialwerten (+0,2 bis +0,25 V). Der Mechanismus der Zunahme der Auflösungsgeschwindigkeit des Stahls auf dem aktiven Abschnitt der Polarisationskurve läßt sich mehr oder weniger gut nur für die an der Anode leicht reduzierbaren Inhibitoren erklären. Was die Inhibitoren anbelangt, die schwerlösliche Verbindungen bilden (NaOH, NH_2, NH_3, Na_3PO_4, Na_2HPO_4, Na_2SiO_3 usw.) oder die Inhibitoren, die sich beim Korrosionspotential in neutralen Medien ($NaNO_2$, Na_2CrO_4, Na_2WO_4, Na_3VO_4 usw.) nicht reduzieren lassen, so ist der Mechanismus dieser Erscheinung nicht klar. Man könnte die Zunahme der Auflösungsgeschwindigkeit auf dem aktiven Abschnitt der Polarisationskurve auf die Hemmung der Anodenreaktion zurück-

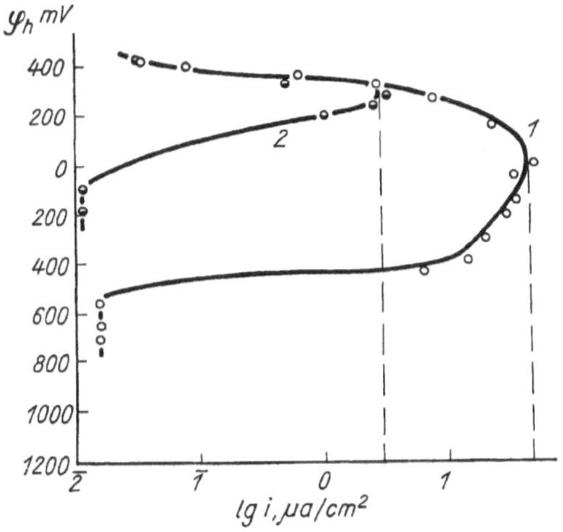

Abb. 4: Polarisationskurven, erhalten am Stahl nach der Methode der Polarisation von dem äußeren Strom (*1*) und nach der Methode der chemischen Passivierung.

1 – 0,1 *n*-Na_2SO_4, potentiostatische Polyrisation (jeder Punkt für 6 Stunden); *2* – 0,1 *n*-Na_2SO_4 + Na_2CrO_4, chemische Passivierung (6 Stunden)

führen, die zur Veredlung des Elektrodenpotentials führt, die entsprechend den Gesetzen der elektrochemischen Kinetik zu einer Beschleunigung der Anodenreaktion führen soll. Die unmittelbaren Experimente zeigen jedoch, daß in dem aktiven Auflösungsbereich die untersuchten Inhibitoren die Geschwindigkeit der Anodenreaktion nicht beeinflussen (Abb. 4). Andererseits hat die Untersuchung der wahren Geschwindigkeit der Kathodenreaktion an der reell korrodierenden Oberfläche, die durch das Vorliegen der Angaben über die Fläche des passiven Teils der Elektrode und die Größe des Korrosionsstromes möglich geworden ist, ergeben, daß die Effektivität des Kathodenprozesses in Gegenwart von ein gemeinsames Ion vom Typ MeO_4^{n-} enthaltenden Inhibitoren nicht zunimmt, was davon zeugt, daß diese Inhibitoren an dem Kathodenprozeß nicht beteiligt sind.

Es soll bemerkt werden, daß nicht alle Inhibitoren eine ebensolche Abhängigkeit der Auflösungsgeschwindigkeit von dem Potential zeigen, die oben gezeigt wurde. In dieser Beziehung ist kennzeichnend das Verhalten von Natriummetavanadat ($NaVO_3$) und Natriumorthovanadat (Na_3VO_4). Diese beiden Verbindungen sind gute Inhibitoren, wirken jedoch verschieden, obwohl die Wertigkeit des Vanadins in diesen gleich ist. Das Metavanadat bildet keinen aktiven Zweig auf der Polarisationskurve. In Gegenwart dieses Inhibitors beobachtet man eine kontinuierliche Abnahme der Korrosionsgeschwindigkeit mit dem Potential. Das Natriumorthovanadat liefert die gleichen Polarisationsdiagramme mit drei charakteristischen Abschnitten wie auch andere Inhibitoren mit dem gemeinsamen Anion vom Typ MeO_4^{n-} (Abb. 5 *a* und 5 *b*).

In anomaler Weise verhält sich auch das Natriumbenzoat. In einem breiten Bereich der Konzentrationen verändert es weder das Potential der Elektrode noch die Korrosionsgeschwindigkeit. Nur bei der Annäherung an die Schutzkonzentration beobachtet man vor dem Eintritt des vollständigen passiven Zustandes eine starke Veredlung des Potentials und das Wachstum der wahren Auflösungsgeschwindigkeit. Sowohl das Natriumorthovanadat als auch das Natriumbenzoat sind anodische Inhibitoren und zeigen in diesem Sinne zum ersten Male Fälle, wo anodische Inhibitoren die Auflösungsgeschwindigkeit bei unvollständigem Schutz nicht erhöhen. Wie weiter unten gezeigt wird, ist dies mit grundsätzlich verschiedenem Wirkungsmechanismus des Natriumorthovanadates und

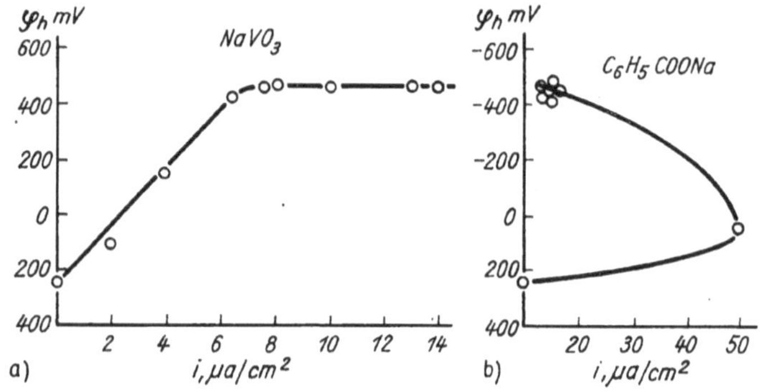

Abb. 5a: Abhängigkeit der Korrosionsgeschwindigkeit von Stahl von dem Potential, das der Elektrode mit Hilfe von Natriummetavanadat und Natriumbenzoat vorgegeben wird.

Natriumbenzoates gegenüber den anderen anodischen Inhibitoren verbunden. Wenn die meisten von uns untersuchten Inhibitoren in dem aktiven Bereich der Auflösung die Korrosion durch teilweise Passivierung der Elektrode verlangsamen, ohne einen wesentlichen Einfluß auf die Anodenreaktion auszuüben, so sind das Natriumorthovanadat und das Natriumbenzoat nicht in der Lage, die Elektrode teilweise zu passivieren, sondern verändern die Auflösungsgeschwindigkeit in einem breiten Bereich der Potentiale durch unmittelbare Verlangsamung der Anodenreaktion.

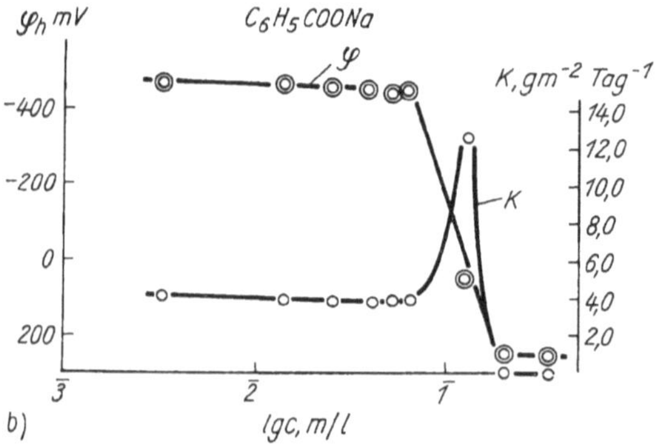

Abb. 5b: Veränderung der Korrosionsgeschwindigkeit von Stahl (g/m² Tag) und des Potentials von Stahl von der Konzentration des Natriumbenzoats.

Es verdient auch beachtet zu werden das Polarisationsdiagramm, das für Alkali erhalten wurde. Es unterscheidet sich von allen anderen dadurch, daß das Potential der teilweisen Passivierung mit dem Potential der vollständigen Passivierung zusammenfällt. Sobald das Stahlpotential durch Alkali auf einen Wert von +2 V verschoben wird, setzt sofort rasche Passivierung ein, es fällt die Auflösungsgeschwindigkeit, während das Potential konstant bleibt. Es tritt die zweifache Rolle der Hydroxylionen in Erscheinung. Auf dem aktiven Abschnitt der Polarisationskurve begünstigen sie bis zu einem bestimmten Potential (+0,2 V) die Beschleunigung des Anodenprozesses anscheinend durch die Bildung des intermediären Komplexes $MeOH^-$ und bei positiveren Potentialen (> +0,2 V) der Passivierung.

Bei der Betrachtung des Wirkungsmechanismus der Korrosionsinhibitoren sind zwei Möglichkeiten zu unterscheiden:

1. Der Inhibitor erhöht die Wirksamkeit des Kathodenprozesses.
2. Der Inhibitor übt keinen unmittelbaren Einfluß auf den Kathodenprozeß aus.

Der erste Mechanismus wird realisiert bei der Verwendung als Inhibitoren von Aminnitrobenzoaten. Diese Verbindungen lassen sich, wie gezeigt wurde, an der Eisenkathode leicht reduzieren, indem sie relativ hohe Stromdichten gewährleisten (Abb. 6), die das Potential auf Werte zu verschieben vermögen, bei denen die Bildung von Oxid höherer Wertigkeit möglich wird. Das Oxid höherer Wertigkeit entsteht dabei dadurch, daß der Inhibitor in der Rolle des Elektronenakzeptors und das niedrigere Oxid in der Rolle des Donators auftritt. Von der Richtigkeit des vorgeschlagenen Mechanismus des Kathodenprozesses kann man sich bei der Betrachtung der Ergebnisse der potentiostatischen

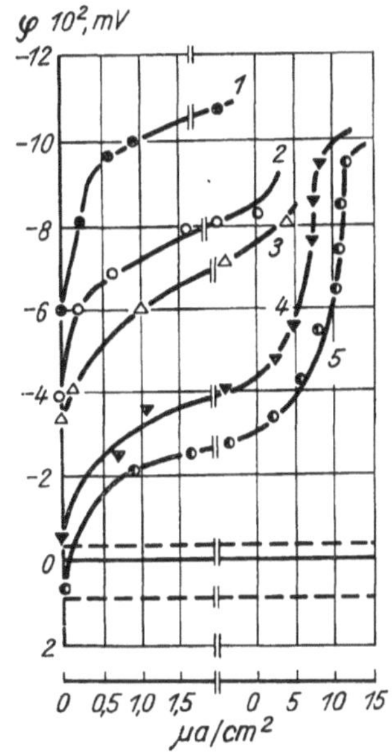

Abb. 6: Reduktion verschiedener Oxydationsmittel am Stahl aus neutralen Elektrolyten.

1 – Luftsauerstoff; *2* – Natriumnitrit; *3* – Dizyklohexylaminnitrit; *4* – Hexamethyleniminnitrobenzoat; *5* – 3,5-Hexamethylenimindinitrobenzoat

Untersuchungen des Anodenprozesses überzeugen [3]. Wenn man dem Boratpuffer 0,01 Mol/*l* Piperidin zugibt, so wächst die Passivierbarkeit des Systems stark an, wobei sich der Bereich des passiven Zustandes weitgehend erweitert und die Passivierungsströme geringer werden. Die gleiche Konzentration von Piperidinbenzoat liefert keinen Effekt. Gibt man jedoch 0,01 Mol/*l* Piperidinnitrobenzoat oder 3,5-Piperidindinitrobenzoat zu, so wird das Anfangspotential des Systems weit über das Potential der vollständigen Passivierung im Boratpuffer hinaus verschoben (Abb. 7). All das deutet auf die Möglichkeit der Bildung in Gegenwart dieser Verbindungen auf dem Eisen von passivierendem Oxid hin. Da jedoch in Gegenwart von Nitrobenzoaten an der Elektrode ein effektiver Kathodenprozeß ablaufen sollte, können die gewöhnlichen potentiostatischen Kurven in den Stromkoordinaten nicht eine vollständige Information über den Anodenprozeß liefern. In diesem Zusammenhang untersuchten wir die Abhängigkeit der Auflösungsgeschwindigkeit von dem Potential nach der Veränderung der Masse (k — φ). Es wurde gefunden, daß in Gegenwart von Aminnitrobenzoaten in der Tat ein starker Unterschied zwischen den Strom- und den gravimetrischen Kennwerten besteht. Bewertet man jedoch die wahre Geschwindigkeit des Anodenprozesses nach der Auflösungsgeschwindigkeit bei gleichen Potentialwerten (Abb. 8), so wird es klar, daß sich die Aminnitrobenzoate in den Anodenprozeß nicht einmischen (die anodischen Kurven im inhibierten und nichtinhibierten Elektrolyten fallen zusammen). Daraus folgt, daß die beobachtete Verschiebung des Potentials über das Potential der vollständigen Passivierung hinaus eine Folge der Steigerung der Wirksamkeit des Kathodenprozesses ist.

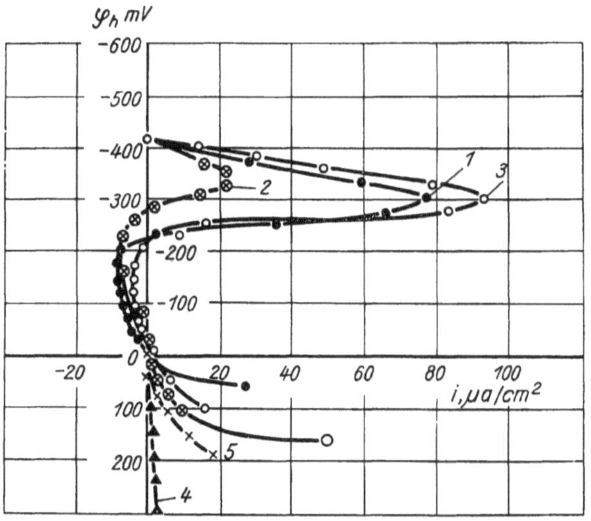

Abb. 7: Einfluß von Piperidin (*2*), Piperidinbenzoat (*3*), Piperidinmetanitrobenzoat (*4*) und 3,5-Piperidindinitrobenzoat (*5*) auf die anodische Auflösung von Stahl in Boratpufferlösung (*1*), die 0,03 N-Na_2SO_4 enthält. (Konzentration des Inhibitors 0,01 Mol/*l*.)

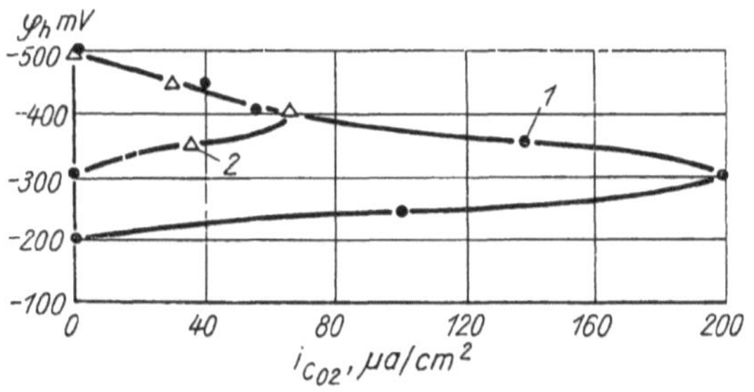

Abb. 8: Einfluß von Piperidin-3,5-dinitrobenzoat auf die wahre Auflösungsgeschwindigkeit von Stahl im Boratpuffer (pH = 7).

1 — Boratpuffer + 0,03 N-Na_2SO_4; *2* — Boratpuffer + 6,75 · 10^{-3} Mol/*l* Piperidin-3,5-dinitrobenzoat.

Im wäßrigen Medium werden die Nitrobenzoat-Ionen der Protonierung unterworfen und wandeln sich in aktiven Komplex um:

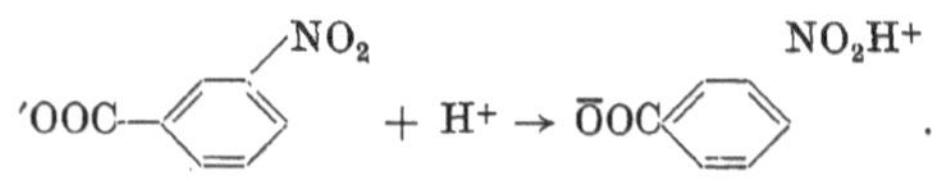

An der Oberfläche des Metalls entreißt das protonierte Anion dem Metall Elektronen, es bildet sich ein Hydroxylamin-Derivat:

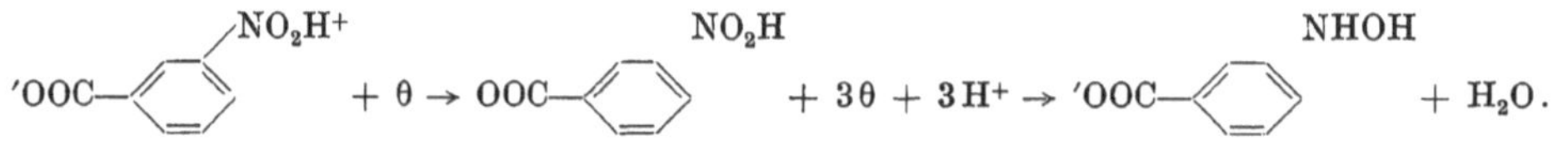

Wie zu ersehen ist, gibt der Inhibitor an das Metall keinen Sauerstoff ab, der den aus Fe_3O_4 bestehenden ursprünglichen Film zu Fe_2O_3 oxydieren könnte. Der Film enthält auch keine Produkte der Reduktion des Inhibitors [4]. Deshalb kann angenommen werden, daß die Oxydation durch den Sauerstoff des Wassers nach der Reaktion vor sich geht:

$$2Fe_3O_4 + H_2O - 2\theta \rightarrow \gamma Fe_2O_3 + 2H^+.$$

Die Entstehung an der Oxidoberfläche einer höheren Wertigkeit erschwert den Übergang der Ion-Atome des Metalls aus dem Gitter in die Lösung, wie dies beim Eintreten der anodischen Passivierung beobachtet wird. Die Funktionen des Inhibitors bestehen nur darin, daß er in der Rolle des Elektronenakzeptors, das heißt der Triebkraft der Anodenreaktion auftritt. Bei einem solchen elektrochemischen Mechanismus wird dem Inhibitor keine spezifische Wirkung zugeschrieben, weil er nur die Fähigkeit besitzen soll, dem korrodierenden Metall ein in dem Bereich des passiven Zustandes liegendes Potential zu erteilen.

Wie soll man sich nun den Wirkungsmechanismus der Inhibitoren vorstellen, die sich in neutralen Elektrolyten mit irgendwie merklicher Geschwindigkeit, die die Abscheidung der passivierenden Schicht ermöglichen könnte, nicht reduzieren lassen?

Hier bestehen mehrere Möglichkeiten, die gegenwärtig in der Literatur lebhaft diskutiert werden. Es geht lediglich darum, zu klären, ob der Schutz ein Resultat der chemischen Umsetzung der Inhibitoren mit der Unterlage ist, die zu einer Veränderung des Valenzzustandes der Ionen führt, oder ein Resultat der Adsorption der nichtreduzierten Inhibitorionen. Wenn beim Schutz eine wichtige Bedeutung das nichtreduzierte Ion hat, so soll das verhältnismäßig niedrige Oxydations-Reduktions-Potential den Schutz begünstigen. Es wird verständlicher, warum kein direkter Zusammenhang zwischen den oxydativen Eigenschaften der Inhibitoren und ihrem Passivierungsvermögen besteht. Was den Wirkungsmechanismus der Amine und Aminbenzoate anbelangt, so ist am wahrscheinlichsten die Adsorption durch das ungeteilte Elektronenpaar der Heteroatome, die nicht zu einem Wechsel des Valenzzustandes der Ionen oder Atome, die im Molekül enthalten sind, führt. Bei einem solchen Mechanismus sollen diese Inhibitoren als Elektronendonatoren auftreten und die Austrittsarbeit vermindern, was auch experimentell bestätigt wird (siehe Tab. 1).

Tabelle 1: Veränderung des Oberflächenpotentials (ΔV_{max}), der Feldstärke ($E_{max}^{O_x}$) und der elektronenchemischen Wirksamkeit der Inhibitormoleküle $\left(\frac{\Delta V}{\Delta N}\right)$

Lfd. Nr.	Bezeichnung	ΔV_{max} [V]	$E_{max}^{O_x}$ [V/cm · 10^6]	Wirksamkeit $\frac{\Delta V}{\Delta N}$ [V/Molekül · cm²]
1.	Diäthylamin (*DÄA*)	+0,51	2,5	$1{,}1 \cdot 10^{-17}$
2.	Hexamethylenimin (*HMI*)	+0,5	2,5	$4{,}3 \cdot 10^{-18}$
3.	Zyklohexylamin (*ZHA*)	+0,7	3,5	$9{,}5 \cdot 10^{-12}$
4.	Dizyklohexylamin (*DZHA*)	+0,7	3,5	$4{,}5 \cdot 10^{-17}$
5.	*DÄA*-Benzoat	+0,3	1,5	$2{,}1 \cdot 10^{-16}$
6.	*HMI*-Benzoat	+0,2	1,0	$2{,}3 \cdot 10^{-16}$
7.	*ZHA*-Benzoat	+0,1	0,5	–
8.	*DZHA*-Benzoat	+0,1	0,5	–

Die Untersuchung hat ergeben, daß die Amine und Aminbenzoate Elektronendonatoren sind, da nach ihrer Adsorption die Austrittsarbeit auf 100 bis 700 mV sinkt. Sie werden alle bei der Adsorption an der Eisenoberfläche positiv geladen. Die an den Inhibitormolekülen lokalisierte positive Adsorptionsladung wird durch die an der inneren Grenze des Oxides $Fe-FeO_x$ liegende negative Ladung ausgeglichen. Die Veränderung des Oberflächenpotentials (ΔV) oder der Austrittsarbeit ($\Delta\varphi = -e\Delta V$) der Menge des an der Elektrode adsorbierten Stoffes direkt proportional.

Angenäherte Berechnungen zeigen, daß die elektrische Feldstärke im Oxid $E_{max}^{O_x}$ dabei etwa $\sim 10^6$ V/cm erreicht, was in der Lage ist, die Wanderung der Kationen an die Oberfläche zu hemmen.

Was den Wirkungsmechanismus der sauerstoffhaltigen Inhibitoren (Na_2CrO_4, Na_2MoO_4, Na_2WO_4) anbelangt, so ist hier die Sachlage weniger klar. Wenn man nämlich die klassischen Positionen der elektrochemischen Passivierung vertritt, so ist es zur Gewährleistung des Schutzes notwendig, daß die Reduktionsgeschwindigkeit des Inhibitors hoch ist. Nur in diesem Falle übersteigt der innere Kathodenstrom den Strom der anodischen Auflösung und das Metall geht in passiven Zustand über. In der Tat aber lassen sich die sauerstoffhaltigen Inhibitoren aus neutralen Elektrolyten nicht reduzieren und in den Fällen, wo die Reduktion teilweise stattfindet, ist die Reaktionsgeschwindigkeit so niedrig, daß die entstehenden Ströme um einige Größenordnungen niedriger liegen als die für die Passivierung notwendigen. So haben wir beispielsweise bei der Untersuchung der Passivierungseigenschaften der Pertechnetat-Ionen an der Metalloberfläche 10^{-8} g/cm² der reduzierten Form des Technetiums registriert [5]. Nach dieser Größe konnte man die entstandene Elektrizitätsmenge berechnen. Diese erwies sich gleich $0,82 \cdot 10^{-8}$ Ah. Wenn man annimmt, daß der ganze Reduktionsprozeß 1 Sekunde dauert, so erweist sich die Stromdichte an der Elektrode etwa 30 μA/cm². Auch in diesem Fall ist die Stromdichte um 3 bis 4 Größenordnungen niedriger als die für die Passivierung von Stahl erforderliche (50 bis 200 mA/cm²). Um einen Ausweg aus dieser schwierigen Lage zu finden, bleibt es anzunehmen, daß der ganze Passivierungsprozeß während 10^{-4} bis 10^{-5} Sekunden abläuft, was an sich wenig wahrscheinlich ist, oder daß ein spezifischer Einfluß der Inhibitoren statthat.

Ein rein elektrochemischer Mechanismus der Inhibierung, bei dem dem Inhibitor kein spezifischer Einfluß zugeschrieben wird, sondern von ihm nur gefordert wird, daß er durch die Reduktions-

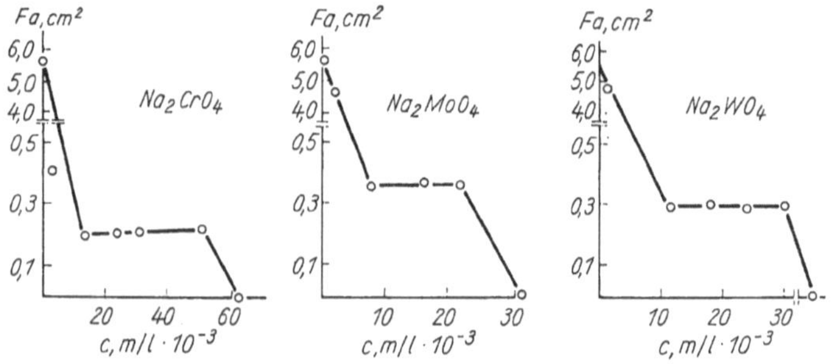

Abb. 9: Veränderung des aktiven Teils der Oberfläche der Elektrode bei der Zugabe zum Elektrolyten (0,1 N-Na_2SO_4 der Korrosionsinhibitoren: Natriumchromat (1), Natriummolybdat (2), Natriumwolframat (3).

reaktion dem Metall ein im Bereich des passiven Zustandes liegendes Potential erteilt, kommt hier nicht zustande. Im Hinblick auf das oben ausgeführte bleibt auch unverständlich der Mechanismus der Verstärkung durch die Inhibitoren ähnlichen Typus der Korrosion bei unvollständigem Schutz.

Es entsteht die Frage, ob überhaupt Passivierung möglich ist, ohne daß die Inhibitoren reduziert werden, und welche Kathodenreaktion in diesem Falle die Entstehung des passivierenden Oxids bewirken kann?

Eine Antwort auf die gestellten Fragen kann man, wie es uns scheint, aus Versuchen zur chemischen Passivierung erhalten. Die Beobachtungen des Verhaltens der Elektrode in inhibierten Elektrolyten ließen eine interessante Besonderheit erkennen, die es möglich macht, einen anderen Auflösungsmechanismus zu begründen, der nicht mit dem unmittelbaren, sondern mit dem indirekten Einfluß des Inhibitors auf den Kathodenprozeß verbunden ist [1]. Es hat sich erwiesen, daß mit der Verschiebung des Potentials nach der positiven Seite hin ein immer größerer Teil der Oberfläche der Elektrode passiviert und aus der Sphäre der Anodenreaktion herausgeleitet wird. Es ist dabei wichtig, darauf die Aufmerksamkeit zu lenken, daß eine solche teilweise Passivierung der Elektrode nur im Bereich der Potentiale zustandekommt, die für die aktive Auflösung kennzeichnend ist, und beim Potential der teilweisen Passivierung endet. Auf dem Abschnitt der für den aktiv-passiven Zustand des Metalls kennzeichnenden Kurve verändert sich der Grad der teilweisen Passivierung der Elektrode in einem breiten Bereich der Potentiale nicht und nur bei der Annäherung an das Potential der vollständigen Passivierung wird die Oberfläche der Elektrode wieder passivert (Abb. 9). Eine solche Gesetzmäßigkeit in der Veränderung des Verhältnisses zwischen dem passiven und dem aktiven Teil der Oberfläche der zusammengesetzten Elektrode macht es möglich, den Mechanismus der Beschleunigung der Korrosion auf dem aktiven Zweig der Polarisationskurve zu erklären. Wenn nämlich auf dem Abschnitt der aktiven Auflösung eine stetige Vergrößerung des Kathodenteils der Oberfläche beobachtet wird, so soll das zu einer Steigerung der Wirksamkeit des für neutrale Elektrolyten gewöhnlichen Kathodenprozesses führen, der in der Reduktion des Wasserstoffes besteht. Somit kann die Steigerung der Auflösungsgeschwindigkeit von Stahl auf dem aktiven Abschnitt der Polarisationskurve mit der Verschiebung des Potentials nach der positiven Seite hin auf den Effekt der inneren Anodenpolarisation ohne Beteiligung des Inhibitors an dem Kathodenprozeß zurückgeführt werden. Die in Tab. 2 angeführten Angaben zeigen, wie sich auf dem aktiven Zweig der Polarisationskurve der Grad der Bedeckung der Elektrode durch das passivierenden Oxid (θ), der gesamte Korrosionsstrom (I) und die wahre Stromdichte (i) auf dem aktiven Teil der Elektrode verändert. Es fällt auf, daß der gesamte Korrosionsstrom nicht proportional dem Bedeckungsgrad der Elektrode durch das passivierende Oxid, sondern in viel geringerem Grade abnimmt, was eine Folge des Zustandekommens der Anodenreaktion auf dem aktiven Teil der Elektrode bei positiveren Potential und der inneren Anodenpolarisation des Metalls ist. Dadurch erhöht sich die Stromdichte auf dem aktiven

Tabelle 2: Veränderung durch die Inhibitoren auf dem aktiven Abschnitt des Polarisationsdiagramms des aktiven Teils der Elektrode (θ), des gesamten Korrosionsstroms (I) und der Stromdichte (i)

Inhibitor	Inhibitorkonzentration [Mol/l]	pH	θ	I	i	Potentielle teilweise Passivierung [mV]
NaOH	0,050	12,5	56,0	1,35	8,72	+200,0
Na_2CO_3	0,056	12,8	26,6	2,74	10,0	−220,0
$NaHCO_3$	0,119	8,8	9,27	1,15	9,72	−425
Na_2SiO_3	0,007	11,4	26,8	4,10	9,30	−245
Na_3PO_4	0,0052	11,4	12,2	1,21	8,0	−290
Na_2HPO_4	0,011	9,0	11,0	1,50	8,42	−245
Na_2CrO_4	0,012	8,0	26,6	4,6	7,2	−310
Na_2WO_4	0,006	6,6	13,3	2,68	6,07	−100
Na_2MoO_4	0,008	6,0	15,0	3,54	6,2	−90
Na_3VO_4	0,04	12,0	20,74	6,76	3,75	−175
$NaNO_2$	0,058	7,4	112	4,4	28,5	0,00
C_6H_5COONa	0,138	7,47	118,5	38,8	3,34	+40,0
$K_2Cr_2O_7$	0,0068	5,8	9,33	2,03	5,0	−150,0
$NaBO_3$	0,039	10,5	3,3	0,64	5,71	−350,0

Teil der Elektrode. Bei Potentialen, die positiver als das Potential der teilweisen Passivierung sind, verändert sich der Mechanismus des Prozesses von Grund aus, indem er von der Kathodenkontrolle zur Anodenkontrolle übergeht. Da sich auf dem für den aktiv-passiven Zustand des Metalls kennzeichnenden Abschnitt der Polarisationskurve das Verhältnis zwischen dem passivierten Teil der Oberfläche und dem im aktiven Zustand befindlichen Teil nicht verändert und folglich von einer Steigerung der Wirksamkeit des Kathodenprozesses nicht gesprochen werden kann, kann die Abnahme der Auflösungsgeschwindigkeit mit der Verschiebung des Potentials nach der positiven Seite hin nur darauf zurückgeführt werden, daß der Inhibitor auf diesem Abschnitt der Polarisationskurve einen unmittelbaren Einfluß auf die Anodenreaktion ausübt, indem er diese verlangsamt. Wie zu ersehen ist, besteht ein prinzipieller Unterschied hinsichtlich der Wirkung der Inhibitoren auf dem aktiven Zweig der Polarisationskurve und auf dem Zweig des aktiv-passiven Zustandes. Auf dem ersteren steigern sie die Wirksamkeit des Kathodenprozesses, ohne sich in den Anodenprozeß einzumischen, während sie auf dem zweiten die Geschwindigkeit des Anodenprozesses verringern, ohne sich in den Kathodenprozeß einzumischen.

Es verdient die Tatsache beachtet zu werden, daß in dem Bereich der aktiven Auflösung sowie an der Grenze des aktiv-passiven Zustandes von Metall spezifische Eigenschaften der Inhibitoren zur Geltung kommen, was in der ungleichen Auflösungsgeschwindigkeit bei dem vorgegebenen Potential zum Ausdruck kommt und davon zeugt, daß die Inhibitoren den Übergang der Kationen aus dem Kristallgitter in die Lösung kontrollieren. Beim Übergang des Metalls in den vollständig passiven Zustand aber verschwinden gleichsam die spezifischen Eigenschaften der einzelnen Inhibitoren, die vollständige Passivierung in Gegenwart aller untersuchten Inhibitoren tritt nämlich bei ein und demselben Potential ein.

Daraus kann man schlußfolgern, daß wenn bei Potentialen, die negativer als das Potential der vollständigen Passivierung liegen, die Inhibitoren einen direkten oder indirekten Einfluß auf die Kinetik der elektrochemischen Reaktionen ausüben und sich an der Bildung der passivierenden Schichten beteiligen, beim Übergang des Metalls in den vollständigen passiven Zustand die Funktionen des Passivierungsmittels ein anderer Passivator, aller Wahrscheinlichkeit nach der Wasserstoff, übernimmt [4, 6]. Da der passive Zustand in Gegenwart der untersuchten Inhibitoren praktisch bei gleichem Potentialwert eintritt, liegt aller Grund zu der Annahme vor, daß die Natur der passivierenden Schicht überall die gleiche ist.

Die Untersuchung der Abhängigkeit des Passivierungspotentials von dem pH-Wert der Lösung in Gegenwart von Inhibitoren (Abb. 10) hat ergeben, daß sie die gleiche bleibt wie auch für die Oxid-

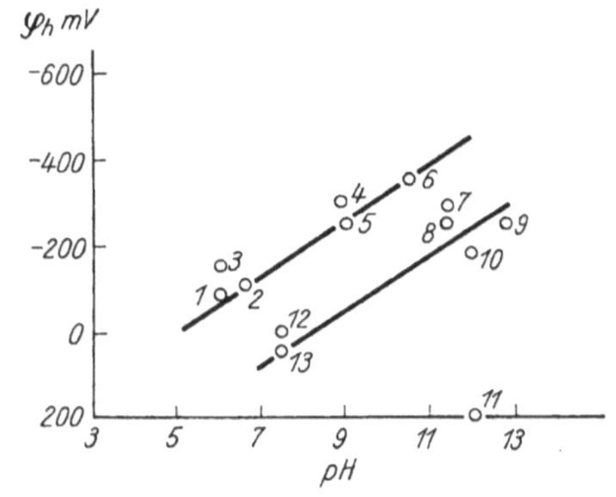

Abb. 10: Veränderung des Potentials der teilweisen Passivierung von Stahl von dem pH-Wert der Lösung.
1 – Na_2MoO_4; *2* – Na_2WO_4; *3* – $K_2Cr_2O_7$; *4* – Na_2CrO_4; *5* – Na_2HPO_4; *6* – $NaBO_3$; *7* – Na_3PO_4; *8* – Na_2SIO_3; *9* – Na_2CO_3; *10* – Na_3VO_4; *11* – NaOH; *12* – $NaNO_2$; *13* – C_6H_5COONa.

elektrode. Daraus folgt, daß in Gegenwart von Inhibitoren die passivierende Schicht ihrer Natur nach eine Oxidschicht ist. Die Funktionen der Inhibitoren bestehen darin, daß ihre Adsorption zur Senkung der freien Gesamtenergie des Systems und Steigerung der Stabilität des passiven Filmes führt. Unter solchen Bedingungen kann die Passivierung auch bei den geringen Stromdichten eintreten, die durch die Kathodenreaktion der Reduktion von Sauerstoff entstehen. In Abhängigkeit von der Natur des adsorbierten Ions, seiner Konzentration an der Oberfläche und der Festigkeit der Bindung verändert sich die Auflösungsgeschwindigkeit, die Polarisierbarkeit und die für die Passivierung notwendige Stromdichte. All das führt zum Schluß, daß bei der chemischen Passivierung das Potential nicht immer die Auflösungsgeschwindigkeit eindeutig bestimmt. In der Arbeit [7], in der zum ersten Male der klassische Fall der chemischen Passivierung behandelt wurde, wurde überzeugend gezeigt, daß unabhängig davon, wodurch der Elektrode das Potential, durch Anodenpolarisation oder chemische Passivierung, erteilt wird, die Auflösungsgeschwindigkeit bei dem vorgegebenen Potential die gleiche ist. Dies dürfte für den Fall zutreffend sein, wenn die chemische Verbindung, mit deren Hilfe das Potential vorgegeben wird, den Anodenprozeß nicht beeinflußt, sondern nur die Wirksamkeit der Kathodenreaktion verändert. In den Fällen, wo der Inhibitor die Kinetik der beiden Reaktionen verändert, ist die anodische Passivierung mit der chemischen bereits nicht identisch und das Potential bestimmt nicht eindeutig die Auflösungsgeschwindigkeit. Beim vorgegebenen Potential kann man in Abhängigkeit von der Natur des adsorbierten Ions verschiedenartige Auflösungsgeschwindigkeiten beobachten und darin kommen die individuellen Eigenschaften der Inhibitoren zur Geltung.

Die von uns für eine Reihe von in adsorbiertem Zustand auf verschiedenen Fragmenten der Oberfläche von Eisen und seiner Oxide befindlichen Inhibitoren ausgeführten quantenchemischen Berechnungen gestatteten es, wertvolle Aussagen über den Wirkungsmechanismus verschiedener Inhibitorklassen zu erhalten. In Tab. 3 und 4 sind Angaben über die Ladungen auf verschiedenen Atomen der Anionen vom Typ MeO_4^{n-} sowohl in freiem als auch in adsorbiertem Zustand, erhalten von uns auf Elektronenrechenmaschinen, angeführt. In der ersten Zeile sind Ladungen auf den Atomen für freie

Anionen, in den nachfolgenden Ladungen derselben Atome in an verschiedenen Eisenfragmenten adsorbiertem Zustand angeführt. Wie zu ersehen ist, verändert sich in adsorbiertem Zustand stark das Verhältnis der Ladungen im Anion MeO_4^{n-}. Es verdient die Tatsache beachtet zu werden, daß die Ladung des zentralen Metallatoms des Anions praktisch konstant und positiv bleibt, während sich die Ladung an den Sauerstoffatomen stark verändert und einen negativeren Wert annimmt. Der Unterschied in den Ladungen des zentralen Atoms in freiem und adsorbiertem Zustand beträgt nicht über 2%, während an den Sauerstoffatomen der Unterschied in den Ladungen 12 bis 178% beträgt. Da der Wert der Ladungen an den Sauerstoffatomen in adsorbiertem Zustand negativer wird, so ist das ein direkter Hinweis darauf, daß die Anionen vom Typ MeO_4^{n-} als Elektronenakzeptoren auftreten, deren Eigenschaften durch die Sauerstoffatome in Erscheinung treten. Dies widerspricht der

Tabelle 3: Ladungen an den Atomen in Anionen vom Typ MeO_4^{n-}

Oberflächen-fragment	Ladungen an den Atomen				
	Cr	O	O	O	O
fr. CrO_4^{-2}	5,387	−1,169	−1,217	−1,500	−1,500
[3Fe]	5,484	−1,614	−1,794	−1,706	−1,710
[2Fe + 0]	5,087	−1,817	−1,849	−0,720	−1,658
[Fe + 20]	5,449	−1,605	−1,606	−1,892	−1,892
	V	O	O	O	O
fr. VO_4^{-3}	4,469	−0,628	−0,633	−0,633	−1,572
[3Fe]	4,423	−1,746	−1,752	1,747	−1,806
[2Fe + 0]	4,777	−1,693	−1,782	−1,776	−1,805
[Fe + 20]	4,492	−1,742	−1,886	−1,741	−1,803

Hypothese von CARTLEDGE [8], die davon ausging, daß die Adsorptionswechselwirkung der Anionen vom Typ MeO_4^{n-} mit dem Eisen durch das die positive Ladung tragende zentrale Metallatom zustandekommt. Die Analyse der algebraischen Summe der Ladungen an den Atomen der betrachteten Anionen zeigt, daß sie alle in adsorbiertem Zustand an beliebigen Eisenfragmenten in Form von Anion bleiben. Dies bestätigt die sich aus den elektrochemischen Untersuchungen ergebende Schlußfolgerung über die Möglichkeit des Schutzes, ohne daß sich das Anion an dem Anodenprozeß beteiligt und folglich sich der Valenzzustand des zentralen Atoms verändert.

Die Amine dagegen treten in adsorbiertem Zustand als Elektronendonatoren auf, die Ladungen an den ein ungeteiltes Elektronenpaar aufweisenden Stickstoffatomen werden, wie dies aus Tab. 4 zu ersehen ist, in adsorbiertem Zustand durch die Abgabe der Elektronen an das Metall positiver. Dies bestätigt die von uns nach der Methode der Kontaktdifferenz der Potentiale erhaltenen Resultate.

Tabelle 4: Ladungen an den Stickstoff- und Kohlenstoffatomen in den Molekülen der Amine und des Ammoniaks

Q Ladung	NR_3	$Fe \cdots NR_3$	NHR_2	$Fe \cdots NHR_2$	NH_2R	$Fe \cdots NH_2R$	NH_3	$Fe \cdots NH_3$
N	−0,815	−0,698	−0,421	0,284	−0,869	−0,185	−0,978	0,323
C	0,097	0,084	−1,086	−0,836	−0,061	0,113	—	—
C	0,037	0,037	−1,086	−0,836	−0,143	0,475	—	—
C	0,273	0,270	−0,197	−0,133	—	—	—	—
C	−0,550	−0,210	−0,197	−0,133	—	—	—	—
C	−0,550	−0,209	—	—	—	—	—	—
C	0,037	0,037	—	—	—	—	—	—

Literatur

[1] I. L. ROSENFELD und L. W. FROLOWA, Ber. d. Akad. d. Wiss. d. UdSSR, **215**, Nr. 6, 1410 (1974)
[2] I. L. ROSENFELD, I. Ja. KERBELEWA u. a., Elektrochimia, Bd. 8, 11, 1700 (1972)
[3] I. L. ROSENFELD, Ju. I. KUSNEZOW, I. Ja. KERBELEWA und W. P. PERSIANZEWA, Metallschutz (im Druck)
[4] I. L. ROSENFELD, „*Flüchtige Korrosionsinhibitoren, Korrosion und Korrosionsschutz, Erkenntnisse der Wissenschaft*", Bd. 1, Verlag WINITI, Moskau 1974, S. 156
[5] W. I. SPIZYN, I. L. ROSENFELD, W. P. PERSIANZEWA, I. I. SAMOSCHNIKOWA und A. F. KUSINA, Sammelband „*Korrosion der Metalle und Legierungen*", Staatsverlag für E- und NE-Metallurgie, Moskau 1963, S. 129
[6] U. K. EVANS, Electroch. acta, **16**, 1825 (1971)
[7] N. Ja. BUNE und Ja. M. KOLOTYRKIN, Zeitschr. für phys. Chemie, Bd. **7**, 35, 1543 (1961)
[8] G. H. CARTLEDGE, J. phys. chem., **60**, Nr. 1 (1956)

Im Rahmen des Symposiums verlieh die Technische Universität KURT SCHWABE am 30. Mai 1975 die Würde eines Ehrendoktors. Im Anschluß an den Festakt hielt KURT SCHWABE folgenden Vortrag:

ZUR PROBLEMATIK DER PASSIVITÄTSFORSCHUNG

Meine sehr verehrten Damen und Herren!

Zweifellos ist der Korrosionsschutz der Metalle durch Passivschichten aus vielen Gründen besonders vorteilhaft und ökonomisch. Einmal regenerieren sich diese Schichten bei Verletzung selbst und schützen so das Metall sehr dauerhaft; zweitens erfordert die Passivschicht keinen hohen Arbeitsaufwand und keinen zusätzlichen Materialverbrauch. Es ist daher nur zu verständlich, daß in hochentwickelten Industrieländern der Korrosionsschutz vorwiegend durch Verwendung passivierbarer Metalle oder Legierungen betrieben wird. Auf diese Weise kann wahrscheinlich am ehesten der hohe Aufwand von 2 bis 3,5% des Sozialbruttoproduktes für Korrosionsverhütung und Korrosionsverluste wesentlich verringert werden. Wenn man dagegen organische Schutzschichten, vor allen Dingen Anstriche verwendet, so hat man nicht nur mit einem sehr hohen Arbeitsaufwand zu rechnen, weil solche Schutzschichten innerhalb kurzer Zeit erneuert werden müssen und die Aufbringung an fertigen Anlagen und Geräten meistens nicht ohne erhebliche manuelle Arbeit möglich ist, sondern man verliert auch in großem Umfange organische Rohstoffe, Lösungsmittel, Bindemittel und auch Füllstoffe. Deswegen schien und scheint mir die Beschäftigung mit der Passivierung der Metalle von vorrangiger Bedeutung; nicht nur als Problem der Ökonomie des Korrosionsschutzes, sondern auch im Interesse der Erhaltung unserer metallischen Rohstoffe, die nicht unerschöpflich sind und nicht in beliebiger Menge, auch nicht aus beliebig großen Tiefen und aus beliebig armen Erzen gewonnen werden können, ohne daß dadurch wieder ein hoher Materialaufwand und eine erhebliche Umweltverschmutzung herbeigeführt wird.

Die Forschung auf dem Gebiet über die Entstehung und Stabilisierung von Passivschichten ist daher sicher eine vorrangige Aufgabe. Kann man aber damit rechnen, daß sie mit Hilfe der Grundlagenforschung so gelöst werden kann, daß von vornherein eine Aussage über die Stabilität von Passivschichten auf neuen Metallen oder Metall-Legierungen möglich ist und auch die Stabilität der Schichten in den verschiedensten Medien vorausberechnet werden kann? Meines Erachtens sind wir von der Lösung der Aufgabe in dieser Form noch außerordentlich weit entfernt. Auf die Gründe dafür möchte ich nun kurz eingehen.

Die Untersuchungsmethoden, deren man sich bei der Aufklärung der Entstehung, Natur und Stabilität von Passivschichten bedienen kann, sind einerseits elektrochemische Methoden. Wenn man solche Methoden anwendet, dann muß man sich im klaren sein, daß diese mindestens nicht voll den praktischen Bedingungen entsprechen, unter denen Passivschichten gebildet werden. Wenn man die elektrochemischen Methoden für die Untersuchung der Auflösung der Metalle oder anderer Elektrodenprozesse einsetzt, dann kann man die Transportvorgänge durch Impulsmethoden oder durch Anwendung der rotierenden Scheibenelektrode ausschalten. Aber im Falle der Passivierungskinetik können Impulsmethoden keine wesentlichen Aussagen liefern, weil die Passivierung in jedem Falle eine Zeit erfordert, innerhalb der auch Transportvorgänge eine Rolle spielen können. Die Methoden der rotierenden Scheibenelektrode gestatten zwar, Aussagen über den Abbau der Passivschichten zu machen, aber nicht so sehr über ihren Bildungsmechanismus. Die coulometrischen Methoden als weitere elektrochemische Methode gestatten wohl, die Passivschicht abzubauen und damit ihre Dicke zu bestimmen, aber andererseits sind sie nicht in der Lage, Aussagen darüber zu machen, unter welchen Bedingungen die Passivschicht entstanden ist.

Die meistens zum Studium der Passivierungsvorgänge verwendeten potentiostatischen Stromspannungskurven müssen bei ihrer Interpretation den OHMschen Spannungsabfall in der Passivschicht und seine zeitliche Veränderung berücksichtigen, das ist vielfach nicht geschehen, dadurch ist aber die Aussagekraft solcher Untersuchungen erheblich eingeschränkt, damit wird auch verständlich, daß oft nur über sehr kleine Spannungsbereiche der Zusammenhang zwischen lg i und dem Potential linear ist. Generell leiden aber die elektrochemischen und praktisch alle Untersuchungsmethoden der Passivierungs- und auch der Korrosionsvorgänge darunter, daß eine Metalloberfläche energetisch inhomogen und daß diese Inhomogenität bei wiederholten Messungen nicht reproduzierbar ist. Darauf wird später noch etwas näher einzugehen sein, zunächst noch einige Bemerkungen zu den optischen Methoden.

Wenn man die elektrochemischen Methoden mit optischen Methoden, z. B. der Ellipsometrie, koppelt, so kann man zwar über die Dicke der Schicht und die Kinetik der Bildung der Schichten verschiedener Dicke Aussagen machen, man kann auch durch ellipsometrische Spektroskopie [1] die Zusammensetzung der Schichten annähernd angeben, aber wo der Sitz der Energiebarriere ist, die die Auflösung des Metalls am stärksten hemmt, kann man damit nicht feststellen.

In den letzten Jahren sind die elektronenoptischen und ionenoptischen Methoden außerordentlich verfeinert und in vielfältiger Hinsicht erweitert worden. Speziell die ionenoptischen Methoden in der Form der Sekundärionenmassenspektroskopie [2] gestatten, Passivschichten auf dem Metall schrittweise systematisch abzubauen und die Änderung der Zusammensetzung zu untersuchen. Alle diese Methoden leiden aber unter dem Nachteil, daß sie im hohen Vakuum angewendet werden müssen und daß daher nicht sicher ist, ob die Schichten, die man dabei untersucht, wirklich mit denen identisch sind, die in Lösungen oder bei einem wäßrigen Oberflächenfilm die wirkliche Energiebarriere darstellen. Dagegen sprechen eine Reihe von Beobachtungen, die einen Wassergehalt der Passivschichten erkennen lassen [3] und auch die Beobachtungen, die zeigen, daß in völlig wasserfreien Medien an den Metallen Eisen und Nickel sogar auch Titan und Aluminium, keine Passivierung auftritt. Die röntgenoptischen Methoden liefern zwar Informationen über die Struktur einer Passivschicht, wenn sie eine kristalline Phase darstellen, aber ob diese für die Passivität verantwortlich ist, gestatten sie nicht zu entscheiden und es gibt ja Argumente dafür, daß amorphe Schichten wegen ihrer geringeren Ionenleitfähigkeit höhere Barrieren für die anodische Auflösung der Metalle darstellen. Die radiochemischen Methoden gestatten keine Aussagen über die Struktur der Passivschicht und nicht einmal über die chemischen Individuen, in denen die Radionuklide enthalten sind. Eines kann man allerdings mit ihnen vielleicht eher beweisen als mit anderen Methoden, nämlich die Inhomogenität solcher Schichten, insbesondere durch die Autoradiographie.

Für die Inhomogenität der Metalloberfläche sind in der letzten Zeit viele experimentelle Beweise mit radiochemischen und elektrochemischen Methoden erbracht worden [4]. Dabei machen die elektrochemischen Untersuchungen von CHAO deutlich, daß selbst an Gold die Oberfläche durch anodische und kathodische Polarisation ständig verändert wird, ohne daß das Metall in Lösung geht. Die Autoren weisen übrigens auf die Problematik optischer Untersuchungsmethoden während der Polarisation der Elektroden insbesondere in bezug auf quantitative Aussagen hin.

Ich halte die Inhomogenität einer Metalloberfläche, auch wenn das Metall extrem rein ist und als Einkristall vorliegt, für ein wesentliches Hindernis, quantitative Aussagen über die Kinetik der Metallauflösung und -passivierung zu machen, vor allem aber die Tatsache, daß es nicht möglich ist, eine Metalloberfläche selbst aus einem Einkristall mit bestimmter Flächenorientierung reproduzierbar herzustellen. Deswegen erhält man für die Elektrodenprozesse der Auflösung des Metalls oder der Bildung von Passivschichten nicht streng die gleichen kinetischen Parameter, wenn man die Messungen unter gleichen Bedingungen aber mit erneuerter Oberfläche wiederholt. Es ist offenbar nicht möglich, durch mechanische, chemische oder elektrochemische Vorbehandlung die Elektrodenoberflächen so zu reproduzieren, daß die inhomogene Energieverteilung, mit der man in jedem Fall infolge von Gitterbaufehlern, Versetzungen usw. rechnen muß, immer wieder die gleiche wird. Dadurch können bei der Auswertung von elektrochemischen Messungen, vor allen Dingen dann, wenn die Elektrode selbst am Prozeß beteiligt ist, ob sie sich löst oder ob sie oxydiert wird, sehr schwer exakte kinetische Aussagen getroffen werden.

Wir haben die Inhomogenität der Metalloberfläche und ihre Nichtreduzierbarkeit an Eisen- und Molybdäneinkristallen sowohl mit elektrochemischen wie radiochemischen und röntgenoptischen Methoden, nämlich der BERG-BARRETT-Methode, nachweisen können [5].

Abb. 1 (Tafel I) zeigt durch BERG-BARRETT-Aufnahmen die Verteilung der Subkorngrenzen auf einem Einkristall (Orientierung 100) nach chemischem Polieren in äthanolischer Schwefelsäure (links), rechts ist die Aufnahme zu sehen, nachdem durch weiteres chemisches Polieren von derselben Oberfläche 0,2 mm abgetragen waren. Man erkennt deutlich die veränderte Verteilung der Subkorngrenzen. Dabei hat es sich gezeigt, daß chemisches oder elektrochemisches Polieren nach vorhergehendem Schleifen Oberflächen ergibt, die in ihrem elektrochemischen und Adsorptionsverhalten relativ am besten übereinstimmen. Immerhin beträgt die Streuung der kinetischen Parameter bei der Aufnahme potentiostatischer Stromdichtespannungskurven oder bei der Adsorption von radioaktiven Ionen auch nach dieser Vorbehandlung mindestens 10%. Daher muß man bei der Zerstörung von Passivschichten durch Halogenidionen auf solchen Oberflächen auch eine nicht reproduzierbare Bildung von Löchern erwarten. Die Lochbildung kann nur auf lokalen Instabilitäten der Passivschicht beruhen, gleichgültig ob man annimmt, daß sie durch eine apriori vorhandene Inhomogenität der Metalloberfläche zustandekommen oder sich erst bei der Bildung der Passivschicht entwickeln. An Eiseneinkristallen (Orientierung 100, geometrische Oberfläche 0,5 cm^2) haben wir zeigen können, daß weder die Lochdichte (Zahl der Löcher pro cm^2) noch die Lage der Löcher nach völlig gleichartiger Oberflächenvorbehandlung desselben Kristalls reproduzierbar ist. Die Kristalloberfläche wurde in äthanolischer Schwefelsäure (3 Mol kg^{-1}) mit einer Stromdichte von 0,5 $A\,cm^{-2}$ elektrochemisch poliert. Nach kurzem Spülen wurde die Elektrode in eine wäßrige Lösung, die 0,065 molar an $Na_2B_4O_7$, 0,35 molar an H_3BO_3 und 0,025 molar an NaCl war, überführt und bei +741 mV (St. H. E.) und 25 °C mit einem Potentiostaten (PS. 2.3. Forschungsinstitut Meinsberg) 60 Minuten polarisiert. Anschließend wurde die Elektrode abgeschliffen bis keine Löcher mehr zu sehen waren und wieder anodisch poliert, dabei wurden insgesamt 0,2 mm abgetragen. Darauf wurde wieder in genau der gleichen Weise 60 Minuten anodisch polarisiert. Diese Verfahren wurden 20mal an der gleichen Elektrode wiederholt und jedes Mal die Lochdichte bestimmt. Wie Tab. 1 zeigt, betrug die Streuung

Tabelle 1: Lochdichten aus 20 Versuchen an einem Eiseneinkristall in einer Lösung aus 0,065 M $Na_2B_4O_7$ + 0,35 M H_3BO_3 + 0,025 M Cl^-; $\varepsilon = 500$ mV/SCE; $t = 60$ min

Lochdichte z/cm^2		
280	161	346
389	275	197
257	208	331
244	402	316
297	198	207
237	291	258

über 100%. Die Verteilung der Löcher auf der Elektrodenoberfläche war nach jeder Polarisation anders, das veranschaulichen einige willkürlich aus den 20 Messungen ausgewählte Aufnahmen der Elektrodenoberfläche in Abb. 2 (Tafel I u. II). Bei großen bzw. zusammengewachsenen Löchern sieht man deutlich die kathodische Schutzwirkung. Wenn man die Elektrodenoberfläche nicht poliert, sondern nur schleift, so daß die Schleifriefen noch mikroskopisch sichtbar sind, dann bilden sich die Löcher bevorzugt längs dieser Riefen, vgl. Abb. 3 (Tafel III). Der Teil 3b des Bildes zeigt dieselbe Stelle nach elektrolytischem Polieren. Auf dem polykristallinen Eisen entstehen wie bekannt die Löcher bevorzugt an den Korngrenzen, aber auch dann ist ihre Zahl und Verteilung nicht reproduzierbar, Abb. 4 (Tafel IV).

Man macht für die Lochbildung vielfach Einschlüsse insbesondere von Sulfiden verantwortlich, Löcher treten allerdings auch an 99,999er Aluminium auf. Wenn Einschlüsse aber die Ursache des Lochfraßes an unseren Fe-Einkristallen sind, dann sind sie mindestens innerhalb verschiedener dünnster Schichten ganz unterschiedlich verteilt. Es bleibt problematisch, wie man aus solchen Untersuchungsergebnissen zu einer quantitativen Theorie gelangen kann. Eine Korrelation zwischen Zahl der Löcher pro cm^2 und Zahl der Einschlüsse besteht ebensowenig wie zwischen Lochzahl und Zahl der Versetzungen pro cm^2. Man macht daher Zusatzannahmen über aktive und inaktive Einschlüsse bzw. Versetzungen [6], wobei aber über die Energieunterschiede, die sicher sehr klein sind, nichts bekannt ist. Eine Riesenzahl von Elektrodenoberflächen müßte untersucht werden, um für *eine* Materialprobe eine statistisch gesicherte, quantitative Aussage machen zu können.

Angesichts dieser Ergebnisse kann man eigentlich kaum erwarten, daß die überaus zahlreichen Untersuchungen über die Bildung und Eigenschaften von Passivschichten sowie die daraus abgeleiteten theoretischen Konzeptionen übereinstimmen. Tatsächlich bestehen sowohl über die Natur und Struktur der Passivschichten wie auch über die Ursachen ihrer unterschiedlichen Stabilität und der lokalisierten Zerstörung verschiedene Auffassungen. Von einer allgemein anerkannten quantitativen Theorie der Passivierung und Passivität, die das Passivverhalten der Metalle und Legierungen voraus zu bestimmen gestattet, kann wohl keine Rede sein. Die nichtreproduzierbare Metalloberfläche bietet auch keine Möglichkeit, eine solche Theorie streng zu prüfen.

Es erhebt sich daher die Frage, ob es sinnvoll ist, nach einer quantitativen Theorie zu suchen und ob wir wenigstens zu gesicherten qualitativen Vorstellungen gelangen können. Ich meine, daß man sich auf einige Fakten stützen kann, die zur qualitativen Interpretation der Passivierung unter realen, praktischen Verhältnissen herangezogen werden können.

Normalerweise erfolgt die Passivierung in Gegenwart von Wasser, ohne Wasser, d. h. in völlig wasserfreien organischen Lösungsmitteln lassen sich die meisten passivierbaren Metalle nicht passivieren, z. B. in sauren Lösungen

Cr in CH_3OH KOLOTYRKIN [7],
Fe und Ni in CH_3CN und DMF sowie DMSO (SCHWABE) [8],
Al in CH_3OH MANSFELD [9],
Ti in CH_3OH MANSFELD [10].

Offenbar ist also bei diesen Metallen und unter diesen Bedingungen das Wasser zur Passivierung nötig. Entweder bildet es die Passivschicht nach

$$Me + OH_2 \rightleftharpoons MeO + 2H^{\cdot} + 2e^- \tag{1}$$

oder es stabilisiert sie, z. B. durch Bildung einer amorphen Hydroxidschicht oder dadurch, daß es immer zur Ausheilung nach Gl. (1) zur Verfügung steht. Die Tatsache, daß schon Wassergehalte $< 0,1\%$ in organischen Lösungsmitteln ausreichen, um die Passivierung hervorzurufen, spricht wohl dafür, daß das Wasser auch an der Bildung der Passivschicht beteiligt ist. Dafür spricht auch, daß solche Metalle zur Passivierung bei Gegenwart von Wasser neigen, bei denen die Reaktion (1) mit einer großen Abnahme der freien Enthalpie verbunden ist (siehe Tab. 2). Allerdings ist das wohl keine sichere Gewähr für eine echte Passivierung, offenbar muß das gebildete Oxid oder Hydroxid auch schwer löslich und fest mit dem Metall verhaftet sein. Beim Molybdän ist ΔG^0 zwar klein, aber es sollte doch passivierbar sein, denn das Ni ist es ja auch. Tatsächlich beobachtet man weder in Säuren noch in alkalischen oder neutralen Lösungen eine wirkliche Passivierung, wenn wir die phänomenologische Definition der Passivität von C. WAGNER akzeptieren (negative Stromspannungscharakteristik bei potentiostatischer Polarisation) und daher auch keinen Lochfraß, Abb. 5.

Neuerdings haben H. BUHL und H. J. RÄTZER-SCHEIBE [11] an einer unter der Oberfläche gehobelten Titanlegierung festgestellt, daß die Repassivierungsgeschwindigkeit nach Abtragen der Passivschicht unabhängig vom in der Lösung vorhandenen Anion ist, solange es sich um wasserhaltige neutrale Lösungen handelt, in wasserfreiem Formamid findet keine Repassivierung statt. Die Stromdichte nimmt nach $\lg i \sim -\lg t$ mit der Zeit ab, die Autoren führen dies auf einen mit der Zeit zunehmenden Bedeckungsgrad mit MeO zurück. Mit einer analogen Konzeption hatten wir 1960 die negative Stromspannungscharakteristik einer passivierbaren Elektrode gedeutet und dafür Berechnungen vorgelegt [12]. Es bleibt allerdings ein Unterschied zwischen der Auffassung von BUHL und unserer. Wir nehmen an, daß die Passivierung, mindestens die primäre, immer eine direkte Umsetzung des Wassers mit der Metalloberfläche ist. BUHL kommt wegen der stark verlangsamten

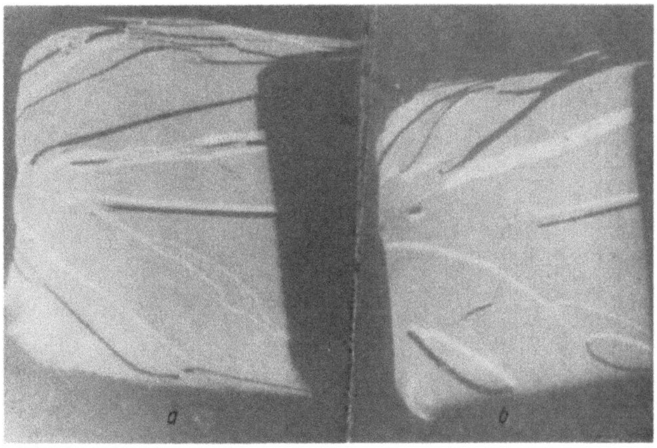

Abb. 1:
Röntgenographische Berg-Barrett-Aufnahmen von einem Eisenkristall der Orientierung (100).
(*a*) Ausgangszustand;
(*b*) nach einem Abtrag von 0,2 mm.

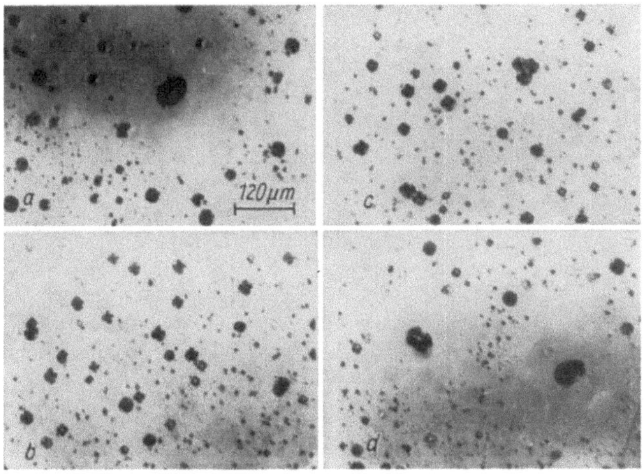

Abb. 2:
Lochverteilung an einem (100)-orientierten Eiseneinkristall nach schrittweisem Abtrag von 0,2 mm

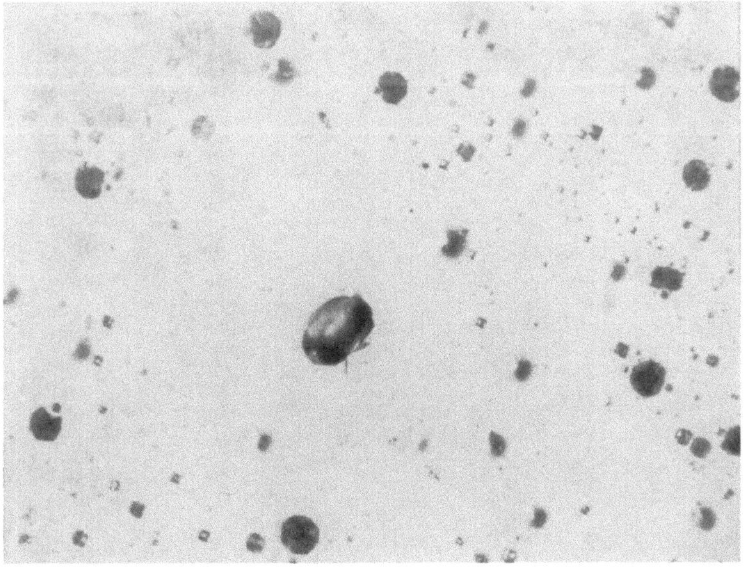

Abb. 2a

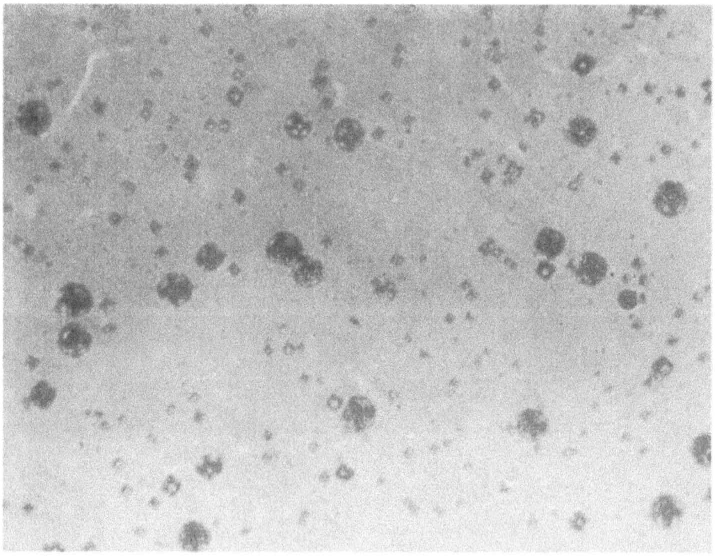

Abb. 2b

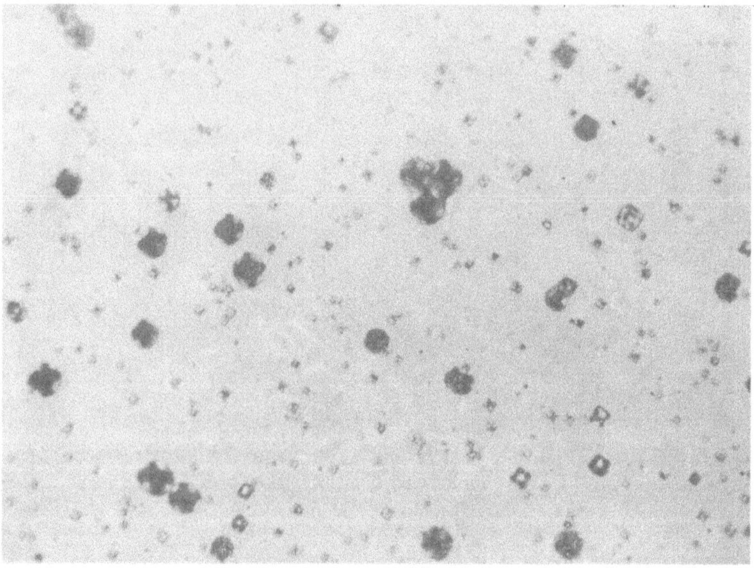

Abb. 2c

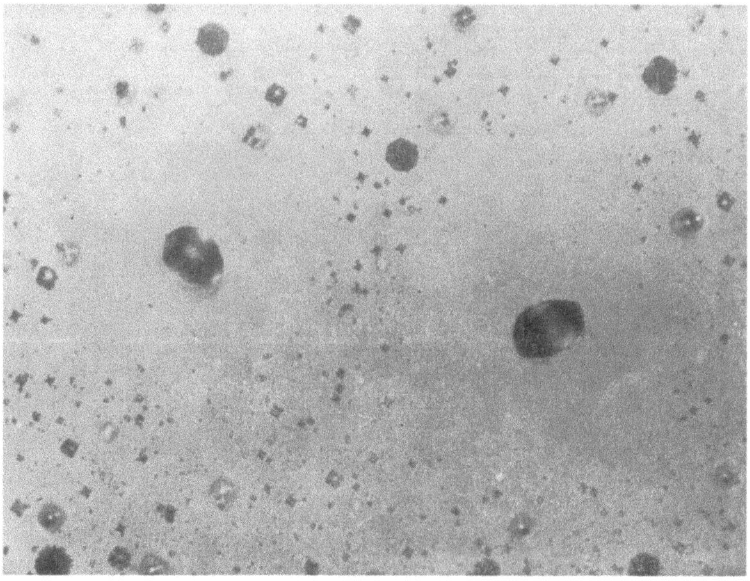

Abb. 2d

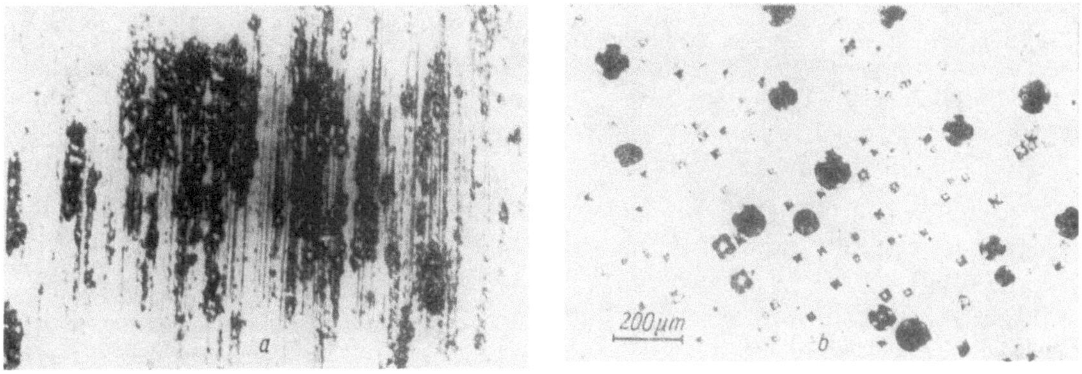

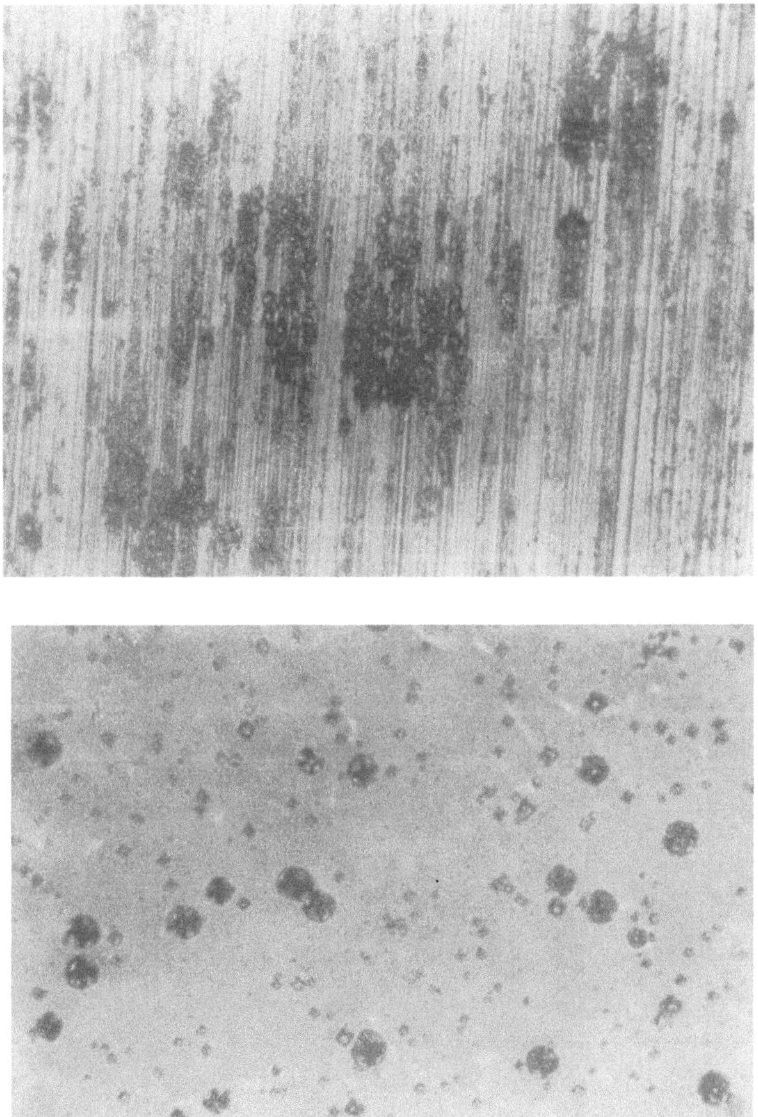

Abb. 3:
Lochanordnung an einem (100)-orientierten Eiseneinkristall
(*a*) nach Naßschleifen mit Papier der Körnung 600,
(*b*) nach elektrolytischem Polieren.

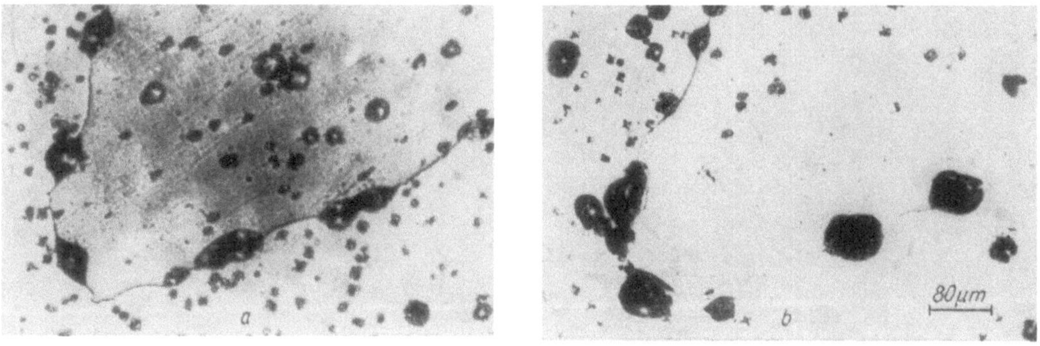

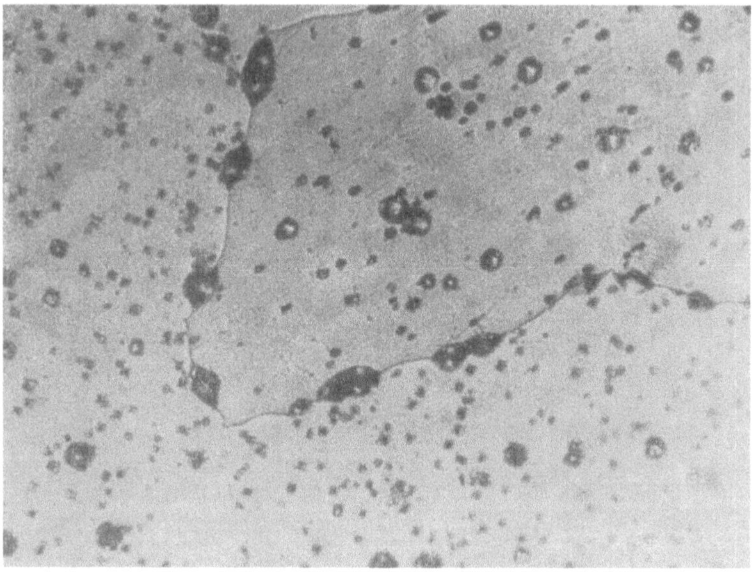

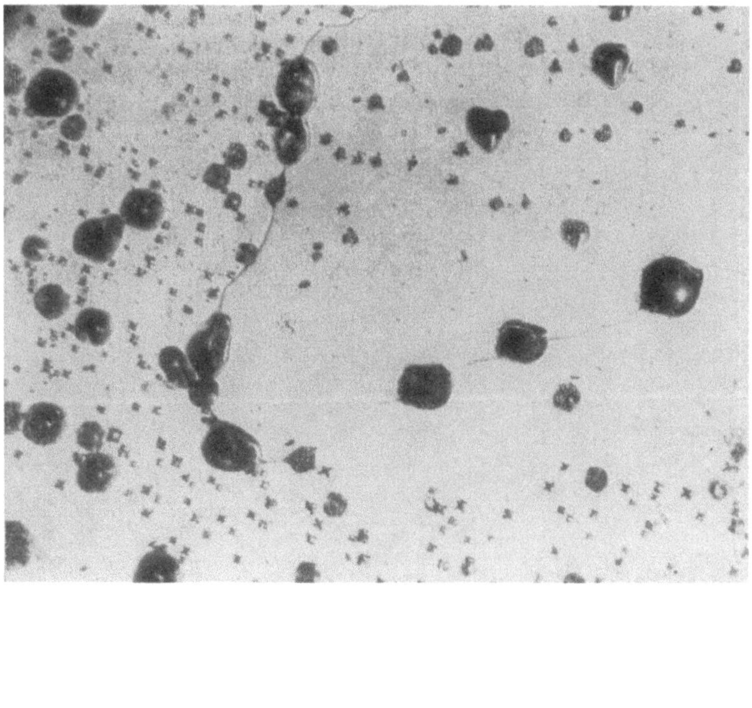

Abb. 4:
Lochfraßkorrosion an polykristallinem Eisen.

(*a*) Ausgangszustand,
(*b*) an der gleichen Stelle wie (*a*) nach einem Abtrag von etwa 0,5 mm.

Repassivierung in hochverdünnten Elektrolytlösungen zu der Auffassung, daß das Ti erst in die Lösung übergehen muß und dieser Vorgang bei niedrigen Konzentrationen und damit geringen Stromdichten eine verzögerte Bildung der Passivschicht aus der übersättigten Lösung hervorruft. Wir glauben, daß diese Auffassung schwer vereinbar mit der spezifischen Adsorption und Reaktion des praktisch immer in großem Überschuß vorhandenen Wassers ist. Die verzögerte Repassivierung

Tabelle 2: Freie Standardenthalpien von Oxidbildungsreaktionen

(1) $nMe + \frac{m}{2} O_2 \rightleftharpoons Me_nO_m$

(2) $nMe + m\,H_2O \rightleftharpoons Me_nO_m + m\,H_2$

Oxid	$\Delta G^0_{298}{}^{(1)}$ kcal mol^{-1}	$\Delta G^0_{298}{}^{(2)}$ kcal mol^{-1}
FeO	−58,3	−1,6
Fe_2O_3	−177,3	−7,2
NiO	−51,4	+5,3
Cr_2O_3	−250,6	−80,5
CrO_2	−128,0	−14,6
CrO_3	−120,2	+49,9
Al_2O_3	−378,0	−207,9
TiO	−118,3	−61,6
Ti_2O_3	−342,4	−172,3
TiO_2	−212,4	−99,0
Ta_2O_5	−456,7	−173,2
ZnO	−76,1	−19,4
MoO_2	−119,0	−5,6
MoO_3	−162,1	+8,0
WO_2	−123,8	−10,4
WO_3	−182,6	−12,5

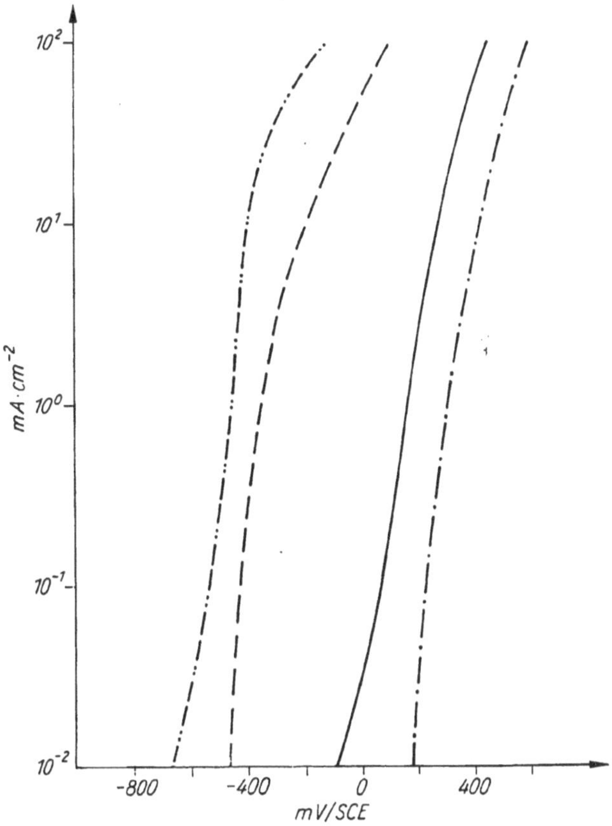

Abb. 5: Anodische Stromdichte-Potential-Kurven von Molybdän in verschiedenen Korrosionsmedien.
---- 1 N HCl; —— 1 N NaCl; -·-·-·- 1 N H_2SO_4; --··- 1 N NaOH.

in sehr verdünnten Elektrolytlösungen läßt sich u. E. auch dadurch deuten, daß wegen der geringen Leitfähigkeit die elektrochemische Gesamtreaktion verzögert ist:

$$\begin{array}{c} Me + OH_2 \rightleftharpoons MeO + 2H^+ + 2e^- \\ \underline{1/2O_2 + H_2O + 2e^- \rightleftharpoons 2OH^-} \\ Me + 1/2O_2 \rightleftharpoons MeO \end{array}$$

Ich möchte allerdings damit nicht behaupten, daß die Passivierung durch das Wasser immer ein elektrochemischer Vorgang sein muß. Damit ist natürlich auch nicht geklärt, ob durch die Passivierungsreaktion mit dem Wasser ein schwerlösliches Oxid bestimmter Struktur gebildet werden muß. Man kann aber feststellen:

Die Gitterparameter der Passivschicht, sofern es sich um oxydische kristalline Schichten handelt, spielen im Verhältnis zu dem Gitterparameter des Metalls *K* eine entscheidende Rolle, wie die Übersicht (Tab. 3) für einige Metalle und ihre stabilen Oxide und Oxidhydrate zeigt. Es sind deswegen die bekannten Oxide und Oxidhydrate aufgeführt, weil nicht eindeutig bekannt ist, welche Verbindungen direkt auf dem Metall aufliegen. Gerade bei den außerordentlich gut schützenden, extrem dünnen

Passivschichten des Cr, Ni oder Ti ist eine röntgenoptische Zuordnung zu einer Oxidphase gar nicht möglich. Eine weitere Feststellung ist in diesem Zusammenhang aufschlußreich:

Die Dicke der Schicht ist offenbar ohne wesentliche Bedeutung für die Schutzwirkung. Wir können dies an der folgenden Tabelle (Tab. 4) erkennen, wo für einige Metalle die Dicke der Schutzschichten, die zur wirksamen Passivierung ausreichen, aufgeführt ist. Die in Tab. 4 zusammengestellten Werte stellen eine relativ willkürliche Auswahl von Meßergebnissen dar, die zum Teil schwer vergleichbar sind. Die Passivierungszeit (Spalte 5) und das Passivierungspotential spielen für die Schichtdicke eine wichtige Rolle, da sie natürlich mit der Zeit wächst und im allgemeinen um so rascher, je höher die

Tabelle 3: Gitterkonstanten einiger Metalle, ihrer Oxide und Oxidhydrate

Metall	Metall-struktur d_M	Oxidstruktur d_O	Oxidhydratstruktur d_{OH}
Fe	$a = 2{,}8604$	FeO $a = 4{,}28-4{,}31$[2]) $a = 4{,}28-4{,}36$ Fe_3O_4 $a = 8{,}397-8{,}394$[2]) $a = 8{,}39$ α-Fe_2O_3 $a = 5{,}43$ γ-Fe_2O_3 $a = 8{,}33-8{,}38$[2]) $a = 8{,}32$ δ-Fe_2O_3 $a = 5{,}09$; $c = 4{,}41$	$Fe(OH)_2$ $a = 3{,}24$; $c = 4{,}47$ α-FeOOH $a = 4{,}64$; $b = 10{,}0$; $c = 3{,}03$ β-FeOOH $a = 10{,}24$; $b = 10{,}56$; $c = 3{,}03$ γ-FeOOH $a = 3{,}87$; $b = 12{,}4$; $c = 3{,}06$ $Fe(OH)_3$ $a = 5{,}70$
Ni	$a = 3{,}5172$	NiO $a = 2{,}95$ Ni_2O_3	$Ni(OH)_2$ $a = 3{,}12$; $c = 4{,}60$ α-NiOOH $a = 2{,}81$; $c = 8$ β-NiOOH $a = 2{,}81$; $c = 4{,}84$ γ-NiOOH $a = 2{,}82$; $c = 20{,}65$
Cr	$a = 2{,}878$	CrO Cr_2O_3 $a = 5{,}35$ CrO_2 $a = 4{,}41$; $c = 2{,}86$ CrO_3 $a = 4{,}79$; $b = 8{,}56$; $c = 5{,}76$	
Al	$a = 4{,}0413$	α-Al_2O_3 $a = 5{,}13$ γ-Al_2O_3 $a = 7{,}9$ γ'-Al_2O_3 $a = 7{,}7-8{,}0$	α-AlO(OH) $a = 4{,}4$; $b = 9{,}4$; $c = 2{,}8$ γ-AlO(OH) $a = 3{,}70$; $b = 12{,}24$; $c = 2{,}85$ $Al(OH)_3$ (Hydrargillit) $a = 8{,}62$; $b = 5{,}06$; $c = 9{,}70$ $Al(OH)_3$ (Bayerit) $a = 5{,}01$; $c = 4{,}76$
Ti	$a = 2{,}95$ $c = 4{,}72$	TiO $a = 4{,}24$ Ti_2O_3 $a = 5{,}42$ Ti_3O_5 $a = 3{,}75$; $b = 9{,}47$; $c = 9{,}72$ TiO_2 (Rutil) $a = 4{,}49$; $c = 2{,}89$ TiO_2 (Anatas) $a = 5{,}36$; $c = 9{,}53$ TiO_2 (Brookit) $a = 9{,}15$; $b = 5{,}44$; $c = 5{,}14$	

Tabelle 3 (Fortsetzung)

Metall	Metall-struktur d_M	Oxidstruktur d_O	Oxidhydratstruktur d_{OH}
Zn	$a = 2{,}659$ $c = 4{,}935$	ZnO $a = 4{,}62$	$Zn(OH)_2$ $a = 3{,}19$; $c = 4{,}65$ $Zn(OH)_2$ $a = 3{,}1$; $c = 7{,}8$ ε-$Zn(OH)_2$ $a = 5{,}16$; $b = 8{,}53$; $c = 4{,}92$
Mo	$a = 3{,}1410$	MoO_2 $a = 5{,}61$; $b = 4{,}84$; $c = 5{,}53$ Mo_4O_{11} $a = 24{,}4$; $b = 5{,}45$; $c = 6{,}70$ Mo_8O_{23} $a = 16{,}8$; $b = 4{,}04$; $c = 13{,}4$ Mo_9O_{26} $a = 16{,}75$; $b = 4{,}03$; $c = 14{,}45$ MoO_3 $a = 3{,}9$; $b = 13{,}8$; $c = 3{,}7$	$MoO_3 \cdot 2H_2O$ $a = 3{,}77$; $b = 6{,}91$; $c = 7{,}34$

Literatur zur Tabelle 3

[1] LANDOLT-BÖRNSTEIN: „*Zahlenwerte und Funktionen*", 6. Auflage, Bd. I: „*Atom- und Molekularphysik*", 4. Teil „*Kristalle*" Springer-Verlag, Berlin—Göttingen—Heidelberg 1955

[2] FASISKA, E. J., Corr. Sci., *7*, 833 (1967)

Tabelle 4: Dicke der Passivschicht und Passivstromdichte von einigen Metallen

Metall	Elektrolyt	Passivschichtdicke d: d [Å]	ε (StHE) [V]	Zeit [min]	Meß-verfahren+)	Passiv-stromdichte i_p [μAcm^{-2}]	Bemerkungen	Lite-ratur
Ni	Boratpuffer pH = 8,4	4 12	−0,36 0,94	60	*E/C*	$< 10^{-1}$	i_p nicht potentialabhängig	[1]
Ni	Na_2SO_4 0,15 N pH = 8,4	9…12	0,17−0,97	60	*C/R*	$< 10^{-2}$	keine Potentialabhängigkeit von d und i_p	[2]
Ni	H_2SO_4 0,01 N + Na_2SO_4 0,5 M pH = 3,15	> 65	0,5		*E*	< 4	wahrscheinlich nicht im Passivbereich eingesetzt	[3]
Ni	H_2SO_4 1 N	1 MS·) 10	0,3 1,04	*t* var.	*K*		Versuchsabbruch bei $i_p = 20\ \mu$Acm^{-2}	[4]
Cr	Na_2SO_4 1 N H_2SO_4 1 N H_2SO_4 0,5 N	25…30 10 10 20	U_R U_R 0,1 (U_R) 1,15		*E* *E*	— — $3 \ldots 5 \cdot 10^{-1}$	Messung am Ruhepotential Cr auf Glas aufgedampft; i_e nicht potentialabhängig	[5]
Fe	$NaNO_2$ 0,1 N	20…30	U_R	1000	*E*	—	Messung am Ruhepotential	[6]
Fe	Boratpuffer pH = 8,4	10 30	−0,46 1,09	60	*C*			[7]
Fe	dto.	17/20 41/46	−0,06 1,04	60	*C/E*			[8]

Tabelle 4 (Fortsetzung)

Metall	Elektrolyt	Passivschichtdicke d				Passiv-stromdichte i_p [μAcm^{-2}]	Bemerkungen	Literatur
		d [Å]	ε (StHE) [V]	Zeit [min]	Meß-verfahren+)			
Fe	Boratpuffer pH = 8,4	15 35	0,0 0,8		C	1	Messung unter quasistationären Bedingungen; i_p nicht potentialabhängig	[9]
Fe	dto.	1 MS·) 35	−0,31 0,9	t var.	K		Versuchsabbruch bei $i_p = 5\ \mu$Acm^{-2}	[10]
Fe	dto.	8 30	−0,16 1,09	60	C	$< 10^{-1}$		[11]
Fe	dto. + As_2O_3 $4 \cdot 10^{-3}$ M	20 30	0,55 0,95	t var.	K		Versuchsabbruch bei $i_p = 5\ \mu$Acm^{-2}	[12]
Fe	H_2SO_4 0,01 N + Na_2SO_4 0,5 M pH = 2,90	30 41	0,9 1,4		E	10^{-1}	stationäre Bedingungen; i_p nicht potentialabhängig	[13]
Fe	H_2SO_4 2 N	8 68	0,83 1,53		C	10^{+1}	i_p nicht potentialabhängig	[14]
Fe	Boratpuffer	20—28	0,45	60	E	6 (Fe)	Schichtdickenabstufung Fe > FeNi > FeCr >	[15]
FeNi 10	pH = 8,4	40—48	1,1			4 (FeNi)	FeNiCr;	
FeCr 10						5 (FeCr)	ip potentiodynam.	
FeNi 10 Cr 10						5 (FeNiCr)	20 mV min^{-1}	
FeCr 24	H_2SO_4 2 N	1…3 MS·)	−0,11	170	C	$2 \cdot 10^{-1}$	Messung am Aktivierungspotential	[16]
Ag	K_2CrO_4 0,25 M	3400	0,8—1,6		C	$3,3 \cdot 10^3$	d und i_p nicht potentialabhängig	[17]

Anmerkung: ·) MS Monoschicht
+) Verfahren der Schichtdickenbestimmung:
C Coulometrie — K Beobachtung von Kernreaktionen
E Ellipsometrie — R Röntgenemissionsspektroskopie

Literatur zu Tabelle 4

[1] N. SATO und K. KUDO, Electrochim. Acta, **19**, 461 (1974)
[2] B. MACDOUGALL und M. COHEN: J. Electrochem. Soc., **121**, 1152 (1974)
[3] J. O'M. BOCKRIS, A. K. N. REDDY und B. RAO, ibid. **113**, 1133 (1966)
[4] J. SIEJKA, C. CHERKI und J. YAHALOM, ibid. **119**, 992 (1972)
[5] V. V. ANDREEVA, Ž. fiz. chim., **38**, 1825 (1964)
[6] J. KRUGER, J. Electrochem. Soc., **110**, 654 (1963)
[7] J. KRUGER und J. P. CALVERT, ibid., **114**, 43 (1967)
[8] K. KUDO, T. SHIBATA, G. OKAMOTO und N. SATO, Corrosion Sci., **8**, 809 (1968)
[9] J. O'M. BOCKRIS, M. A. GENSHAW, V. BRUSIC und H. WROBLOWA, Electrochim. Acta, **16**, 1859 (1971)
[10] B. AGIUS und J. SIEJKA, J. Electrochem. Soc., **120**, 1019 (1970)
[11] M. COHEN, ibid., **121**, 191 C (1974)
[12] B. AGIUS und J. SIEJKA, ibid., **122**, 723 (1975)
[13] K. J. VETTER und F. GORN, Z. physik. Chem. N.F., **86**, 113 (1973)
[14] J. L. ORD und J. H. BARTLETT, J. Electrochem. Soc., **112**, 160 (1965)
[15] K. N. GOSWAMI und R. W. STAEHLE, Electrochim. Acta, **16**, 1895 (1971)
[16] R. P. FRANKENTHAL, J. Electrochem. Soc., **116**, 580 (1969)
[17] M. PRAŽÁK, Coll. czechoslov. chem. Commun., **29**, 1, 1773 (1964)

anodische Polarisation ist. Auch die Elektrolytzusammensetzung beeinflußt die Schichtdicke, weil von ihr die Abbaugeschwindigkeit der Passivschicht abhängt. Es soll durch die Tabelle lediglich nachzuweisen versucht werden, daß schon eine oder wenige Monoschichten ausreichen, um einen sehr niedrigen anodischen Strom zu erzielen, der durch das Wachsen der Passivschicht nicht wesentlich verstärkt wird. Andererseits vermögen dicke Schichten wie Ag_2CrO_4 auf Ag die anodische Auflösung relativ wenig zu behindern, wie die um drei Größenordnungen höhere „Passivstromdichte" zeigt, wenn keine primäre Passivierung vorliegt.

Aus diesen experimentellen Befunden kann man vielleicht einige Regeln ableiten, aber sicher keine quantitativen Gesetzmäßigkeiten. Die lange umstrittene Frage: Ist für die Passivierung nur eine Monoschicht adsorbierten Sauerstoffs oder ein Phasenoxid erforderlich? kann man meiner Auffassung nach so beantworten: Der Primärschritt der Passivierung erfordert nur eine Monoschicht Sauerstoff, wie auch nur eine Monoschicht eines Inhibitors zur Hemmung der anodischen Metallauflösung genügt [13].

Der primären Passivität folgt die sekundäre Passivität durch Aufwachsen eines Phasenoxids mit geringer Stromdichte. Aber ohne primäre Passivität gibt es auch keine sekundäre Passivität, andererseits ist natürlich die Chemisorptionsschicht ohne Oxidphase leicht verletzbar. Viele Autoren haben wohl in der letzten Zeit ähnliche Auffassungen ausgesprochen und durch neue Argumente und Experimente gestützt.

Die wesentliche Barriere für die anodische Auflösung des Metalls liegt an der Nahtstelle zwischen Metall und Oxid, was auch aus den theoretischen Arbeiten von N. Cabrera und N. F. Mott [14] hervorzugehen scheint. Das ist aber gleichzeitig ein weiterer Grund, der die experimentelle Forschung über die Natur der Passivschicht sehr erschwert, denn diese Nahtstelle ist mit optischen und elektronenoptischen Mitteln kaum zu fassen. Ich hatte vor Jahren einmal angeregt zu prüfen, ob man mit Hilfe der Mössbauer-Spektroskopie an oxydiertem Fe-57-Pulver, neben dem Metall und seinen Oxiden einen Zwischenzustand feststellen könne, bisher ist mir aber in dieser Richtung nichts bekannt geworden.

Die doppelte Rolle des Wassers einerseits bei der anodischen Auflösung des Metalls, andererseits bei seiner Passivierung ist inzwischen von verschiedenen Seiten zur Grundlage der Berechnung von Stromspannungskurven beim Passivierungsvorgang gemacht worden [15], nachdem wir diese Konzeption (1960) entwickelt und Dr. Ebersbach [16] in ihrer Dissertation gezeigt hatte, daß man damit Kurven berechnen kann, die dem experimentellen Stromspannungsverlauf entsprechen. Das Metall kann also in Gegenwart von Wasser zwei Simultanreaktionen unterliegen; nämlich einmal der anodischen Auflösung unter Bildung hydratisierter Ionen und zum anderen der chemischen oder elektrochemischen Umsetzung mit dem Wasser unter Bildung einer oxydischen Passivschicht. Ob eine Passivierung eintritt, hängt nun nicht nur von der Schwerlöslichkeit des Oxids ab, sondern auch von der Geschwindigkeit, mit der diese beiden in Konkurrenz stehenden Reaktionen ablaufen. Passivität wird um so eher eintreten, je größer die Geschwindigkeit der Reaktion mit dem Wasser zu *Me*O und je geringer die Geschwindigkeit der anodischen Auflösung ist. Das wird deutlich, wenn man die Austauschstromdichten für die Gebrauchsmetalle, an denen die Passivität besonders eingehend untersucht worden ist, betrachtet (Tab. 5).

Metalle mit hoher Austauschstromdichte wie Pb oder Ag zeigen keine echte Passivierung. Sie bedürfen einer Schicht, die sich aus der übersättigten Lösung abscheidet [17]. Die verschiedenen prinzipiellen Reaktionsmöglichkeiten des Wassers mit dem Metall sind in Abb. 6 dargestellt. Über die Passivierungsgeschwindigkeit können wir auf Grund der bisherigen experimentellen Möglichkeiten wenig aussagen. Es ist aber naheliegend, daß solche Metalle, bei denen die Änderung der freien Enthalpie bei Umsetzung des Metalls mit dem Wasser besonders stark negativ ist, auch eine hohe Geschwindigkeit der Passivierungsreaktion zeigen werden. Und tatsächlich sind ja die stark zur Passivität neigenden Metalle wie Aluminium, Titan, Tantal sehr elektronegative Metalle und haben eine große Affinität für die Reaktion mit dem Wasser.

Die außerordentlich niedrigen Austauschstromdichten, z. B. des Nickels im Verhältnis etwa zum Silber, machen es verständlich, daß dieses Metall ebenso wie das Chrom eine sehr starke Tendenz hat, oxidische Passivschichten zu bilden. Natürlich ist dies nur eine sehr qualitative Feststellung und ich

Tabelle 5: Standardelektrodenpotentiale $\overset{\circ}{\varepsilon}_H$ und Austauschstromdichten i_0 (bezogen auf $C_{Me^{z+}} = 1\ M$) von Gleichgewichtsreaktionen $Me \rightleftharpoons Me^{z+} + ze^-$

Metall	$\overset{\circ}{\varepsilon}_H$ [V] niedrigste Oxydationsstufe	Literatur	i_0 (Me/Me^{z+}) [A cm^{-2}]	Literatur
Fe	Fe/Fe^{2+} −0,44	[1]	Fe/Fe^{2+} $10^{-4} \dots 10^{-8}$	[2], [6]
Ni	Ni/Ni^{2+} −0,25	[1]	Ni/Ni^{2+} $10^{-5} \dots 10^{-9}$	[3], [6], [8]
Cr	Cr/Cr^{2+} −0,86 Cr/Cr^{3+} −0,71	[1]	Cr/Cr^{3+} $< 10^{-10}$	[4]
Al	Al/Al^{3+} −1,67	[1]	Al/Al^{3+} $10^{-7} \dots 10^{-8}$	[5]
Ti	Ti/Ti^{2+} −1,75	[1]		
Mo	Mo/Mo^{3+} −0,20	[7]	Mo/Mo^{3+} $\approx 10^{-4}$	
Pb	Pb/Pb^{2+} −0,126	[1]	Pb/Pb^{2+} $10^{2} \dots 10^{0}$	[6]
Ag	Ag/Ag^{+} +0,80	[1]	Ag/Ag^{+} $10^{2} \dots 10$	[6]

Literatur zu Tabelle 5

[1] *„Tabellenbuch Chemie"*, VEB Dt. Verlag f. Grundstoffindustrie, Leipzig 1968
[2] V. A. ROITER, V. A. JUCA und E. S. POLUJAN, Acta physicochim. USSR, **10**, 389 (1939)
[3] M. HOLLNAGEL und R. LANDSBERG, Z. physik. Chem. **212**, 94 (1959);
M. HOLLNAGEL, Wiss. Z. Univ. Greifswald, Math.-nat. R. **8**, 172 (1958/59)
[4] K. SCHWABE, Electrochim. Acta, **3**, 186 (1960)
[5] TAMAMUSHI R., Supplement to Electrochim. Acta, **9**, 963 (1964), "*Kinetic Parameters of Electrode Reactions*", 1972
[6] N. TANAKA und R. TAMAMUSHI, Electrochim. Acta **9**, 963 (1964)
[7] A. A. POZDEEVA et al., Corr. Sci., **6**, 149 (1966)
[8] V. A. JUCA und L. D. KOPYL, Ž. fiz. chim., **14**, 1074 (1941)

bin weit davon entfernt zu glauben, daß man daraus schon sichere Schlußfolgerungen für die Herstellung von Legierungen ableiten kann, die bei Gegenwart von Wasser stabile Passivschichten bilden.

Wie schwierig es ist, darüber hinausgehende quantitative Aussagen zu machen, zeigt die Tatsache, daß bei Veränderung der Zusammensetzung von Legierungen Effekte auftreten, die sich kaum deuten lassen. Das kann man an dem Verhalten der drei Metalle — Eisen, Nickel, Chrom — bei Zusatz von Molybdän belegen. Man erkennt hier, daß der Zusatz von 8% Molybdän zu einem dieser drei Metalle ganz unterschiedlich auf das Lochfraßverhalten wirkt.

Während beim Eisen und Nickel der Molybdänzusatz die Lochfraßbildung eher begünstigt als zurückdrängt, ist beim Chrom durch den Molybdänzusatz das Metall praktisch unangreifbar geworden, auch in Salzsäure, vgl. Abb. 7 (Taf. V—VII). Die Veränderung der potentiostatischen Stromspannungskurven durch 8% Molybdänzusatz gegenüber denen der reinen Metalle in 1 N H_2SO_4 zeigen die Abb. 8, 9 und 10 (Taf. VII u. VIII). Wie man sieht, wird die Cr—Mo-Legierung überhaupt nicht aktiv, während Eisen durch den Mo-Zusatz eine höhere Passivstromdichte hat. Selbst eine qualitative

Deutung der Wirkung des Molybdänzusatzes zu Fe, Ni und Cr ist vorerst sehr spekulativ. Wenn man mit HOAR [18], KRUGER [19] und anderen annimmt, daß die Stabilität der Passivschicht des Chroms von ihrem amorphen Zustand herrührt, der eine sehr geringe Ionenleitfähigkeit hat, dann sollte das Molybdän bzw. seine Oxidphase diese Ionenleitfähigkeit noch weiter verringern, dabei könnte auch eine Stabilisierung des Wassergehaltes in der Passivschicht eine Rolle spielen. Für $Cr(OH)_3$ wurde

Die aktive Auflösung (I) und die Passivierung (II) der Eisenmetalle

I

$$Me + H_2O \rightleftharpoons MeOH + H^+ + e^- \quad (I_1)$$
$$MeOH \rightarrow MeOH^+ + e^- \quad (I_2)$$

$$Me + H_2O \rightarrow MeOH^+ + H^+ + 2e^- \quad (I_3)$$
$$MeOH^+ + H^+ \rightleftharpoons Me^{++} + H_2O \quad (I_4)$$

$$Me \rightarrow Me^{++} + 2e^- \quad (I)$$

II

$$Me + H_2O \rightleftharpoons MeOH + H^+ + e^- \quad (II_1)$$
$$MeOH + H_2O \rightarrow Me(OH)_2 + H^+ + e^- \quad (II_2)$$
$$Me(OH)_2 \rightleftharpoons MeO + H_2O \quad (II_3)$$

$$Me + H_2O \rightleftharpoons MeO + 2H^+ + 2e^- \quad (II)$$

Abb. 6: Reaktionsmöglichkeiten von Metall und Wasser.

kürzlich nachgewiesen, daß es die Wasserabgabe von FeO(OH) Goethit ($\approx Fe_2O_3 \cdot H_2O$) stark hemmt [20], eine entsprechende Wirkung könnte ein MO-Oxidhydrat auf das Chromoxid selbst haben. Die Gegenwart des Wassers könnte die sofortige Repassivierung des relativ unedlen Chroms in bekannter Weise hervorrufen, wenn die primäre Passivschicht z. B. durch Cl^- lokal zerstört wird.

Ich darf also zusammenfassend folgendes feststellen: Die Methoden zur Untersuchung der Bildung von Passivschichten, ihrer Struktur im Zusammenhang mit ihrer Stabilität und korrosionshemmenden Wirkung reichen nicht aus, um sichere Prognosen für die Herstellung solcher stark korrosionshemmenden Passivschichten zu treffen.

Unserer Meinung nach machen zwei Probleme die quantitative Deutung der Passivierungsvorgänge außerordentlich schwierig, wenn nicht unmöglich.

1. Die Metalloberfläche ist nicht reproduzierbar, die kinetischen Parameter, die aus elektrochemischen oder anderen Messungen erhalten werden, zeigen deswegen eine große Streubreite.
2. Alle bekannten Untersuchungsmethoden gestatten entweder nur indirekte Aussagen, insbesondere die elektrochemischen, oder wir wissen nicht, ob wir wirklich den Sitz der Energiebarriere beobachten, die die anodische Auflösung des Metalls unterbindet.

Natürlich ist es denkbar, neue Methoden zu entwickeln oder zu kombinieren, die diesen Mangel beseitigen. Fraglich erscheint es mir in jedem Fall, ob es gelingt, Metalloberflächen streng reduzierbar herzustellen. Wenn es aber gelingen sollte, dann sind sie so völlig verschieden von der Oberfläche eines realen, metallischen Werkstoffs, daß man aus solchen Untersuchungen nicht auf das Korrosions- und Passivverhalten dieses Werkstoffs schließen kann. Es ist dies, meine ich, einer der Fälle, wo der verantwortungsbewußte Wissenschaftler sorgfältig zu prüfen hat, welcher Erkenntnisgewinn generell zu erwarten ist und welchen Wert er für die Volkswirtschaft hat, bevor er ein Forschungsprogramm aufstellt. Danach muß sich auch der Aufwand für die Aufgaben der Grundlagenforschung richten. Tatsächlich sind die wichtigsten Erkenntnisse über die Bildung solcher Passivschichten auf Legierungen, soweit ich weiß, in erster Linie empirisch gewonnen worden. Natürlich auf Grund eines vorliegenden Erfahrungsschatzes, aber doch immerhin mehr oder weniger empirisch. Das gilt sowohl für die Chrom-Nickel-Stähle, als auch für den Molybdänzusatz zur Verhütung des Lochfraßes bei diesen Metallen. Ich halte es zur Zeit für aussichtsreicher, unter Verwendung aller bisherigen Erfahrungen dieses pragmatische Vorgehen fortzusetzen. Allerdings ist es dann notwendig, alte und neue Erfahrungen über das Korrosions- und Passivverhalten der Metalle und Legierungen interdisziplinär und international systematisch zur Auffindung neuer passivierbarer Werkstoffe zu nutzen. Ich halte diesen Weg deswegen für richtig — bei aller Anerkennung der Bedeutung der Grundlagenforschung auf diesem Gebiet — weil er rascher zu konkreten Ergebnissen führt und die Korrosion ja nicht darauf wartet und weil andererseits der Aufwand an metallischen Werkstoffen ständig steigt trotz Einsatz von organischen Hochpolymeren und anorganischen Werkstoffen, vor allen Dingen, wenn wir die Entwicklungsländer mit berücksichtigen, und weil andererseits die Rohstoffvorräte für metallische Werkstoffe begrenzt sind.

Gestatten Sie mir, in diesem Zusammenhang noch eine allgemeine Bemerkung über Forschungen im Hinblick auf die Zukunft, die heute ja durch die Bevölkerungsentwicklung und die Probleme des Umweltschutzes zum Teil mit Sorge betrachtet werden muß. Moderne, insbesondere naturwissenschaftliche Forschung erfordert einen enormen materiellen Aufwand an Großgeräten und Wissenschaftlern, zum Teil auch an Zeit und Material für die Vorbereitung. Die Verantwortung des Wissenschaftlers verlangt, daß er immer sorgfältig prüft, ob die geplanten Untersuchungen eine echte Chance haben, zu wirklich neuen Erkenntnissen zu gelangen. Das gilt sowohl für die experimentelle Forschung, hier muß die Untersuchungstechnik zur Beantwortung der Frage geeignet und das Modell der Untersuchungsmethode optimal angepaßt sein, andererseits aber mit dem realen Objekt in den wesentlichen Eigenschaften übereinstimmen. Bei Beachtung dieser primitiven Regeln ließe sich viel unnötiger Aufwand vermeiden. Das gilt aber auch für die theoretische Forschung in bezug auf den Rechenaufwand und die Zeit. Prof. CASIMIR hat einmal darauf hingewiesen, daß nach seinen Erfahrungen mindestens die Hälfte aller Computerstunden überflüssig sei, wenn vorher sorgfältig überlegt würde, ob die Rechnung wirklich sinnvoll ist und neue Erkenntnisse erwarten läßt.

Wir sollten ferner in allen Bereichen der wissenschaftlichen Forschung berücksichtigen, daß die Ergebnisse der Grundlagenforschung für ihre technische Realisierung einen enorm viel größeren Aufwand erfordern als die Grundlagenforschung selbst, nicht nur einen sehr hohen materiellen Aufwand, sondern auch einen enormen Aufwand an menschlicher Arbeitskraft, und deswegen ist nur so viel Grundlagenforschung vertretbar, daß die praktische Realisierung ihrer Ergebnisse gesichert werden kann. Da der dafür erforderliche Investitionsbedarf und die notwendige Zahl von Arbeitskräften die Forderung für die Grundlagenforschung oft um Größenordnungen übersteigt, müssen die wirtschaftlichen Erfolgschancen sorgfältig geprüft werden, und zwar möglichst schon beim Beginn gezielter Grundlagenforschung. Die Bedeutung, die diese Frage hat, wird daraus sichtbar, daß die Zahl der Wissenschaftler, die sich um die Grundlagenforschung bemühen, in einem außerordentlich starken Anwachsen begriffen ist — ich glaube, das gilt für alle hochentwickelten Länder — und daß andererseits für die technische Realisierung schon jetzt häufig nicht genügend technisches Personal zur Verfügung steht. Diese Disproportion wird um so gefährlicher, je rascher die Probleme der Bevölkerungsentwicklung auf uns zukommen; und die Weltkonferenz über die Bevölkerungsentwicklung hat ja gezeigt, daß in dieser Hinsicht in den nächsten Jahrzehnten keine Veränderung der Zuwachsrate zu erwarten ist, daß andererseits aber die Ernährungs- und Versorgungsprobleme dieser Bevölkerung dadurch außerordentlich rasch wachsen, und man wird also darauf achten müssen, daß alle Forschungsergebnisse diesen Aufgaben für die Menschheitsversorgung untergeordnet werden, auch im Hinblick auf die Frage, ob Ergebnisse der Grundlagenforschung in absehbarer Zeit für solche Aufgaben wirklich mit einem vertretbaren Aufwand eingesetzt werden können.

Ich habe mir erlaubt, diese Bemerkungen auf Grund der Erfahrungen, die ich in verschiedenen Bereichen im Laufe meines langen Lebens gemacht habe, hier vorzubringen, weil ich glaube, daß keine Zeit zu verlieren ist, und weil m. E. ein Teil der immer wieder diskutierten Überführungsschwierigkeiten darauf zurückzuführen ist, daß einerseits der Aussagewert der Grundlagenforschung für die Praxis überschätzt wird, zumal wenn die Auswahl der Modelle für die Forschung sich in prinzipiellen Eigenschaften vom realen Objekt unterscheiden, und weil andererseits der Aufwand der zusätzlich nach Abschluß der Grundlagenforschung noch für Entwicklungsarbeiten zu leisten ist, oft völlig unterschätzt wird.

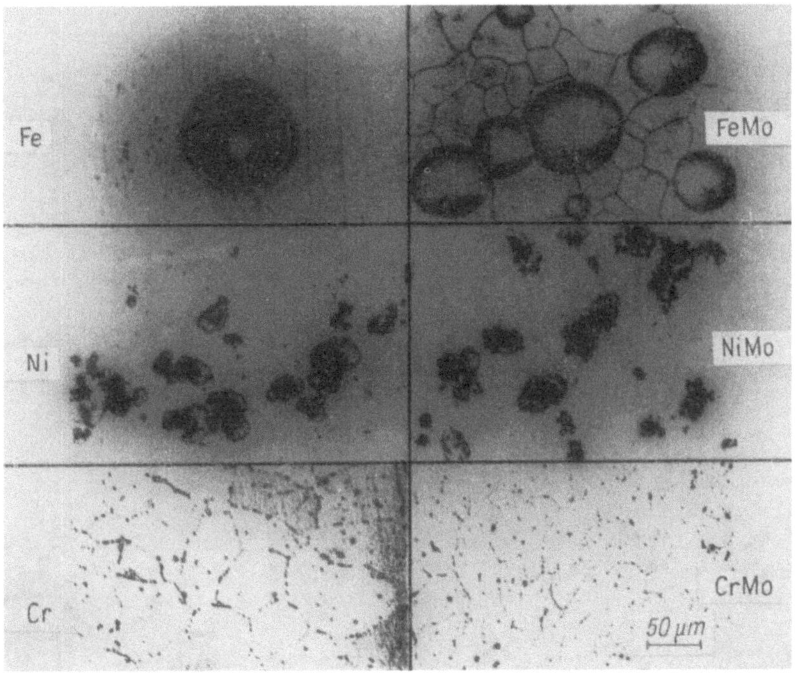

Abb. 7:
Lochfraßkorrosion an Fe sowie FeMo, an Ni sowie NiMo und an Cr sowie CrMo.

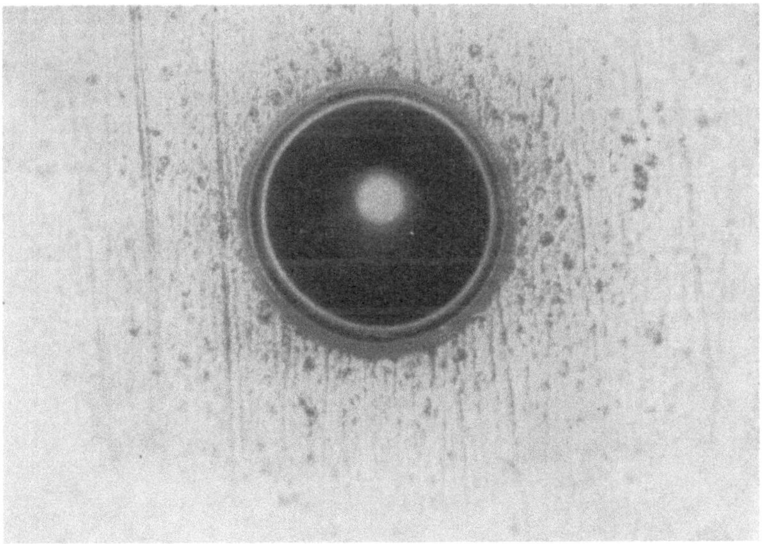

Abb. 7a

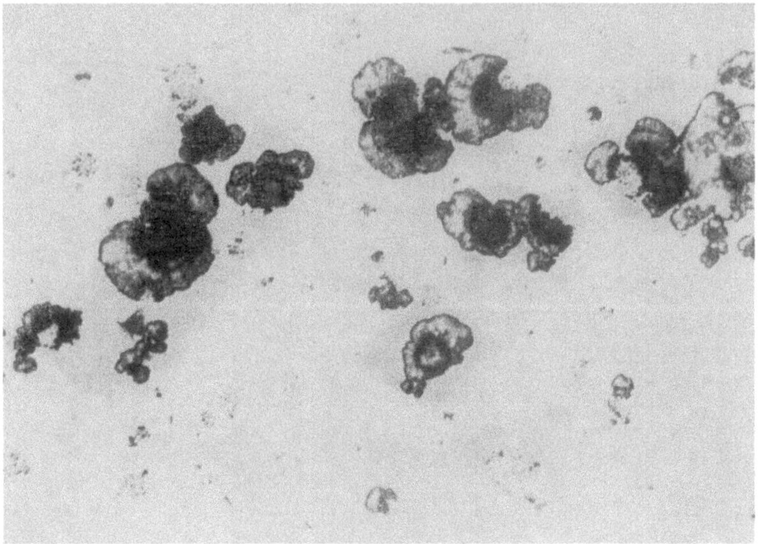

Abb. 7b

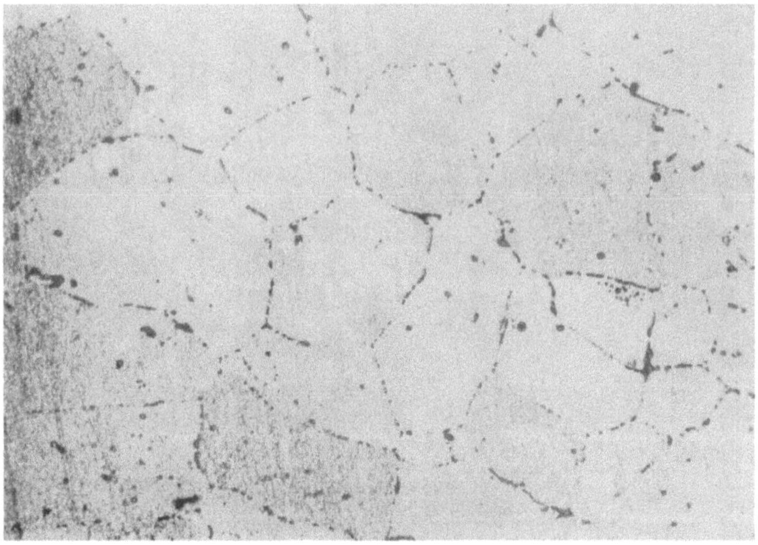

Abb. 7c

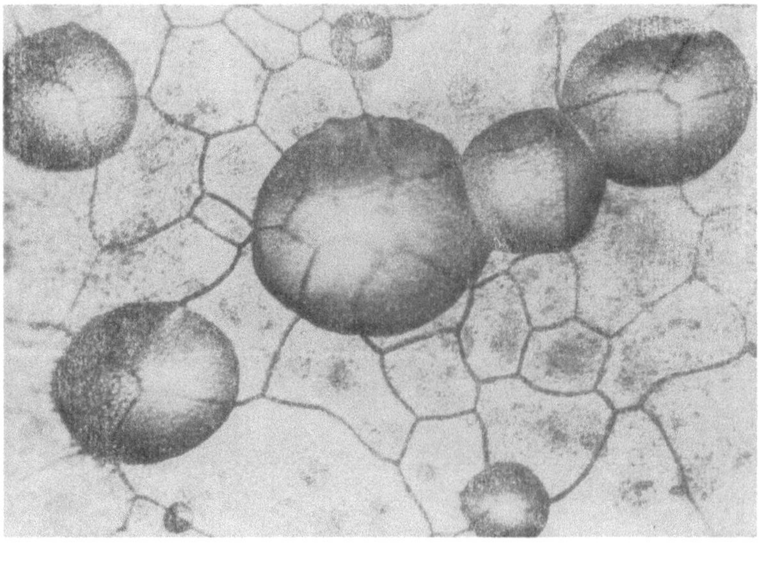

Abb. 7d

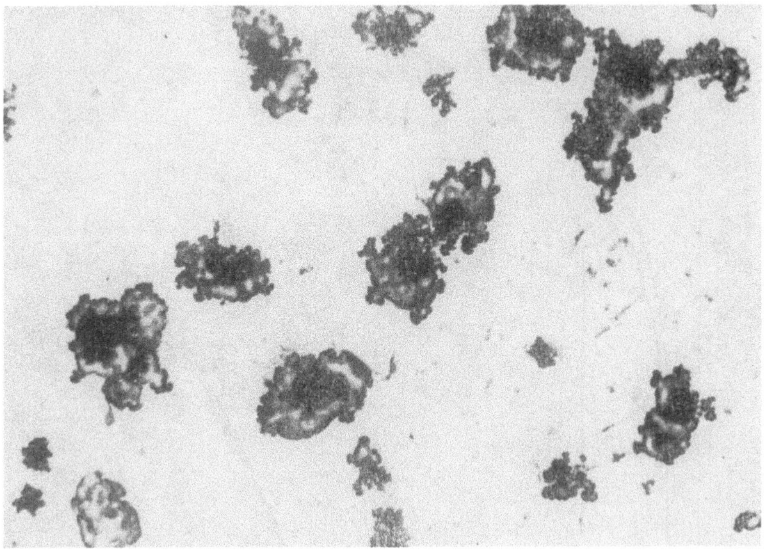

Abb. 7e

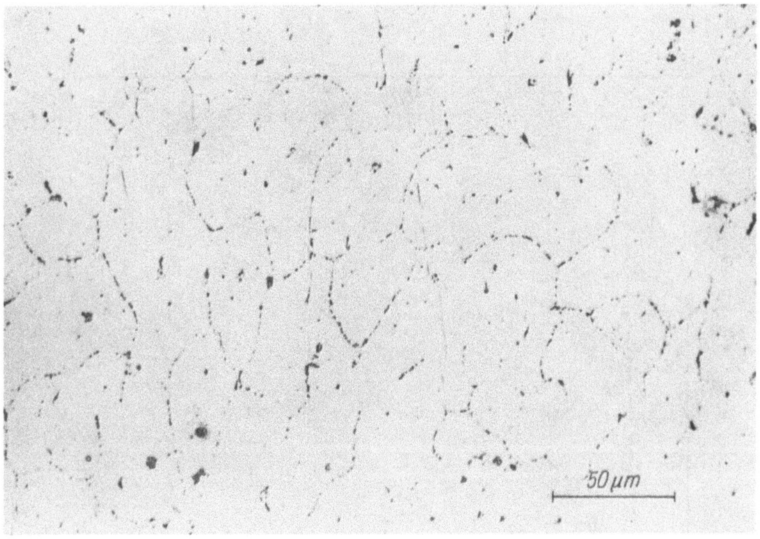

Abb. 7f

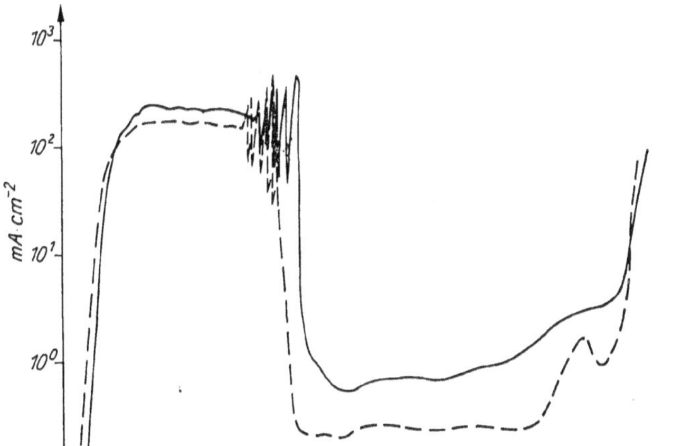

Abb. 8:
Anodische Stromdichte-Potential-Kurven von Fe und FeMo.
- - - - Fe;
—— - FeMo (8% Mo);
Elektrolyt: 1 N H_2SO_4.

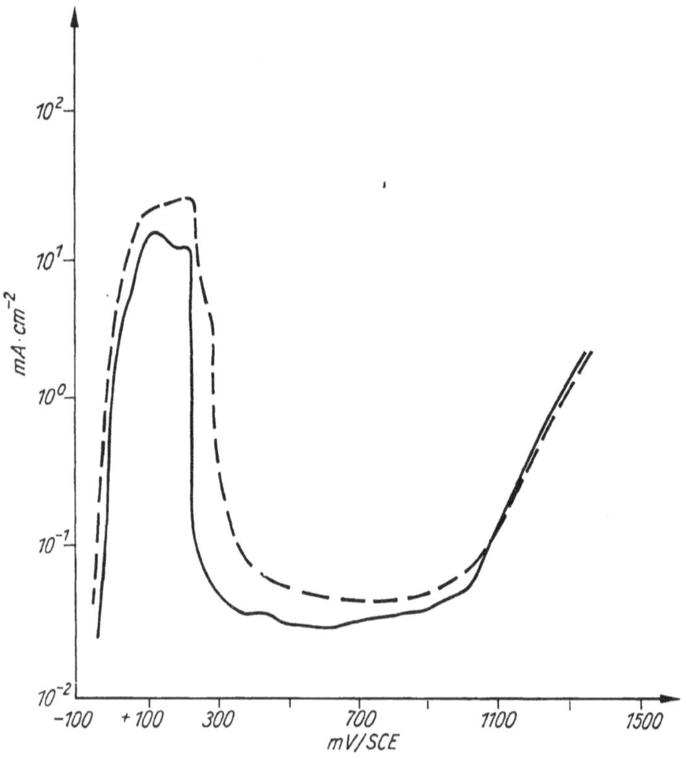

Abb. 9:
Anodische Stromdichte-Potential-Kurven von Ni und NiMo.
- - - - Ni;
—— - NiMo (8% Mo);
Elektrolyt: 1 N H_2SO_4.

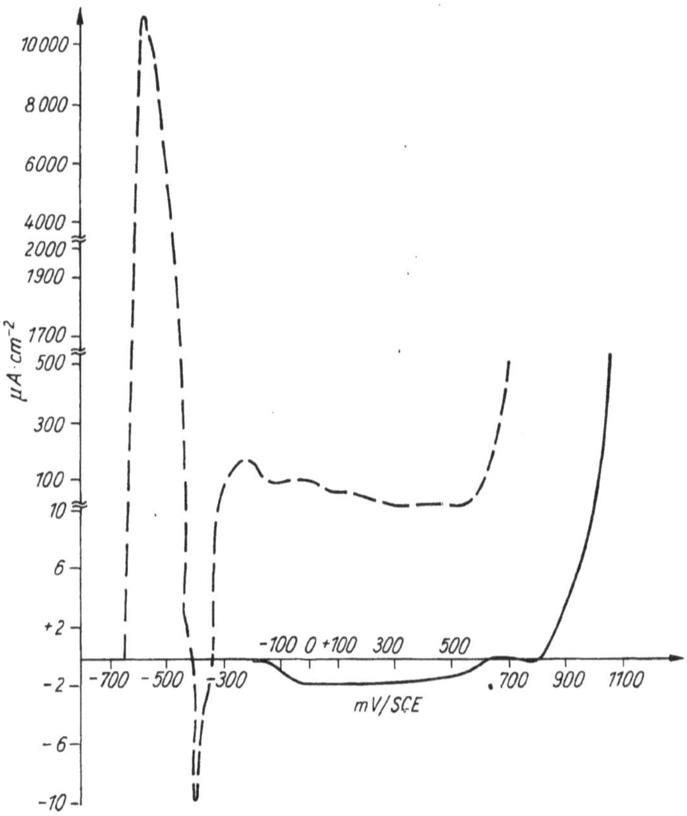

Abb. 10:
Anodische Stromdichte-Potential-Kurven von Cr und CrMo.
- - - - Cr;
—— - CrMo (8% Mo);
Elektrolyt: 1 N HCl.

Meine Mitarbeiter haben mir zum 65. Geburtstag einen Wandspruch geschenkt, der lautet: „*Genieße* jede Stunde, denn es ist später, als du denkst."

Ich glaube, wir sollten ihn aus der Verantwortung gegenüber der menschlichen Gesellschaft und ihrer Zukunft, alle, ob jung oder alt, in der Abwandlung beherzigen: *Nütze* jede Stunde, denn es ist später, als du denkst, — und wenn die vielen Worte des Lobes und der Anerkennung, die mir in diesen Tagen zuteil wurden, meinem Verhalten in diesem Sinne galten, dann fühlte ich mich besonders geehrt.

Literatur

[1] C. L. McBee und I. Kruger, Nature Phys. Sci., **230**, 194 (1971); Surface Sci., **16**, 340 (1969)
[2] H. Seiler, Chem. Ing. Techn., **46**, 805 (1974)
[3] E. Heitz: in M. G. Fontana und R. W. Staehle, "*Advances in Corrosion Science and Technology*", Vol. 4, S. 149, Plenum Press, (1974)
J. M. Kolotyrkin und G. G. Kossyj, Zascita metallov, **1**, 272 (1965)
K. Schwabe und W. Schmidt, Corrosion Science, **10**, 143 (1970)
M. Jänchen und K. Schwabe, veröffentl. in Physik. Chem. (Leipzig) im Druck
F. Mansfeld, J. Electrochem. Soc , **120**, 2, 188 (1973)
F. Mansfeld, J. Electrochem. Soc., **118**, 1412 (1971)
[4] V. E. Kazarinov und V. N. Andreev, Elektrochimija, **10**, 196 (1974)
H. D. Suschke, K. Schwabe und J. Broers, Oberfläche-surface, **15**, 295 (1974)
F. Chao, M. Costa und A. Tadjeddine, Surface Science, **46**, 265 (1974); s. auch J. Gonzales-Velasco, J. Heitbaum und W. Vielstich, Metalloberfläche, **29**, 168 (1975)
F. I. Nagirnjak, G. F. Naumova, E. A. Chopynov und N. M. Schuravleva, Zav. Laborat. **41**, 292 (1975)
[5] H. D. Suschke, K. Schwabe und J. Broers, Oberfläche-surface, **15**, 295 (1974)
H. Worch und K. Schwabe, „*II. Internationale Konferenz der RGW-Länder über Korrosion und Korrosionsschutz*", Prag 1975
[6] B. G. Wranglen, Corrosion, **30**, 462 (1974); Corrosion Science, **14**, 331 (1974)
[7] J. M. Kolotyrkin und G. G. Kossyj, Zascita metallov, **1**, 272 (1965)
[8] K. Schwabe und W. Schmidt, Corrosion Science, **10**, 143 (1970)
M. Jänchen und K. Schwabe, veröffentl. in Physik. Chem. (Leipzig), im Druck
[9] F. Mansfeld, J. Electrochem. Soc., **120**, 2, 188 (1973)
[10] F. Mansfeld, J. Electrochem. Soc., **118**, 1412 (1971)
[11] H. Buhl und H. J. Rätzer-Scheibe, Werkstoffe und Korrosion, **26**, 2 (1975)
[12] K. Schwabe, Electrochim. Acta, **3**, 186 (1960); Werkstoffe und Korrosion, 18, 961 (1967)
[13] K. Schwabe, Chem. Technik, **13**, 268, 275 (1961)
Dechema-Monographie, Bd. **45**, 273 (1963), Verlag Chemie
Amsel u. a., J. Phys. Chem. Solids, **36**, 883 (1971); s. auch P. Schubert, C. Herold und W. D. Schulz, Isotopenpraxis, **8**, 321 (1972).
[14] N. Cabrera und N. F. Mott, Repts. Prog. Phys., **12**, 163 (1948/49)
[15] R. P. Frankenthal, J. Electrochem. Soc., **114**, 542 (1967)
Novoselskij u. a., Elektrochimija (russ.), **7**, 421 (1971) u. folgende Arbeiten
Vgl. auch D. Geana, A. A. Elmiligy und W. J. Lorenz, Corrosion Science, **13**, 505 (1973)
[16] U. Ebersbach, Dissertation TU Dresden, Sektion Physik 1969
K. Schwabe, Electrochim. Acta, **3**, 186 (1960); Angew. Chemie, **78**, 253 (1966)
[17] K. Schwabe, Habilitationsschrift, TU Dresden, 1933
[18] T. P. Hoar, J. Electrochem. Soc., **117**, 17C (1970)
[19] C. L. McBee und J. Kruger, Electrochim. Acta, **17**, 1337 (1972)
[20] S. C. Maiti und S. K. Adhya, Technology. Sindri (Indien), **11**, 82 (1974)

J. M. Kolotyrkin

Forschungsinstitut für physikalische Chemie „L. J. Karpov“, Moskau

EINIGE GESETZMÄSSIGKEITEN DER SELEKTIVEN AUFLÖSUNG VON EISEN- UND NICKELLEGIERUNGEN

In der Literatur wird wiederholt die Vermutung geäußert, daß viele Besonderheiten des Korrosionsverhaltens von Legierungen mit der selektiven Auflösung einzelner Legierungskomponenten zusammenhängen können. Deshalb konzentrierte sich in letzter Zeit die Aufmerksamkeit der Wissenschaftler auf die Untersuchung von Gesetzmäßigkeiten der Auflösung einzelner Legierungskomponenten und ebenso auch auf die Untersuchung von Gesetzmäßigkeiten bei der Bildung von Oberflächenschichten, die mit einer unter den gegebenen Bedingungen korrosionsbeständigeren Komponente angereichert sind. Allerdings war die Durchführung derartiger Untersuchungen bis vor einigen Jahren oftmals schwierig, weil zuverlässige Methoden zur direkten Bestimmung von sehr geringen Auflösungsgeschwindigkeiten ebenso wie auch zuverlässige Methoden zur Messung der Änderungen in der chemischen Zusammensetzung sehr dünner Oberflächenschichten fehlten. Durch das Fehlen solcher Methoden war es praktisch unmöglich, die Kinetik der selektiven Auflösung von Legierungskomponenten unter Bedingungen zu ermitteln, bei denen sich die Komponenten mit sehr geringer Geschwindigkeit auflösen. Aus diesem Grunde fehlten praktisch bis vor kurzem Literaturangaben über Gesetzmäßigkeiten zur selektiven Auflösung von Legierungskomponenten im Passivzustand.

In den letzten Jahren hat sich die Lage völlig verändert. Große Möglichkeiten zur Bestimmung geringer Korrosionsgeschwindigkeiten von Legierungen und insbesondere zur Bestimmung geringer Auflösungsgeschwindigkeiten einzelner Komponenten wurden durch die in unserem Institut entwickelten γ-spektroskopischen Verfahren [1—6] geschaffen, die eine hohe Empfindlichkeit und Selektivität aufweisen. Diese Verfahren gestatten es, in den Auflösungsprodukten (Korrosionsprodukten) gleichzeitig 5 bis 6 Legierungskomponenten zu bestimmen. Für einige Komponenten reichte die Empfindlichkeit aus, um die Auflösung von 1% und manchmal auch 0,1% einer monoatomaren Schicht zuverlässig zu registrieren.

Zur Bestimmung von Änderungen in der chemischen Zusammensetzung dünner Oberflächenschichten ($\geqq 1\ \mu m$) hat man in den letzten Jahren erfolgreich begonnen, Röntgen-Mikroanalysatoren einzusetzen. Für die Analyse noch dünnerer Schichten (bis hin zu monoatomaren Schichten) erscheinen offenbar die Auger-Spektroskopie und die Elektronenspektroskopie als sehr aussichtsreich.

Bereits die ersten Untersuchungen, die mit den genannten Verfahren durchgeführt wurden, haben gezeigt, daß diese Entwicklungsrichtung außerordentlich nützlich ist. Das kann anhand von Ergebnissen demonstriert werden, die kürzlich in unserem Labor bei Untersuchungen zur anodischen Auflösung der binären Legierung Fe—28%Cr erzielt wurden. Die Versuche wurden mit Proben durchgeführt, die eine homogenisierende Wärmebehandlung durchlaufen hatten. Dadurch wurde die Existenz solcher Zonen ausgeschlossen, deren chemische Zusammensetzung sich von der Mischkristallzusammensetzung unterscheiden könnte. Auch der Gehalt an Phasen-Gefügekomponenten wurde damit auf ein Minimum reduziert. Mit dem γ-spektroskopischen Verfahren wurden die Auflösungsgeschwindigkeiten der Legierung und jeder Legierungskomponente bei verschiedenen Potentialen ermittelt. Diese Potentiale entsprachen folgenden Bereichen: Übergang vom Aktiv- zum Passivzustand; Passivzustand und Transpassivität.

Die ermittelten Ergebnisse sind in Abb. 1 dargestellt. In Abhängigkeit von der Zeit sind im oberen Teil die Auflösungsgeschwindigkeiten der Legierungen und ihrer Komponenten Eisen und Chrom aufgetragen, im unteren Teil dahingegen der Selektivitätskoeffizient für Chrom, der angibt, um wievielmal sich das Konzentrationsverhältnis von Chrom zum Eisen in den Auflösungsprodukten vom Konzentrationsverhältnis dieser Komponenten in den Legierungen unterscheidet.

Es ist zu erkennen, daß bei einem konstanten Anodenpotential*) gewöhnlich sowohl die Auflösungsgeschwindigkeit der gesamten Legierung als auch die partiellen Auflösungsgeschwindigkeiten der Legierungskomponenten mit der Zeit stark abnehmen. Das ist besonders zu Beginn der Polarisation in der betreffenden Lösung festzustellen.

Im Passivbereich läuft die Legierungsauflösung mit einem bevorzugten Übergang von Eisen in die Lösung ab, wobei die partielle Auflösungsgeschwindigkeit des Eisens ungefähr um eine Größen-

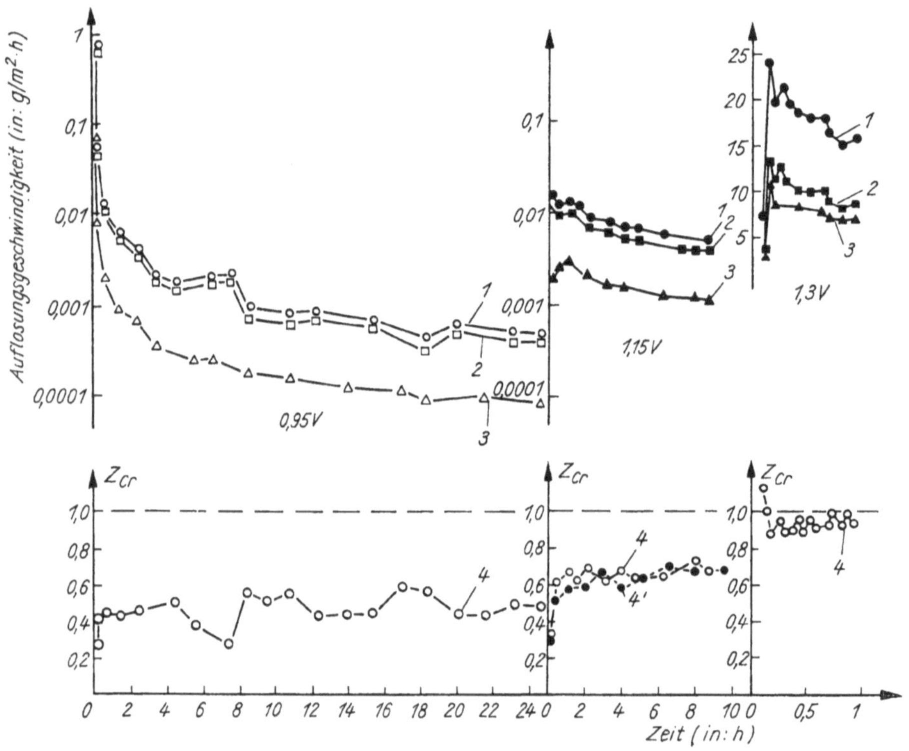

Abb. 1: Zeitabhängigkeit der summaren Auflösungsgeschwindigkeit (*1*) und der partiellen Auflösungsgeschwindigkeiten von Eisen (*2*) und Chrom (*3*) sowie Zeitabhängigkeit des Koeffizienten Z_{Cr} (*4*) für die Legierung Fe—28%Cr in 1 n H_2SO_4 bei 40°C. Schrittweise Potentialeinstellung: *a* — 0,95 V, *b* — 1,15 V, *c* — 1,3 V. Kurven *1'*—*4'* wurden bei 1,15 V an einer neuen Probe (ohne „Vorgeschichte") aufgenommen.

ordnung höher als die Auflösungsgeschwindigkeit des Chroms ist und damit die Gesamtauflösungsgeschwindigkeit fast vollständig durch die Eisenauflösung bestimmt wird. Das bedingt einen niedrigeren Wert des Selektivitätskoeffizienten Z_{Cr} für Chrom. Dieser Koeffizient hat einen besonders niedrigen Wert in der Anfangsperiode, die durch hohe instationäre Auflösungsgeschwindigkeiten der Legierung gekennzeichnet ist. Der anschließende Übergang zu stationären Auflösungsbedingungen wird von einer Zunahme des Koeffizienten Z_{Cr} begleitet. Allerdings bleibt auch unter diesen Bedingungen, entgegen der Meinung einiger Wissenschaftler [7], der Wert dieses Koeffizienten wesentlich unter Eins. Die Größe Z_{Cr} hat ebenso wie die Auflösungsgeschwindigkeit der Legierung keinen exakt konstanten Wert, wobei die beachteten Abweichungen vom Mittelwert — sowohl in Richtung geringerer als auch in Richtung höherer Werte — nicht auf Ungenauigkeiten bei den Messungen zurückgeführt werden können.

Aus Ergebnissen solcher Messungen wurde eine Kurve gezeichnet (Abb. 2), die die Potentialabhängigkeit der Mittelwerte des Selektivitätskoeffizienten für Chrom unter den sich einstellenden stationären

*) Die Potentialangaben sind auf das Potential der Standard-Wasserstoffelektrode bezogen.

Auflösungsbedingungen wiedergibt. Es ist ersichtlich, daß die Größe Z_{Cr} nur im Bereich des Aktivzustandes den Wert Eins etwas überschreitet. Beim Übergang der Legierung in den Passivzustand nimmt Z_{Cr} zunächst stark ab, verringert sich dann anschließend stetig und erreicht bei 0,75 V (dieses Potential entspricht dem Ende des Passivbereiches) mit 0,4 seinen Minimalwert. Bei weiterer Potentialerhöhung steigt Z_{Cr} an und nimmt bei 1,3 V (dieses Potential entspricht den bereits recht hohen Auflösungsgeschwindigkeiten der Legierung) einen Wert nahe Eins an.

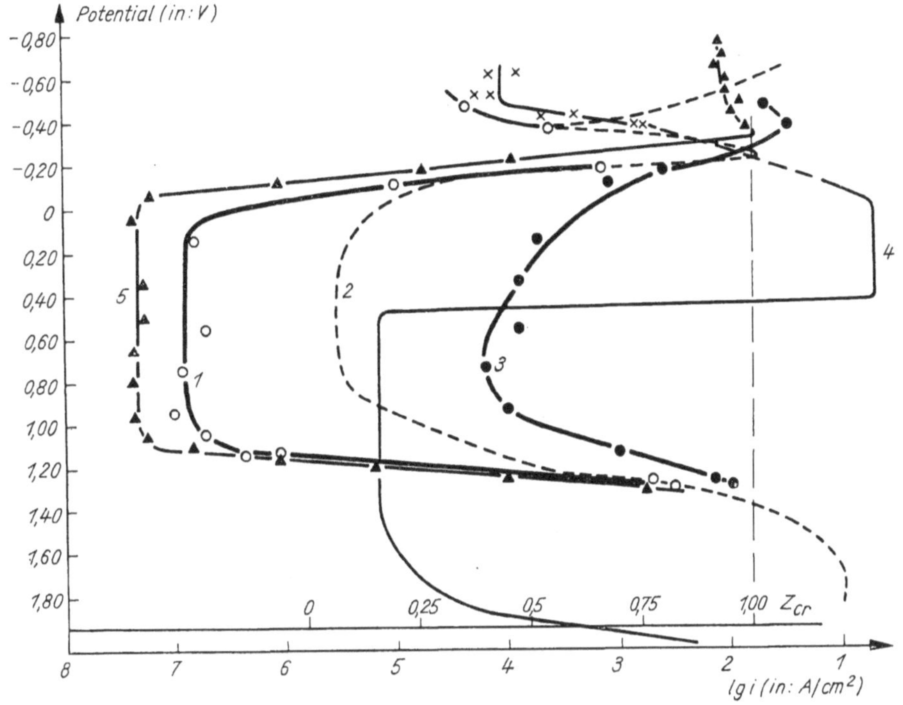

Abb. 2: Gegenüberstellung der elektrochemischen Korrosionseigenschaften der Legierung Fe-28% Cr (*1, 2, 3*), Eisen (*4*) [15] und Chrom (*5*) [15] in 1n Schwefelsäure.

1, 4, 5 — Potentialabhängigkeit der Auflösungsgeschwindigkeit.

2 — kathodische und anodische potentiodynamische Kurven (1 V/h); *3* — Abhängigkeit des Koeffizienten Z_{Cr} vom Potential (1—3: bei 40°C; 4, 5: Raumtemperatur).

Diese Angaben zeigen, daß in einem breiten Potentialbereich, das dem Passivbereich, dem Übergangsbereich vom Aktiv- zum Passivzustand und teilweise dem Bereich der Transpassivität entspricht, die Legierungsauflösung bei vorwiegendem Übergang von Eisen in die Lösung abläuft und daß nur im Bereich der aktiven Auflösung eine selektive Auflösung des Chroms vorliegt.

Die Gegenüberstellung der Kurven 4 und 5 mit den Kurven 1 und 3 in Abb. 6 weist auf eine gewisse Übereinstimmung zwischen selektiver Auflösung der Mischkristalle und den elektrochemischen Korrosionseigenschaften der Mischkristallkomponenten hin. Selektiv aufgelöst wird diejenige Komponente, deren Auflösungsgeschwindigkeit im individuellen Zustand beim vorgegebenen Potential wesentlich höher als die Auflösungsgeschwindigkeit der anderen Komponente ist. Die Richtigkeit dieser Schlußfolgerung steht für Mischkristalle mit verhältnismäßig hohem Gehalt an Komponenten außer Zweifel. Die Ausweitung dieser Schlußfolgerung auf den Bereich sehr geringer Konzentration (Verunreinigungen, Begleitelemente) bleibt allerdings ungeklärt und erfordert eine spezielle Überprüfung.

Prinzipiell analoge Ergebnisse wurden auch bei Untersuchungen zum Verhalten einiger austenitischer nichtrostender Stähle im Passivzustand erzielt. Diese Ergebnisse gestatten es auch, eine Schlußfolgerung über die Chromanreicherung an der Stahloberfläche während des Auflösungsvorganges zu ziehen. Die Intensität der Anreicherung hängt vom Potential ab und ändert sich im Laufe der Zeit merklich. Aus Abb. 3 ist z. B. ersichtlich, daß in der Anfangsperiode Z_{Cr} einen sehr niedrigen Wert hat. Dann nimmt die Größe Z_{Cr} zu und strebt nach einem Maximum einem Wert zu, der wesentlich kleiner als Eins ist. Dieser Endwert verringert sich bei Potentialerhöhung und nimmt mit steigender Temperatur zu.

Es sei hervorgehoben, daß auch im betrachteten Fall die durch Eisenverarmung und Chromanreicherung der Oberflächenschicht hervorgerufene anfängliche Zunahme von Z_{Cr} einer wesentlichen Verringerung der Auflösungsgeschwindigkeit entspricht. Anschließend beobachtet man, daß Z_{Cr}

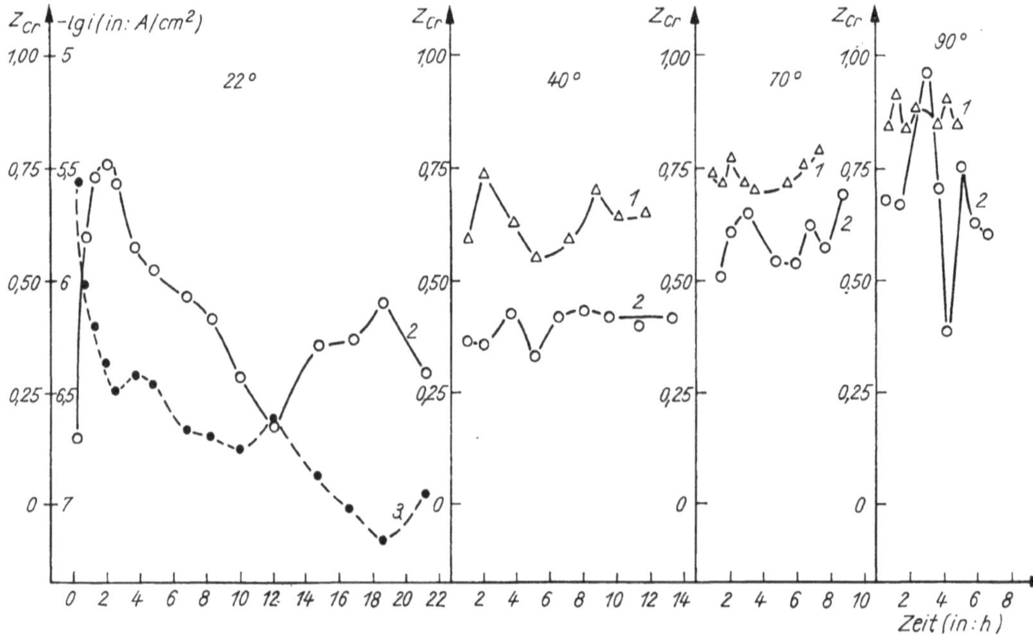

Abb. 3: Zeitabhängigkeit des Koeffizienten Z_{Cr} bei Auflösung des Stahls Cr20Ni20 in 1n Schwefelsäure; schrittweise Temperaturerhöhung.

1 — bei 0,25 V; *2* — bei 0,6 V; *3* — Zeitabhängigkeit der Stahlauflösung bei 0,6 V.

hintereinander ein Minimum und ein Maximum durchläuft, während für die Auflösungsgeschwindigkeit entsprechend ein Maximum bzw. Minimum vorliegt. Folglich liegt auch im Falle des Chrom-Nickel-Stahls die passive Oberfläche in einem dynamischen Zustand vor.

Weitere Untersuchungen haben gezeigt, daß unsere Schlußfolgerung über die Existenz einer Übereinstimmung zwischen den individuellen elektrochemischen Korrosionseigenschaften der Legierungskomponenten und deren Anfälligkeit zur selektiven Auflösung auch für die Legierung Ni70Mo28 gerechtfertigt ist. Das wurde bei der Auflösung dieser Legierung in siedender 21%HCl festgestellt. Aus Abb. 4 geht klar hervor, daß sich auch im betrachteten Falle Z_{Mo} bei jedem Potential auf einen gewissen Wert einpendelt, um den herum periodische Schwankungen auftreten, die deutlicher als bei der Legierung Fe—28%Cr ausgeprägt sind.

Aus der Gegenüberstellung der Kurven 1 und 5 folgt, daß zwischen der zeitlichen Änderung der Gesamtauflösungsgeschwindigkeit der Legierung Ni70Mo28 und dem Selektivitätskoeffizienten ebenso wie auch bei Legierungen auf Eisenbasis eine recht klare Korrelation zu beobachten ist.

Wie aus Abb. 5 ersichtlich ist, hängt der sich unter stationären Auflösungsbedingungen einstellende gemittelte Wert Z_{Mo} wesentlich vom Potential ab. Beim negativsten Potentialwert nimmt Z_{Mo} den Wert 0,66 an, was ungefähr einem 20%igen Molybdängehalt in den Auflösungsprodukten gegenüber einem 28%igen Gehalt in der Legierung entspricht. Die Potentialverschiebung in positiver Richtung wird von einer Zunahme der Größe Z_{Mo} begleitet. Bei 0,19 V beträgt der Wert Z_{Mo} bereits 1,5 und entspricht damit einem 38%igen Molybdängehalt in den Anodenprodukten. Aus der Analyse der in Abb. 5 dargestellten Ergebnisse geht hervor, daß dieser Wert Z_{Mo} in der Nähe des Maximalwertes

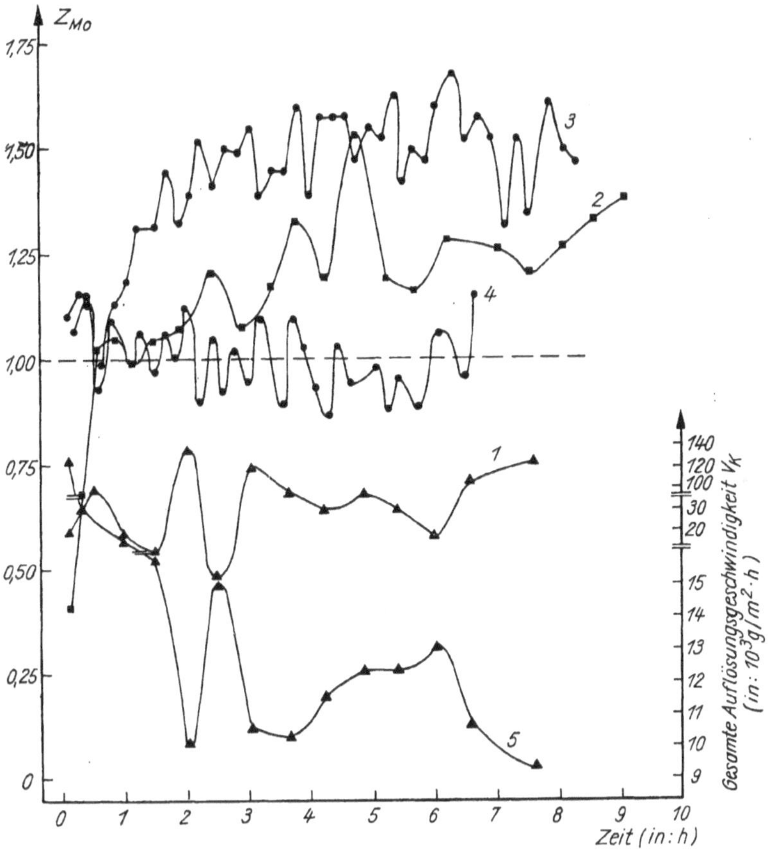

Abb. 4: Zeitabhängigkeit des Koeffizienten Z_{Mo} (*1–4*) und der summaren Auflösungsgeschwindigkeit (*5*) der Legierung Ni70Mo28 in siedender 21%-HCl bei folgenden Potentialen: *1* und *5* — 0,11 V; *2* — 0,15 V; *3* — 0,19 V; *4* — 0,22 V.

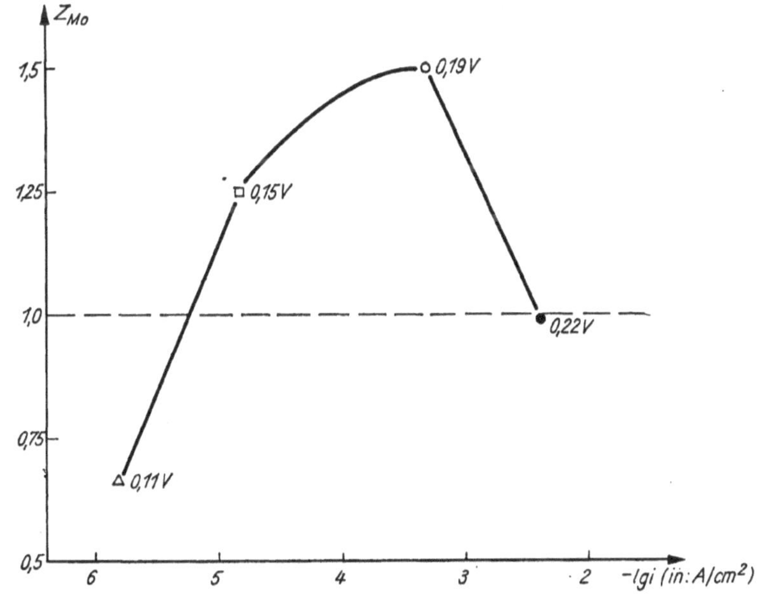

Abb. 5: Abhängigkeit der gemittelten Werte Z_{Mo} von der summaren Auflösungsgeschwindigkeit der Legierung Ni70Mo28 in siedender 21%-HCl unter potentiostatischen Bedingungen (anhand der Ergebnisse aus Abb. 4).

liegt. Bei einer anschließenden Potentialverschiebung um 30 mV, die von einem Ansteigen der Auflösungsgeschwindigkeit bis auf den Wert $4 \cdot 10^{-3}$ A/cm² begleitet ist, verringert sich Z_{Mo} stark und nimmt einen Wert nahe Eins an.

Auf diese Weise können sich bereits geringe Potentialänderungen wesentlich auf die Selektivität der Auflösung von Komponenten der Nickel-Molybdän-Legierung auswirken, und zwar nicht nur quantitativ, sondern auch qualitativ. Dieser Potentialeinfluß auf die Größe Z_{Mo} befindet sich in befriedigender Übereinstimmung mit den elektrochemischen Korrosionseigenschaften der Legierungskomponenten. In Abb. 6 ist die Potentialabhängigkeit der Auflösungsgeschwindigkeiten für die Legierung und ihre Komponenten dargestellt. Wie daraus ersichtlich ist, befindet sich das Molybdän

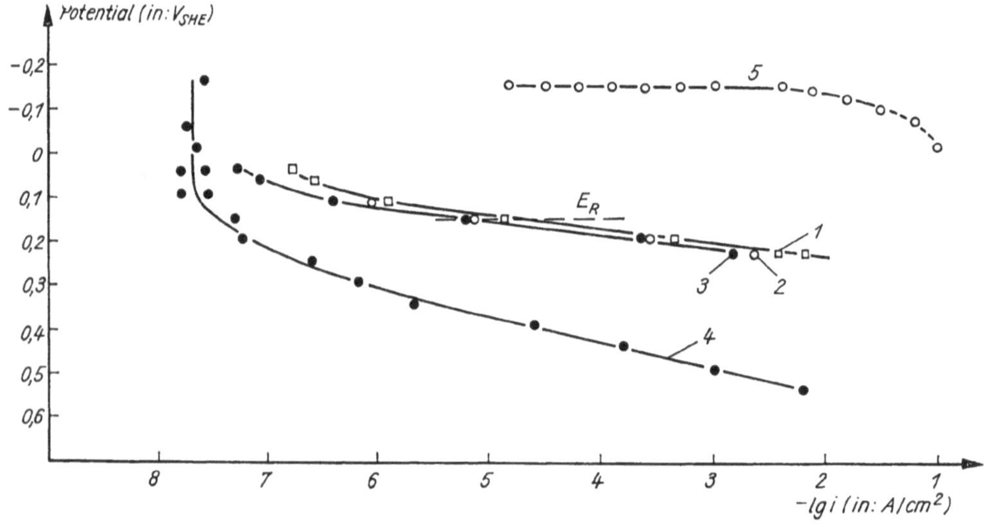

Abb. 6: Potentialabhängigkeit der Auflösungsgeschwindigkeit des Molybdäns (*4*) und der Legierung Ni70Mo28 (*1*), sowie der partiellen Auflösungsgeschwindigkeiten für Nickel (*2*) und Molybdän (*3*) unter stationären Bedingungen in siedender 21%-HCl; *5* — stationäre anodische Polarisationskurve für Nickel unter den gleichen Bedingungen.

bei Potentialen unterhalb 0,12—0,14 V im Passivzustand und löst sich mit einer sehr geringen Geschwindigkeit auf, die nicht vom Potential abhängt. Gerade bei diesen Potentialwerten ist Z_{Mo} kleiner 1; dieser Sachverhalt zeugt von einer Anreicherung des Molybdäns auf der Legierungsoberfläche. Die Situation verändert sich beim Übergang zu positiveren Potentialen. Die bei diesem Übergang beobachtete Zunahme der Größe Z_{Mo} bis auf Werte größer 1 hängt zweifellos mit der Transpassivität des Molybdäns zusammen. Es ist dabei von wesentlicher Bedeutung, daß die Transpassivität des Molybdäns in der Legierung beträchtlich erleichtert ist.

Aus Abb. 6 ist ersichtlich, daß die totalen und auch die partiellen Auflösungsgeschwindigkeiten der Legierung Zwischenwerte im Vergleich mit den Auflösungsgeschwindigkeiten der Komponenten in ihrer individuellen Form annehmen. Das ist ein Hinweis darauf, daß sich die Komponenten in der Legierung gegenseitig beeinflussen, und zwar derartig, daß die Auflösung der weniger korrosionsbeständigeren Komponenten erschwert wird, während die Auflösung der korrosionsbeständigeren Komponente erleichtert wird. Wesentlich ist allerdings, daß die beständigere Komponente einen stärkeren Einfluß hat.

Auch Untersuchungen zur Auflösung von Komponenten des Stahls Cr23Ni28Mo3Cu3Ti im Aktivzustand [3] befinden sich in guter Übereinstimmung mit den bereits genannten Ergebnissen. Es wurde festgestellt, daß bei der Auflösung des Stahls unter diesen Bedingungen (Potential: +0,03 V) die Werte für Z_{Cu} und Z_{Mo} kleiner Eins sind. Dieses Ergebnis stimmt gut mit der hohen Korrosionsbeständigkeit des reinen Kupfers und Molybdäns im angegebenen Potentialbereich überein und weist auf die Anreicherung dieser Elemente auf der Stahloberfläche hin. Die Richtigkeit dieser Schlußfolgerung wurde durch die radiochemische Analyse der Auflösungsprodukte bei anschließender Abätzung der Oberflächenschichten im Passivbereich (Potential: 0,6 V) bestätigt. Dieser Potentialwert entspricht dem Transpassivitätsbereich des Molybdäns und der intensiven elektrochemischen Auflösung des Kupfers.

Für die unmittelbare Beobachtung von Änderungen der chemischen Zusammensetzung an der Legierungsoberfläche während der Auflösung wurde die Röntgen-Mikrosonde MS-46 der Firma „Cameca" eingesetzt. Die Eindringtiefe des Elektronenstrahls betrug 1 μm. Folglich steht dieses Verfahren in der Empfindlichkeit wesentlich hinter den γ-spektroskopischen Methoden zurück, weil es nur gestattet, die Zusammensetzungsänderung von verhältnismäßig dicken Oberflächenschichten zu verfolgen, die man z. B. bei der Auflösung einer Legierung im Passivzustand in einer vertretbaren Zeit nicht aufbauen kann. Im Zusammenhang damit wurden die Legierungsproben längere Zeit bei

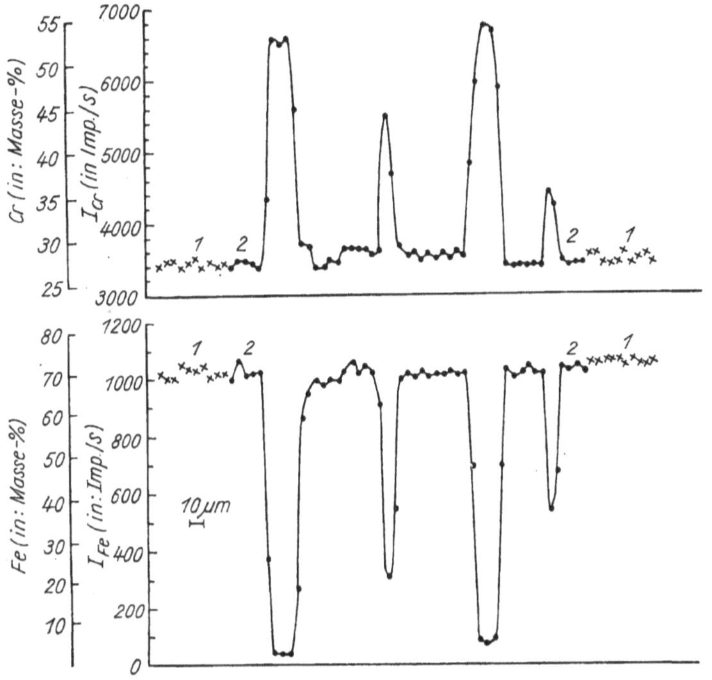

Abb. 7: Chrom- und Eisenverteilung an der nicht angeätzten Oberfläche der Legierung Fe—28% Cr (*1*) und auf einzelnen Oberflächenbezirken nach 28,5 h Anätzen in 1n H_2SO_4 bei 1,15 V; 40°C (*2*). Die Messungen wurden mit dem Mikroanalysator MS-46 durchgeführt.

Potentialen im Anfangsbereich der Transpassivität gehalten. Unter diesen Bedingungen ist bereits eine beträchtliche Auflösungsgeschwindigkeit der Legierung mit noch bemerkbaren Selektivitätseffekten feststellbar.

Die ermittelten Ergebnisse sind in Form von charakteristischen Röntgenstrahlungsspektren für zwei Elemente, die an ein und derselben Stelle der Oberfläche bei der Abtastung mit dem Elektronenstrahl entlang einer ausgewählten Richtung bestimmt wurden, in Abb. 7 und 8 dargestellt. Dort ist auch eine Prozentskala für den Elementgehalt angegeben; diese Skala wurde anhand einer Vergleichsprobe des Elementes berechnet.

Aus den Angaben der Abb. 7 ist ersichtlich, daß die langzeitige Polarisation der Legierung unter den angegebenen Bedingungen wirklich zu einer Chromanreicherung und Eisenverarmung der Oberfläche führt. Allerdings findet eine derartige Zusammensetzungsänderung nicht auf der gesamten Oberfläche statt, sondern nur auf einigen verhältnismäßig kleinen Oberflächenbezirken. Dabei ist es sehr wesentlich, daß jedem Intensitätsmaximum für Chrom ein Intensitätsmaximum für Sauerstoff entspricht (Abb. 8). Das zeugt davon, daß das auf der Oberfläche festgestellte überschüssige Chrom unmittelbar an den Sauerstoff gebunden ist.

Auf der übrigen Oberfläche der Legierung konnte nur in einigen Fällen eine geringe Erhöhung des Chromgehaltes beobachtet werden. Diese Erscheinung hängt offenbar damit zusammen, daß die sich auf der Oberfläche ausbildende Schicht mit veränderter chemischer Zusammensetzung eine Dicke aufwies, die beträchtlich geringer als die Eindringtiefe des Elektronenstrahls war.

Bei der Deutung des Mechanismus der selektiven Auflösung wird häufig die Vorstellung geäußert, daß sich in Wirklichkeit beide Legierungskomponenten mit Geschwindigkeiten auflösen, die proportional der jeweiligen Konzentration in der Legierung sind und daß die Selektivität das Ergebnis einer elektrochemischen Abscheidung der elektropositiveren Komponente auf der Legierungsoberfläche ist.

Im betrachteten Fall ist dieser Mechanismus thermodynamisch nicht möglich, weil die Versuche bei Potentialen durchgeführt wurden, die bedeutend positiver als die Potentiale für eine Reduktion von Chrom- und Eisenionen waren.

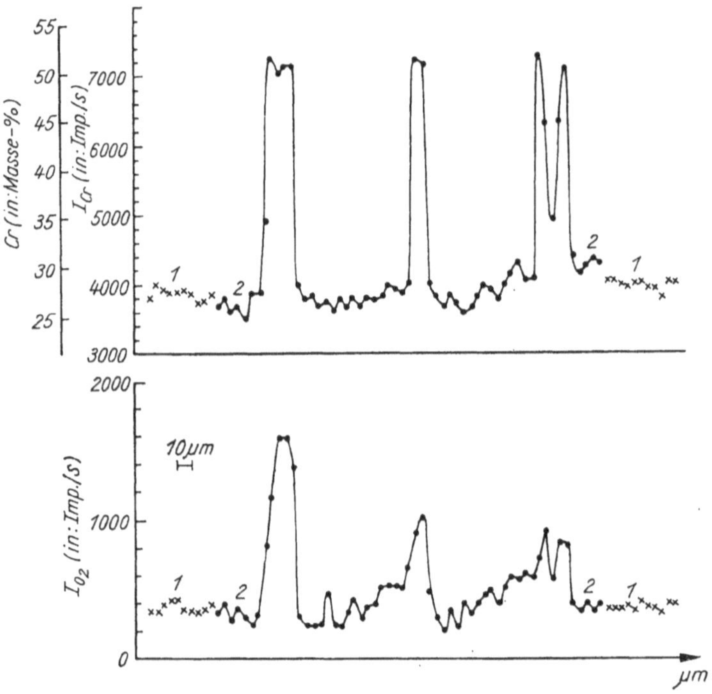

Abb. 8: Chrom- und Sauerstoffverteilung auf der nichtgeätzten (*1*) und geätzten Oberfläche der Legierung Fe—28%Cr. Die Messungen wurden mit dem Mikroanalysator MS-46 an der gleichen Probe wie in Abb. 7 durchgeführt.

Die genannten Ergebnisse stimmen gut mit einem anderen Mechanismus überein, der unter der Bezeichnung Volumendiffusions-Mechanismus bekannt ist [8]. In diesem Fall wird angenommen, daß sich nur (oder bevorzugt) die unbeständigere Komponente auflöst und daß sich in der Oberflächenschicht der Legierung nicht nur die beständigere Komponente anreichert, sondern auch Leerstellen im Kristallgitter anhäufen, wodurch der Diffusionskoeffizient in den Oberflächenschichten beträchtlich größer sein kann als der Diffusionskoeffizient im Innern der Legierung. Beim Erreichen des stationären Zustandes kann die selektive Auflösung durch die Diffusion der unbeständigeren Komponente aus dem Innern zur Oberfläche der Legierung (und Diffusion der beständigeren Komponente in entgegengesetzter Richtung) aufrechterhalten werden. In den Veröffentlichungen von PICKERING und WAGNER [8] wurde gezeigt, daß bei ununterbrochener Entstehung von Leerstellen durch diesen Mechanismus sogar bei Normaltemperatur beträchtliche Auflösungsgeschwindigkeiten auftreten können.

Wir haben festgestellt, daß bei hohen Auflösungsgeschwindigkeiten der Legierungen oberhalb $1 \cdot 10^{-3}$ A/cm² die Größe Z gewöhnlich Werte annimmt, die im Bereich um 1 liegen. Das kann man besonders gut in den Abb. 4 und 5 erkennen. Offenbar ist die Zerstörungsgeschwindigkeit des Kristallgitters der Legierung unter diesen Bedingungen mit der Diffusionsgeschwindigkeit der selektiv aufgelösten Komponenten über Leerstellen in oberflächennahen Schichten vergleichbar oder größer als die letztgenannte Diffusionsgeschwindigkeit.

Aus der Gesamtheit der genannten Angaben geht hervor, daß die selektive Auflösung von Legierungen durch einige Besonderheiten gekennzeichnet ist, die große Bedeutung für den Einfluß der selektiven Auflösung auf die Korrosionsbeständigkeit der Legierungen haben.

Die erste Besonderheit besteht in der weiter oben festgestellten starken Verringerung der Auflösungsgeschwindigkeit der Legierung und der partiellen Auflösungsgeschwindigkeiten einzelner Legierungskomponenten im Anfangsstadium der Probenexposition bei konstantem Potential. Es gibt Grund zur Annahme, daß dieser Effekt in beträchtlichem Maße durch die Anreicherung der beständigeren Komponente an der Oberfläche hervorgerufen wird. Wenn das der Fall ist, dann folgt aus den genannten Angaben, daß sich durch die selektive Auflösung während des Korrosionsvorgangs auf der Legierungsoberfläche eine Schicht ausbilden kann, die in ihrer Beständigkeit wesentlich die Korrosionsbeständigkeit der Oberflächenschicht mit der ursprünglichen Zusammensetzung übertrifft.

Die zweite Besonderheit besteht in einer ungleichmäßigen Chromanreicherung der Oberfläche. Das zeugt zweifellos von einer ungleichmäßigen Auflösung der Oberfläche bei der Herausbildung des stationären Zustandes. Diese Schlußfolgerung wurde nach der Auflösung durch eine unmittelbare Untersuchung der Proben mit Licht- und Rasterelektronenmikroskopie bestätigt. Es wurde festgestellt, daß die Auflösung gewöhnlich neben Einschlüssen beginnt und sich dann auf die Korngrenzen und einzelne Bereiche der Kornflächen ausbreitet. Wie in unserem Labor gezeigt wurde, kann unter bestimmten Bedingungen die Geschwindigkeit der lokalen Auflösung nichtrostender Stähle an Einschlüssen derartig hoch sein, daß sich an diesen Stellen Vertiefungen in der Art des Lochfraßes bilden [10]. Dieses Ergebnis stimmt gut mit Angaben überein, die in den letzten Jahren von verschiedenen Wissenschaftlern bei Untersuchungen zur Art der Lochfraßbildungszentren in Chloridlösungen und anderen Medien mit depassivierend wirkenden Anionen ermittelt wurden [11]. In diesen Arbeiten wurde gezeigt, daß an unsensibilisierten nichtrostenden Stählen sowie an Eisen und Nickel die Grenzbereiche zwischen Metall und nichtmetallischen Einschlüssen die bevorzugtesten Stellen für eine Lochfraßentstehung sind. Kürzlich wurde in unserem Institut durch Einsatz der Röntgenmikroanalyse und der Rasterelektronenmikroskopie gezeigt, daß nichtmetallische Einschlüsse nicht nur bei unsensibilisierten austenitischen Stählen unterschiedlicher Zusammensetzung (Chrom-Nickel-, Chrom-Nickel-Silizium-, Chrom-Nickel-Molybdän-Stähle), sondern auch Nickel-Chrom-Legierungen, Titan und sogar beim Platin bevorzugte Zentren für die Lochfraßbildung sind [12, 13].

Aus all diesen Angaben sind zwei wichtige Schlußfolgerungen zu ziehen. Die erste Schlußfolgerung besteht darin, daß durch die selektive Auflösung während des Korrosionsvorganges eine Ausheilung der Schwachstellen erfolgt, auf die — bildhaft ausgedrückt — „verstärkte Flicken“ aufgesetzt werden. Die zweite, allgemeinere Schlußfolgerung besteht darin, daß als Schwachstellen an der Oberfläche die Grenzbereiche zwischen Metall und nichtmetallischen Einschlüssen und möglicherweise die Schnittpunkte dieser Bereiche mit den Korngrenzen in Betracht kommen. Die hohe Korrosionsaktivität der Einschlußgrenzbereiche wird offenbar durch die hohe Schadhaftigkeit und die Verschmutzung des Kristallgitters neben den Einschlüssen hervorgerufen. Bei sulfidischen Einschlüssen kommt noch die katalytische Wirkung des Schwefelwasserstoffs oder der sich bei Auflösung dieser Einschlüsse bildenden Sulfidionen hinzu.

Wie die in un erem Institut durchgeführten Untersuchungen gezeigt haben, kann unter gewissen Bedingungen der Einfluß nichtmetallischer Einschlüsse den Einfluß der Legierungselemente überdecken. Es ist beispielsweise bekannt, daß ein Legieren mit im Austenit löslichem Silizium zur Erhöhung der Lochfraßbeständigkeit des Stahls führt. Es wurde allerdings festgestellt, daß durch Zugabe von Silizium bei gleichzeitiger Verunreinigung durch SiO_2-Einschlüsse unter bestimmten Bedingungen an diesen Einschlüssen starke Narbenkorrosion eintritt, wobei die Beständigkeit des Chrom-Nickel-Silizium-Stahls beträchtlich schlechter als die Beständigkeit des Chrom-Nickel-Stahls ist [14].

Die dritte Besonderheit beinhaltet die periodischen zeitlichen Schwankungen der Auflösungsgeschwindigkeiten der Legierung und einzelner Legierungskomponenten. Eine Analyse aller Ergebnisse zeigte, daß diese Schwankungen nicht mit Meßfehlern zusammenhängen, sondern in Wirklichkeit den Auflösungsvorgang der Legierung charakterisieren. Es gibt Grund zu der Annahme, daß als Ursache der Schwankungen die selbständige lokale Aktivierung einzelner Mikrobereiche der Legierungsoberfläche und deren anschließende Repassivierung nach dem Mechanismus der selektiven

Auflösung in Erscheinung treten. Die Aktivierung kann unterschiedliche Ursachen haben, dazu gehört die Freilegung neuer Einschlüsse mit Leerstellenanhäufung in den angrenzenden Metallschichten sowie auch die Zerstörung der Passivschicht infolge innerer Spannungen in dieser Schicht.

Diese Ergebnisse und einige andere Resultate, die in den letzten Jahren ermittelt wurden, zeigen überzeugend, daß die früher üblichen Vorstellungen über die Passivschicht als statisches Gebilde falsch sind. In Wirklichkeit stellt diese Schicht ein außerordentlich dynamisches Gebilde dar, in dem fast ununterbrochen lokale Zerstörungen auftreten, die ein „Aufflackern" der Auflösungsgeschwindigkeit hervorrufen. Das wurde von SUCHOTIN [9] gut demonstriert, der mit einem trägheitsarmen Gerät gezeigt hat, daß die stationäre Auflösungsgeschwindigkeit des passiven Eisens bei gleichbleibendem Potential keine streng konstante Größe darstellt, wie das aus der Anzeige eines gewöhnlichen Gerätes hervorgeht, sondern ununterbrochen um einen bestimmten Betrag schwankt.

In Abhängigkeit von den jeweiligen Bedingungen können die aktiven Oberflächenbereiche entweder erneut repassiviert oder Zentren für die Lochfraßbildung werden.

Literatur

[1] J. M. KOLOTYRKIN, A. N. CHEMODANOV, M. A. DEMBROVSKI, W. M. KNAJAZEVA, N. N. RODIN, Korrosion (DDR) Nr. 4, 45 (1970)

[2] M. A. DEMBROVSKIJ, Zaščita metallov **8**, 387 (1972)

[3] N. N. RODIN, A. V. PLASKEEV, M. A. DEMBROVSKIJ, J. M. KOLOTYRKIN, W. M. KNJAZEVA, Zaščita metallov, **9**, 387 (1973)

[4] J. M. KOLOTYRKIN, Electrochim. Acta, **18**, 593 (1973)

[5] J. M. KOLOTYRKIN, W. M. KNJAZEVA, „*Isvestija Severo-Kavkazkogo Naučnogo centra vysšej školy*", estestvennye nauki, 1974, Nr. 2. 11

[6] A. P. PČELNIKOV, L. I. KRASINSKAJA, A. D. SITNIKOV, V. V. LOSEV, Elektrochimija, **11**, 37 (1975)

[7] N. D. TOMAŠOV, „*Tr. 3-go Mezdunarodnogo kongressa po korrozii metallov*", 1966, M., „Mir", t. I, 1968, S. 39

[8] H. W. PICKERING, C. WAGNER, J. Electrochem. Soc., **114**, 698 (1976)

[9] A. M. SUCHOTIN, M. V. MICHALEVA, V sb. „*Passivnost i korrozija metallov*", L., „Chimija", 1971, S. 51

[10] J. M. KOLOTYRKIN, L. I. FREIMAN, G. S. RASKIN, Ž. O. GOJNACKAJA, Dokl. AN SSSR **220**, Nr. 1, 156 (1975)

[11] Z. SZKLARSKA-SMIAŁOWSKA, Corrosion, **28**, 388 (1972)
G. WRANGLEN, Corr. Sci., **9**, 585 (1969); **14**, 331 (1974)
H. WORCH, I. GARZ, W. SCHATT, Werkstoffe und Korrosion, **24**, 872 (1974)

[12] L. I. FREIMAN, G. S. RASKIN, J. M. KOLOTYRKIN, L. J. CHARITONOVA, „*II-aja Meždunarodnaja Naučno-techničeskaja konferencija po probleme SEV „Razrabotka mer zaščity metallov ot korrozii*". Praga, Apr. 1975, Sekcija I, S. 440

[13] J. I. MIRONOV u. a., Zaščita metallov, **10**, 401 (1974)

[14] S. A. GLAZKOVA, L. I. FREIMAN, G. S. RASKIN, G. L. ŠVARC, Zaščita metallov, **8**, 960 (1972)

[15] J. M. KOLOTYRKIN, W. M. KNJAZEVA, „*Itogi nauki i techniki*", *ser. Chimija*, „*Korrozija i zaščita ot korrozii*", t. 3, M., 1974, S. 5

N. D. Tomaschow, G. P. Tschernowa, L. A. Tschigirinskaja und M. F. German

Laboratorium für Korrosion der Legierungen des Instituts für physikalische Chemie
der Akademie der Wissenschaften der UdSSR, Moskau

DER EINFLUSS DER LEGIERUNGSELEMENTE AUF PASSIVIERUNG UND KORROSIONSBESTÄNDIGKEIT DER LEGIERUNGEN AUF EISEN-CHROM-BASIS

Zusammenfassung

Es wurde der Einfluß von Kathodenlegierung (Pd 0,1—0,5%) auf Passivierung und Korrosionsbeständigkeit der Legierungen Fe—Cr (25—100%) und Fe-25Cr-Stähle, die Mn, Ni, Mo und N enthalten, in heißen konzentrierten Lösungen der H_2SO_4 und verdünnten Lösungen der HCl untersucht. Festgestellt wurde, daß Pd-Kathodenzusätze die Neigung zur Selbstpassivierung und Korrosionsbeständigkeit der erwähnten Legierungen in den untersuchten Bedingungen bedeutend erhöhen. Es wurde bemerkt, daß das Vorhandensein des Mn die Selbstpassivierung der Cr—Ni- und Cr—Ni—Mo-Stähle der austenitischen und austenitisch-ferritischen Klasse sich unter dem Einfluß der kathodischen Modifikation verbessert. Es wurde demonstriert, daß die mit Pd kathodischmodifizierten Cr-Stähle Korrosionsbeständigkeit in verdünnten Lösungen der HCl bekommen.

Eine besonders starke Hemmung der Korrosionsgeschwindigkeit bei Metallen und Legierungen ist bei deren Übergang in den Passivzustand zu verzeichnen. In der Feststellung von Gesetzmäßigkeiten der Selbstpassivierbarkeit von Legierungen und der Erhöhung der Stabilität deren Passivzustands besteht der wichtigste Weg, um Legierungen mit erhöhter Korrosionsbeständigkeit zu finden [1, 2]. Die Selbstpassivierbarkeit der Legierungen kann unter Einhaltung folgender Bedingungen erzielt werden:

- Verschiebung des Passivierungspotentials zur negativen Seite, und des stationären Korrosionspotentials zur positiven Seite hin auf solch eine Weise, daß das Korrosionspotential positiver als das Passivierungspotential wird;
- Herabsetzung des Stroms der anodischen Auflösung einer Legierung im Passivzustand.

Wie früher festgestellt [1]—[3], kann die erste Bedingung nicht nur durch unmittelbare Steigerung der Passivierbarkeit, z. B. bei Chromgehaltserhöhung, sondern auch durch Erleichterung des Kathodenvorgangs durch Zusatz unbedeutender Mengen an elektrisch positiven Elementen mit niedriger Wasserstoffüberspannung (Pd, Pt, Rn, Re u. a.) erfüllt werden, d. h. nach dem Kathodenlegierungsverfahren. Das Legieren von Titan, nichtrostenden Stählen und Chrom durch kleinere Kathodenzusätze erhöht bekanntlich deren Korrosionsbeständigkeit als Ergebnis der Selbstpassivierung beträchtlich [1]—[8]. Die Selbstpassivierung einer Legierung mit Kathodenzusatz erfolgt infolge deren Ansammlung auf der Legierungsoberfläche im Zeitraum der aktiven Auflösung, woraus eine wesentliche Verringerung der Wasserstoffüberspannung und die Verschiebung des Legierungspotentials zur positiven Seite hin, in den passiven Bereich resultiert. Die Selbstpassivierung der kathodenmodifizierten Legierungen ist sowohl vom Charakter der Anreicherung des Kathodenzusatzes auf der Oberfläche zum Zeitpunkt des Korrosionsbeginns als auch vom Passivierungsvorgang bei der Legierungsbasis abhängig.

Es wurde der Einfluß von Pd (0,1···0,5%) auf das elektrochemische und Korrosionsverhalten der Legierungen des Fe—Cr-Systems (25···100% Cr) (siehe Tab. 1) und der Legierungen Fe—25%Cr mit Mo (0,5···3%), Ni (6%), Mn (2···8%) gewöhnlichem (0,03···0,04%) und erhöhtem Gehalt des Stickstoffs (0,05···1,0%) (siehe Tab. 2) untersucht.

Das Mikrogefüge der Legierungen mit und ohne Pd war praktisch nicht zu unterscheiden.

Die Prüfungen wurden in H_2SO_4-Lösungen (10···50%) bei 100 °C sowie in HCl-Lösungen (1···3%) bei 20···100 °C durchgeführt. Die Proben wurden vor dem Versuch kathodisch aktiviert. Die nicht Pd-modifizierten Fe-Cr-Legierungen sowie Cr lassen sich unter diesen Verhältnissen nicht passivieren. Die Korrosionsgeschwindigkeit ist hoch (Abb. 1 *a*) und steigt mit der Erhöhung des Cr-Gehalts auf 50% in 1% HCl und auf 60% in H_2SO_4-Lösungen; die Korrosionsgeschwindigkeit des reinen

Tabelle 1: Verfahren zur Herstellung und die chemische Zusammensetzung von Legierungen des Fe—Cr-Systems

Legierung*)	Charakteristik der Legierungen			Hauptkomponente**) [Gew.-%]	
	Schmelzführung	Verformung	Wärmebehandlung	Cr	Pd
25Cr	Vakuum von 10^{-4} Torr	Schmieden und Walzen	Glühen unter Vakuum	28,0	—
25CrPd	Dreifaches Umschmelzen	in der Luft bei 800 °C	von 10^{-4} Torr bei 780 °C	25,8	0,15
32Cr	unter Heliumschutz		1,5 h lang	31,3	—
32CrPd				33,0	0,2
40Cr				38,2	—
40CrPd				38,2	0,17
50Cr	Dreifaches Umschmelzen	Nicht verformt,	Nicht behandelt	48,5	—
50CrPd	unter Argonschutz	gegossen		46,5	0,1
60Cr				58,6	—
60CrPd				58,8	0,18
Cr				100	—
Cr—Pd				~100	0,15

*) Alle Legierungen wurden im Induktionsofen erschmolzen und danach dem Homogenisierungsglühen unter Vakuum von 10^{-4} Torr bei 1250 °C im Verlaufe von 5 Stunden unterzogen.

**) Gehalt an Fremdelementen (in %): 0,009 C, 0,05 W, 0,025 Si, 0,005 Al.

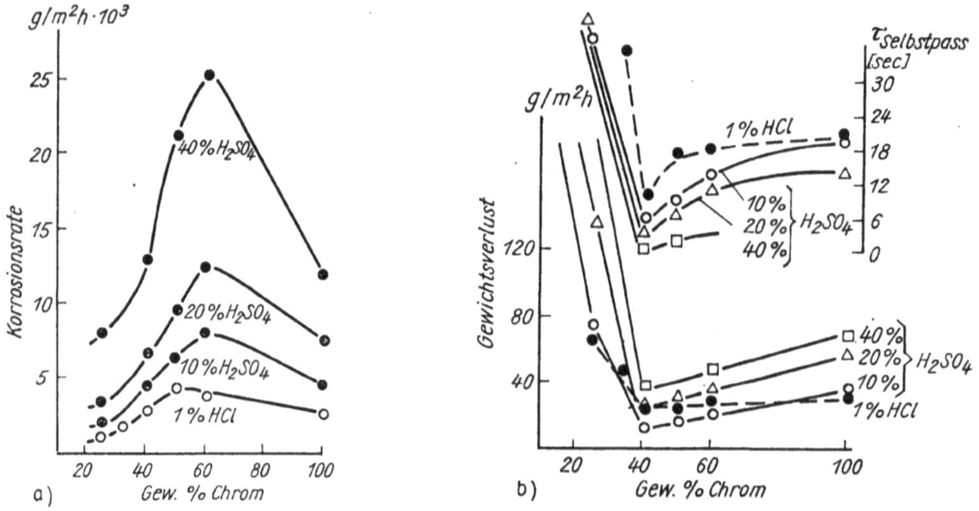

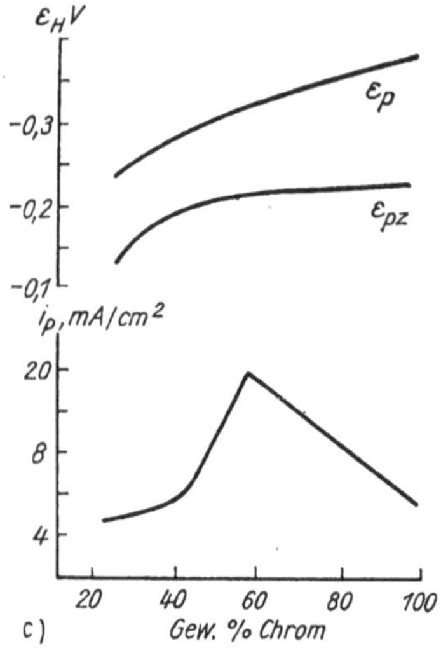

Abb. 1: Chromgehaltsabhängigkeit

a) der Korrosionsgeschwindigkeit der Stähle ohne Pd in H_2SO_4 verschiedener Konzentration bei 100 °C und in 1% HCl beim Sieden;

b) der Selbstpassivierungszeit der Stähle mit Pd ($\tau_{s\text{-}pas.}$) und der Korrosionsverluste (Q) in der gleichen Zeit;

c) des Stroms (i_p) und des Anfangs- (ε_p) und Vollpassivierungspotentials (ε_{pZ}) in 10% H_2SO_4 bei 25 °C.

Tabelle 2: Verfahren zur Herstellung, chemische Zusammensetzung und Struktur hochlegierter nichtrostender Chromstähle

Legierung	Charakteristik der Legierung			Hauptkomponente Gew.- %*)						Struktur der Legierungen***)
	Schmelzführung	Verformung	Wärmebehandlung	Cr	Ni	Mo	Mn	Pd	N	
25Cr	Unterschlackeschmelzen	Stahlblockschmieden in	Glühen bei 770 °C	25,2	—	—	0,6	—	**)	*F*
25Cr0,2Pd	im Hochfrequenzofen	der Luft und Blech-	30 min lang	25,6	—	—	0,6	0,2	**)	*F*
25Cr0,5Pd		walzen nach der	Wasserkühlung	25,6	—	—	0,6	0,5	**)	*F*
25Cr6Ni		Erhitzung auf 1100 °C	Härten ab 1000 °C	24,7	6,1	—	0,6	—	**)	*A* + 40% *F*
25Cr6Ni0,2Pd			(Dauer 10 min)	24,4	6,2	—	0,6	0,2	**)	*A* + 40% *F*
25Cr6Ni0,5Pd			Wasserkühlung	24,4	6,2	—	0,6	0,5	**)	*A* + 40% *F*
25Cr8MnN	Einsatzgutschmelzen	Stahlblockschmieden in	Härten ab 1150 °C	23,0	—	—	8,4	—	0,94	*A* + *F* (Spuren)
25Cr8MnN0,2Pd	in der Argon-	der Luft nach der	(Dauer 30 min)	22,2	—	—	7,7	0,2	0,69	*A* + 8% *F*
25Cr8MnN0,5Pd	atmosphäre.	Erhitzung auf 1200 °C	im Wasser	22,0	—	—	8,3	0,5	0,91	*A* + 50% *F*
25Cr8Mn6NiN	Blockerschmelzen in der			25,0	5,4	—	7,0	—	0,77	*A* + 9% *F*
25Cr8Mn6NiN0,2Pd	Stickstoffatmosphäre			26,5	5,6	—	7,1	0,2	0,97	*A*
25Cr8Mn6NiN0,5Pd	beim Partialdruck			26,5	5,1	—	9,0	0,5	0,83	*A* + 10% *F*
25Cr2Mn6Ni2MoN	von 4 atm			24,3	6,8	2,1	2,2	—	0,53	*A* + 11% *F*
25Cr2Mn6Ni2MoN0,2Pd				27,0	8,1	2,0	0,8	0,2	0,92	*A* + 13% *F*
25Cr8Mn2MoN				18,0	—	1,9	10,0	—	0,77	*A* + 40% *F*
25Cr8Mn2MoN0,2Pd				23,5	—	1,7	9,5	0,2	0,77	*A* + 50% *F*
25Cr8Mn2MoN0,5Pd				24,5	—	2,0	9,3	0,5	0,97	*A* + 36% *F*

*) C = 0,06%.
**) N entspricht dem Normalgehalt in industriellen Stählen und ist gleich 0,03—0,04%.
***) *A* — Austenit, *F* — Ferrit.

Anmerkung: Die Stähle, die eine erhöhte Stickstoffmenge enthalten, besitzen hohe Festigkeitseigenschaften ($\sigma_B \sim 100$ kg/mm²) bei einem ausreichend hohen Stand der Plastizitätseigenschaften ($\delta \sim 50$—58%). Die Stähle wurden im Patoninstitut elektrischer Schweißung (UdSSR, Kiew) verfertigt und in Zusammenarbeit mit diesem Institut untersucht [9].

Chrom liegt etwas darunter. Nach der anfänglichen Auflösung im Aktivzustand (im Verlaufe von einigen Sekunden) werden die Legierungen mit Pd selbstpassiviert, dabei verschiebt sich das Potential sprungartig zu positiveren Werten (+0,03 V). Bei Fe—Cr-Legierungen mit Pd ist die Mindestzeit der Passivierung ($\tau_{Spas.}$) und der Mindestwert der Korrosion in der Passivierungszeit (Q) bei etwa 40% Chromgehalt in der Legierung zu verzeichnen (Abb. 1 *b*).

Beim Einsatz der passivierbaren Legierungen hat die Korrosionsgeschwindigkeit im Passivzustand eine entscheidende Bedeutung. Deshalb wurden spezielle Versuche zur Bestimmung der Korrosionsgeschwindigkeit im Passivzustand bei der Legierung Fe—40CrO,2Pd (Tab. 3) durchgeführt. Diese Legierung weist eine hohe Korrosionsbeständigkeit in 40— und 50%iger H_2SO_4 bei 100 °C und in 20%iger H_2SO_4 beim Siedepunkt auf. In der siedenden 30%iger H_2SO_4 und der mit noch höheren Konzentrationen wird diese unzureichend beständig.

Tabelle 3: Korrosionsgeschwindigkeit (g/m²h) bei der Legierung 40CrO, 2Pd im Passivzustand in Schwefelsäurelösungen (Versuchsdauer 10 Stunden)

% H_2SO_4	20	30	40	50
100 °C	0,05	0,05	0,05	0,06
Siedepunkt	0,08	20	20	20

Die höchste Korrosionsbeständigkeit der Legierung Fe40Cr0,2Pd unter Fe—Cr-Legierungen kann auf die beste Kombination von Passivierungscharakteristiken zurückgeführt werden. Wie aus Abb. 1 *c* ersichtlich, verschieben sich Passivierungs- und Vollpassivierungspotentiale in der negativen Richtung bei der Erhöhung des Chromgehalts. Die größte Verschiebung des Vollpassivierungspotentials ist bei der Legierung Fe40Cr zu verzeichnen und verändert sich bei der weiteren Erhöhung des Chromgehalts wenig. Gleichzeitig ändert sich der Passivierungsstrom bei der Erhöhung des Cr-Gehalts auf 40% unbedeutend, nimmt aber beim Übergang zur Legierung Fe—60%Cr stark zu.

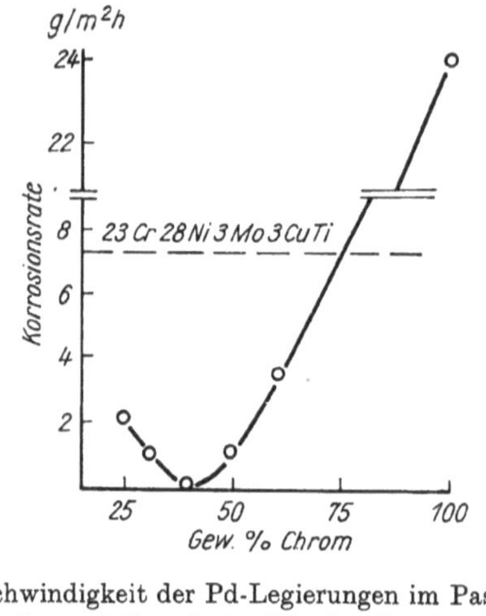

Abb. 2: Abhängigkeit der Korrosionsgeschwindigkeit der Pd-Legierungen im Passivzustand in 1% HCl beim Sieden vom Chromgehalt in der Legierung (Versuchsdauer 6 Stunden). (Mit der Strichlinie ist die Korrosionsgeschwindigkeit des Stahls 23Cr28Ni3Mo3CuTi gekennzeichnet.)

Nichtrostende Stähle, darunter auch Chromstähle, sind bekanntlich in Salzsäurelösungen unbeständig. Die durchgeführten Prüfungen der Fe—Cr-Legierungen mit Pd in siedender 1%iger HCl-Lösung ergaben, daß sich die Legierung 40Cr0,2Pd als besonders beständig erweist (Abb. 2). Die Korrosionsgeschwindigkeit beim Stahl 23Cr28Ni3Mo3Cu ist unter diesen Verhältnissen viel größer. In Salzsäurelösungen mit einer höheren Konzentration (1,5% und 2%) steigert sich die Korrosionsgeschwindigkeit bei der Legierung 40Cr0,2Pd. Der Vergleich der Korrosionsgeschwindigkeit in 40% H_2SO_4 bei 100 °C bei Fe—Cr-Legierungen mit und ohne Pd und bei einigen säurefesten Legierungen (Tab. 4) zeigt, daß die Legierung Fe40Cr0,2Pd eine höhere Beständigkeit als genannte säurefeste Legierungen unter gegebenen Verhältnissen aufweist.

In den letzten Jahren hat es sich gezeigt, daß die industrielle Anwendung hochlegierter Chromstähle (20···28% Cr) aussichtsreich ist [10, 11]. Im vorliegenden Beitrag wird der Einfluß verschiedener Legierungszusätze auf die Selbstpassivierbarkeit und Korrosionsbeständigkeit der Stähle untersucht, die 25% Cr und (0,2···0,5%) Pd enthalten (Tab. 2).

Die Ergebnisse der Korrosionsprüfungen in Schwefelsäurelösungen (Abb. 3) zeigen, daß durch Legieren des Stahls 25Cr8MnN mit Ni (Kurve *2*) oder mit Mo (Kurve *3*) die Korrosionsgeschwindigkeit

Tabelle 4: Korrosionsgeschwindigkeit bei säurefesten Legierungen in 40% H_2SO_4 bei 100 °C

Legierung	25Cr, 40Cr, 60Cr, Cr (ohne Pd)	23Cr28Ni 3Mo3CuTi	Hastelloy		Ti30Mo	40Cr0,2Pd
			„*A*“ u. „*B*“	„*C*“		
K, mm/Jahr	∼10000	3,0	0,2	3,0	0,18	0,05

verringert wird. Die Pd-Modifikation der zu untersuchenden Stähle (0,1···0,5%) erhöht deren Korrosionsbeständigkeit beträchtlich. Dabei ist hervorzuheben, daß die Pd-Stähle, die gleichzeitig Mn und Ni enthalten, die geringste Korrosionsgeschwindigkeit aufweisen. Die Selbstpassivierung der Mo-Stähle wird bei Erhöhung der Aggressivität des Mediums erschwert. Die in Salzsäurelösungen (1—3%) bei 20—100 °C durchgeführten Untersuchungen (Abb. 4) ergaben, daß eine Selbstpassivierung der zu untersuchenden Pd-modifizierten Stähle möglich ist. Dabei ist hervorzuheben, daß Molybdän auf die

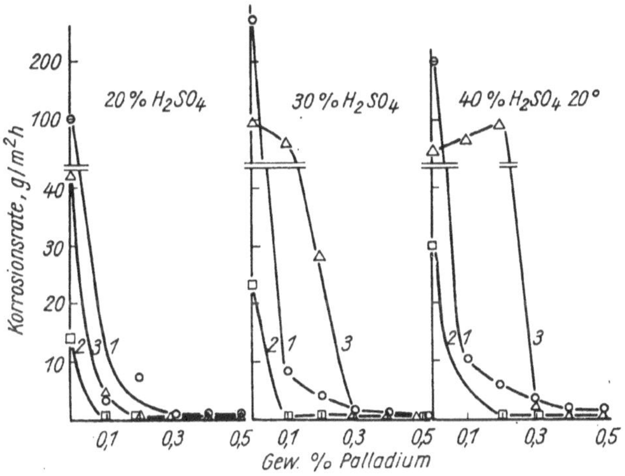

Abb. 3: Pd-Einfluß auf Korrosionsgeschwindigkeit der Legierungen 25Cr8MnN (Kurve *1*), 25Cr8Mn6NiN (Kurve *2*), 25Cr8Mn2MoN (Kurve *3*), in 20-, 30- und 40%iger H_2SO_4 bei 20 °C. Versuchsdauer 24 Stunden.

Selbstpassivierung der Pd-Stähle in Salzsäurelösungen einen günstigen Einfluß ausübt. Bei der Erhöhung der HCl-Konzentration auf 2% läßt sich zum Beispiel der Stahl Cr—Mo—0,2Pd und Cr—Mn—Ni—0,2Pd in 3% HCl nicht mehr passivieren und weist eine niedrige Korrosionsbeständigkeit auf. Gleichzeitiges Legieren der Cr—Mn-Stähle mit Mo und Pd (0,2%) sowie mit Ni, Mo und Pd (0,2%) läßt deren Korrosionsbeständigkeit in der Salzsäure beträchtlich erhöhen. Wie auch früher gezeigt [12], besitzt der Stahl mit Molybdän und Palladium 25Cr3MoTi0,5Pd eine hohe Korrosionsbeständigkeit in der Salzsäure bei 100 °C bis zu 1,65% Konzentration, der Stahl mit 0,2% aber bis zu 1,1% Konzentration. Der gleiche Stahl ohne Pd ist unter diesen Verhältnissen vollkommen unbeständig.

Zur Klärung des Einflusses der Legierungskomponente auf das anodische Verhalten der Stähle wurden potentiokinetische Kurven für Stähle ohne Palladium sowie Potential-Korrosionskurven von Abb. 5 aufgenommen. Das Mo- und Ni-Legieren des Cr—Mn-Stahls hat eine Verringerung der Überspannung beim Kathodenvorgang der Wasserstoffentwicklung und des Anodenstroms im Bereich der aktiven Auflösung zur Folge, das Passivierungspotential verschiebt sich dabei zum Teil zur positiven Seite hin, das Vollpassivierungspotential ändert sich praktisch nicht (Abb. 5). Früher wurde

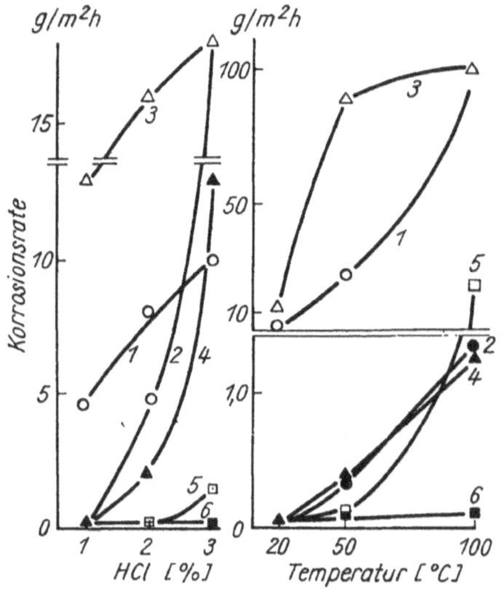

Abb. 4: Einfluß von HCl-Konzentration (*a*) und Temperatur (1% HCl) (*b*) auf die Korrosionsgeschwindigkeit folgender Stähle: 25Cr8Mn6NiN (Kurve *1*), 25Cr8Mn6NiN0,2Pd (Kurve *2*), 25Cr8Mn2MoN (Kurve *3*), 25Cr8Mn2MoN0,2Pd (Kurve *4*), 25Cr2Mn2Mo6NiN (Kurve *5*) und 25Cr2Mn2Mo6NiN0,2Pd (Kurve 6).

festgestellt [13, 14], daß beim Nickel-Legieren hochlegierter Chromstähle die Passivierungs- und Vollpassivierungspotentiale zur positiven Seite hin wesentlich verschoben werden. Deshalb wurde der Einfluß von Mangan auf das Anodenverhalten der Chrom- und Chrom-Nickel-Stähle detaillierter untersucht. Dazu wurden Potential-Korrosionskurven für reinen Chrom-, Chrom-Nickel-, Chrom-Mangan- und Chrom-Nickel-Mangan-Stähle im unmittelbaren Bereich nach dem Übergang aus dem Aktiv- in den Passivzustand aufgenommen, d. h. in dem Bereich, wo sich Korrosionspotentiale der

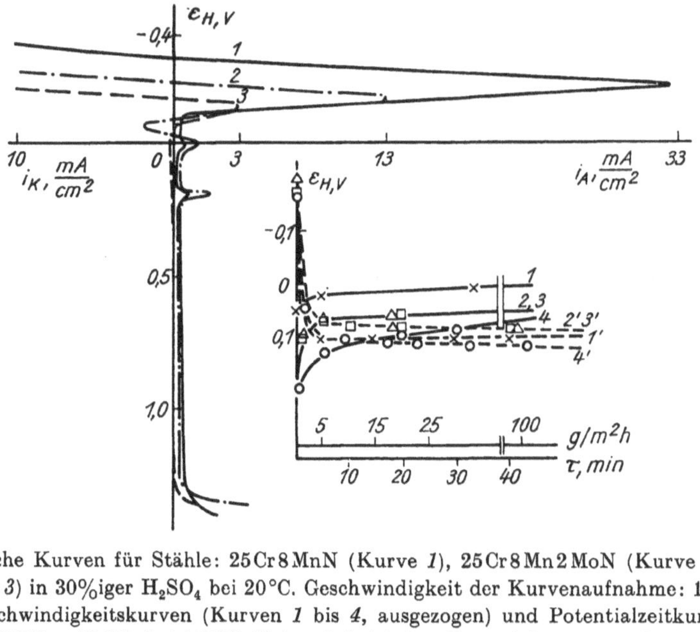

Abb. 5: Potentiokinetische Kurven für Stähle: 25Cr8MnN (Kurve *1*), 25Cr8Mn2MoN (Kurve *2*) und 25Cr8Mn6NiN (Kurve *3*) in 30%iger H_2SO_4 bei 20°C. Geschwindigkeit der Kurvenaufnahme: 1,2 V/h.
Potential-Korrosionsgeschwindigkeitskurven (Kurven *1* bis *4*, ausgezogen) und Potentialzeitkurven (Kurven *1'* bis *4'*, gestrichelt) in 20%iger H_2SO_4 bei 100°C. Haltezeit bei jedem Potentialwert von 10 min bis 2 Stunden.
1 — 25Cr, *2* — 25Cr8MnN, *3* — 25Cr8Mn6NiN, *4* — 25Cr6Ni; *1'* bis *4'* sind die gleichen, mit 0,5% Pd legierten Stähle.

Pd-Stähle nach deren Selbstpassivierung einstellen (Abb. 5 *a*). Aus Ergebnissen dieser Versuche folgt, daß das Vollpassivierungspotential des Chromstahls durch Mn zu positiveren Werten hin (Kurve *2*), weniger jedoch als durch Nickel (Kurve *4*) verschoben wird. Der Chromstahl, der gleichzeitig Mn und Ni enthält (Kurve *3*), verhält sich wie Chrom-Mangan-Stahl. Aus den Potentialzeitkurven der gleichen, mit 0,5% Pd legierten Stähle ist es ersichtlich, daß sich die Potentiale der Cr-, Cr—Mn- und Cr—Mn—Ni-Stähle mit Pd (jeweils die Kurven *1'*, *2'* und *3'*) positiver als Vollpassivierungspotentiale dieser Stähle einstellen und die Stähle mit Pd eine hohe Korrosionsbeständigkeit besitzen. Das Potential des Chrom-Nickelstahls mit Pd (Kurve *4'*) befindet sich noch im Bereich nicht voller Stahlpassivierung, und die Stahlkorrosionsgeschwindigkeit bleibt deshalb relativ hoch.

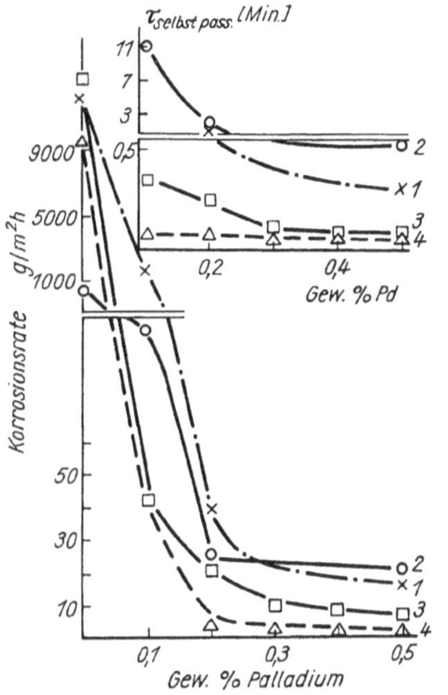

Abb. 6: Pd-Einfluß auf die Korrosionsgeschwindigkeit und auf die Selbstpassivierungszeit der Stähle in 20%iger H_2SO_4 bei 100°C. Versuchsdauer 5 Stunden.

1 — 25 Cr, *2* — 25 Cr 6 Ni, *3* — 25 Cr 8 MnN, *4* — 25 Cr 8 Mn 6 NiN.

Aus dem Verlauf der Selbstpassivierungszeit und der Korrosionsgeschwindigkeit obengenannter Stähle mit Pd ist es ersichtlich (Abb. 6), daß sich Chrom-Mangan- und Chrom-Mangan-Nickel-Stähle am schnellsten, Chrom-Nickel-Stähle jedoch am langsamsten selbst passivieren. Dementsprechend liegen die mittleren Korrosionsgeschwindigkeiten bei Cr—Ni-Stählen höher, bei Cr—Mn—Ni- und Cr—Mn-Stählen aber am niedrigsten.

Gleichzeitiges Ni- und Mn-Legieren des stickstoffreichen Chromstahls erweist sich als günstig für die Erhöhung dessen Korrosionsbeständigkeit durch Einführung des Kathodenzusatzes in den Stahl.

Damit können die untersuchten austenitischen nichtrostenden Stähle mit erhöhtem Stickstoffgehalt, die über eine hohe Festigkeit und Formbarkeit sowie hohe Korrosionsbeständigkeit bei deren Pd-Legieren verfügen, zur industriellen Anwendung kommen, im Falle, daß eine Kombination dieser Eigenschaften gefordert wird.

Literatur

[1] N. D. Tomaschow und G. P. Tschernowa, „*Passiwnost i sastschita metallow ot korrosii*" (Passivität und Korrosionsschutz der Metalle), Verlag Nauka, Moskau 1965

[2] N. D. Tomaschov, Sastschita metallow, **3**, 3 (1967)

[3] N. D. Tomaschow und G. P. Tschernowa, „*Korrosija i korrosionnostoikije splawy* (Korrosion und korrosionsbeständige Legierungen), Verlag Metallurgija, Moskau 1973

[4] M. STERN und H. WISSENBERG, J. Electrochem. Soc., **106**, 755, 759 (1959)
[5] N. D. GREEN, C. BISHOP und M. STERN, J. Electrochem. Soc., **108**, 836 (1961)
[6] W. R. FISCHER, C. ILSCHNER-GENSCH und C. KNORR, Werkstoffe und Korrosion, **12**, 597, (1961)
[7] N. D. TOMASCHOW und G. P. TSCHERNOWA, Sastschita metallow, **11**, (1975) (im Druck)
[8] M. STERN, Patent US. 3246980, Cl 75-134, April, 19, 1966
[9] N. D. TOMASCHOW, W. J. LAKOMSKIY, G. P. TSCHERNOWA, G. F. TORCHOW, L. A. TSCHIGIRINSKAJA und W. A. SLISCHANKOWA, Sastschita metallow (1975) (im Druck)
[10] R. I. KNOTH, G. E. LASKO und W. A. MATEJKA, Chem. Eng., **77**, 574, (1970)
[11] M. A. STREICHER, Corrosion, **30**, 77, 115, (1974)
[12] N. D. TOMASCHOW, G. P. TSCHERNOWA und O. N. MARKOWA, Sastschita metallow, **9**, 696, (1973)
[13] G. P. TSCHERNOWA, L. N. WOLKOW und N. D. TOMASCHOW; im Sammelband „*Korrosija i sastschita metallow*". Verlag Nauka, Moskau, 14, (1970)
[14] A. A. BABAKOW, L. I. POSSYSSAJWA, N. D. TOMASCHOW und G. P. TSCHERNOWA, im Sammelband „*Korrosija i sastschita metallow*", Verlag Nauka, Moskau, 5, (1970)

I. Garz

Technische Hochschule „Otto von Guericke" Magdeburg, Sektion Technologie der metallverarbeitenden Industrie, Wissenschaftsbereich Werkstofftechnik

DER EINFLUSS DER KALTVERFORMUNG AUF DIE PASSIVIERUNG VON NICKEL UND DAS LOCHFRASSPOTENTIAL VON Fe-Ni-EINKRISTALLEN

Zusammenfassung

1. An Ni(321)-Proben in 1 *n* H_2SO_4 wurden folgende Auswirkungen einer Zugverformung festgestellt:
   - Das Passivierungspotential (Potential bei i_{max}) wird in kathodischer Richtung verschoben, dabei besteht bei Verformungsgraden > 10% Proportionalität zur Gleitliniendichte.
   - Der Steilabfall der Stromdichte wird bei hohen Verformungsgraden verzögert und der Aktiv-Passiv-Übergangsbereich mit steigendem Verformungsgrad verbreitert.
   - Die Maximalstromdichten steigen mit dem Verformungsgrad und bei > 10% proportional der Gleitliniendichte an.
   - Die Passivstromdichten werden durch kleine Kaltverformungen gesenkt.
2. Das Lochfraßpotential vom FeNi(210) wird durch kleine Zugverformungen negativer, bei > 22% steigt E_L jedoch wieder an.
3. Die Ergebnisse werden unter Heranziehung der Strukturmerkmale der Proben sowie auf der Basis adsorptionstheoretischer Vorstellungen diskutiert.

1. Einführung

Die Beeinflussung des Korrosions- bzw. Passivierungsverhaltens durch eine Kaltverformung ist wegen ihrer gezielten Anwendung zur Festigkeitssteigerung und ihrem oft unbeabsichtigtem Auftreten bei der Herstellung metallischer Bauteile nicht nur von theoretischem Interesse. Die Ergebnisse der bisher vorliegenden Untersuchungen unterscheiden sich jedoch z. T. beträchtlich in ihren grundlegenden Aussagen, wozu allerdings auch die Verschiedenheit der untersuchten Systeme, d. h. des korrodierenden Metalls, des Korrosionsmediums, des elektrochemischen Meßverfahrens und nicht zuletzt Unterschiede hinsichtlich der Art der Verformung (Zug, Druck, Schockverformung, Walzen, Torsion) beitragen.

Im Aktivbereich wird überwiegend ein Ansteigen der anodischen Stromdichte gefunden, z. B. von Bühler und Schwenk [1], Kunze [2] sowie von Garz und Häfke [3] an Ni. Eichkorn, Lorenz u. a. [4, 5] sowie Kunze [6] konnten zeigen, daß der Mechanismus der aktiven Eisenauflösung in sauren Medien stark vom Strukturzustand des Metalls abhängt.

Das Passivierungspotential E_{p1} als Potential des Stromdichtemaximums i_{max} wird nach Finley und Myers [7] bei Fe und nach Elajaperumal und Balachandra [8] bei austenitischem Stahl durch Kaltverformung nicht verändert, jedoch wird im letzten Fall ein Ansteigen von i_{max} gefunden und auf die verformungsbedingte Bildung von Martensit zurückgeführt. Bei Nickel wird unabhängig von der Verformungsart eine Verschiebung von E_{p1} in kathodischer Richtung und damit ein günstiger Einfluß der Verformung auf die Passivierung festgestellt [9]—[11], [3]. Dagegen scheinen sich Unterschiede in der Verformungsart und der Ausgangsstruktur besonders auf i_{max} auszuwirken. Die Verbreiterung des Aktiv-Passiv-Übergangsbereichs und damit eine Verkürzung des Passivbereichs wird von Finley und Myers [7] für Eisen als Hauptkriterium der Kaltverformung angesehen, sie wurde unter bestimmten Bedingungen auch für Ni [12] gefunden.

Überwiegend ungünstig wird der Verformungseinfluß auf die Passivstromdichte [13] sowie auf das Durchbruchspotential [3, 14] und das Lochfraßpotential [15, 16] beschrieben.

Mazza u. Mitarb. [12] wiesen nach, daß bei Zugverformung die Lage der Prüffläche zur Zugachse die Ergebnisse beeinflußt.

In der vorliegenden Arbeit soll an einkristallinem Nickel der Einfluß einer Zugverformung auf das Passivierungsverhalten unter Berücksichtigung der Gleitlinienbilder als einem relativ einfach zugänglichem Kriterium des Strukturzustandes untersucht werden.

2. Versuchsdurchführung und Ergebnisse

Aus einem Ni-Einkristall der Reinheit 99,9% wurden Flachzugproben herausgearbeitet, deren Seiten- und spätere Arbeitsflächen die Orientierung [*321*] aufwiesen.

Die Orientierung der Stabachse des Einkristalls und damit gleichzeitig der Zugachse der Proben wurde zu [*320*] bestimmt. Nach einem Spannungsarmglühen und elektrolytischer Entfernung der Bearbeitungsschicht wurden die Proben bei konstanter Verformungsgeschwindigkeit um 3, 10, 22 bzw. 50% verformt. Diese Verformungsgrade entsprechen der wahren Dehnung. Aus den Versuchsdaten einer bis zum Bruch verformten Probe wurde die Schubspannungs-Abgleitungskurve unter Verwendung der für Einfachgleitung geltenden Umrechnungsbeziehungen [17] berechnet (Abb. 1). Nach der Lage des Einkristalls im Orientierungsdreieck (Stabachsenorientierung [*320*] nahe der Symmetralen

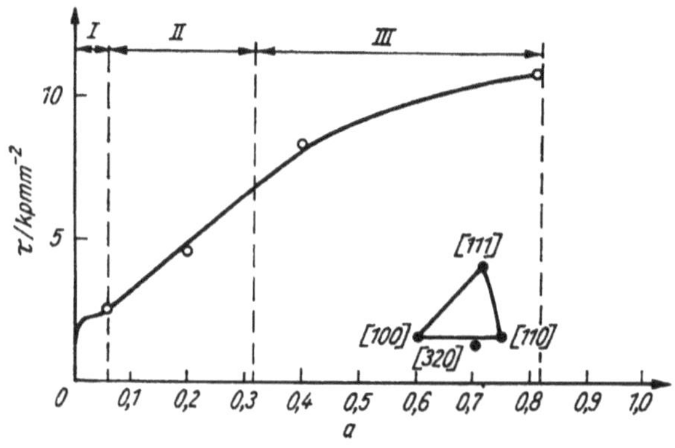

Abb. 1: Schubspannungs-Ableitungskurve des Nickeleinkristalls

[*100*]—[*111*]) war das Auftreten von Doppelgleitung zu erwarten. Die Schubspannungs-Abgleitungs-Kurve, deren Form außer von der Kristallorientierung von der Probenform abhängt, zeigt noch Andeutungen der für mittelorientierte Kristalle typischen Dreiteilung in die Bereiche Einfachgleitung, Verfestigung und dynamische Erholung, was wahrscheinlich auf die auch aus den Gleitlinienbildern ersichtliche dominierende Rolle des primären Gleitsystems zurückzuführen sein dürfte. Die Bereichseinteilung ist jedoch nicht willkürfrei. Die Zuordnung der um 3 und 10% verformten Proben zum Bereich *II* und der übrigen zum Bereich *III* erfolgte daher außer nach der Lage der Proben auf der $\tau-a$-Kurve unter Berücksichtigung der Gleitlinienbilder (Abb. 2 bis 5 auf Taf. IX). Der Bereich *II* ist durch eine strukturierte Feingleitung gekennzeichnet. Beim Übergang von $\eta = 3$ (Abb. 2) zu 10% Verformung (Abb. 3) ist bei prinzipiell gleichem Gleitlinientyp ein Ansteigen der Gleitstufenhöhen zu erkennen.

Für $\eta = 22\%$ (Abb. 4) ist vor allem die Fragmentierung der Gleitlinien und das Vorliegen zahlreicher Quergleitungen auffällig, die als Charakteristikum des Bereichs *III* der $\tau-a$-Kurve anzusehen sind. Bei genauer Betrachtung erscheinen die Gleitlinien in sich strukturiert, was auf eine Gleitbandbildung hindeuten könnte. Bei $\eta = 50\%$ (Abb. 5) treten die für den Bereich *III* genannten Strukturmerkmale, besonders die Fragmentierung, verstärkt in Erscheinung.

Die Unterschiede im Oberflächenprofil sind auch bei REM-Aufnahmen mit y-Modulation (Abb. 6 bis 8 auf Taf. X) gut erkennbar. Bei dieser Technik wird das Integral der Intensität der Sekundärelektronen erfaßt. Dabei ist allerdings zu beachten, daß selbst bei konstant gehaltener Proben-Detektor-Geometrie die Höhenprofilverhältnisse nur qualitativ wiedergegeben werden.

Zur quantitativen Charakterisierung der Gleitlinienbilder wurden die den Laufweg der Verletzungen charakterisierenden Gleitlinienlängen, die mittleren Gleitliniendichten und die Gleitstufenhöhen bestimmt.

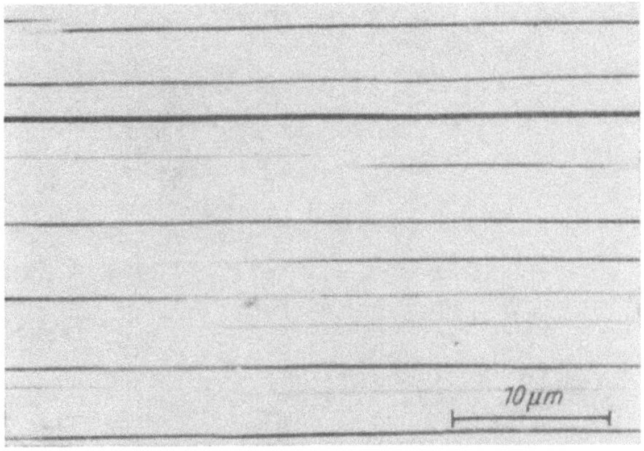

Abb. 2:
Lichtoptische Gleitlinienbilder der verformten Proben, $\eta = 3\%$.

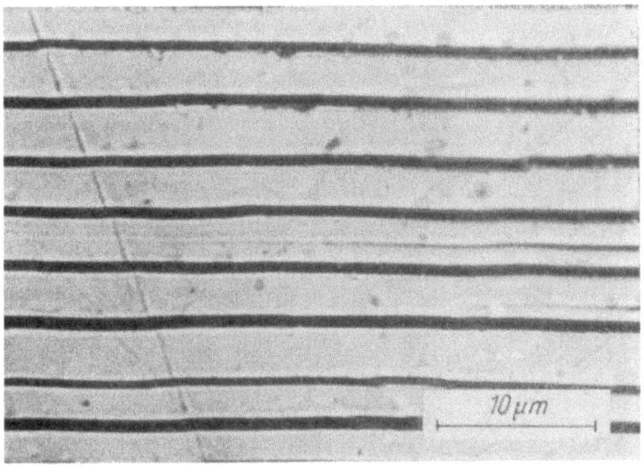

Abb. 3:
Lichtoptische Gleitlinienbilder der verformten Proben, $\eta = 10\%$.

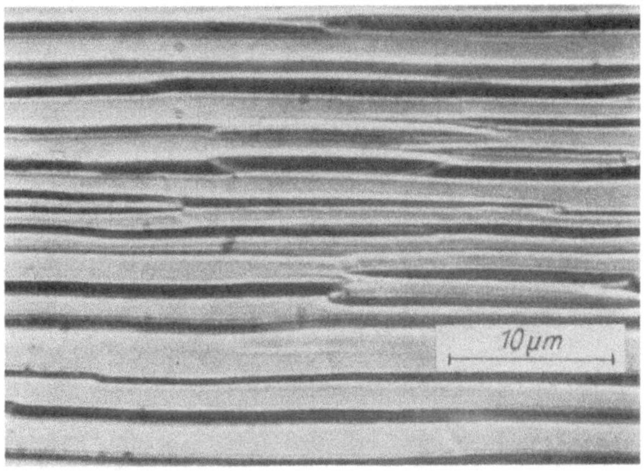

Abb. 4:
Lichtoptische Gleitlinienbilder der verformten Proben, $\eta = 22\%$.

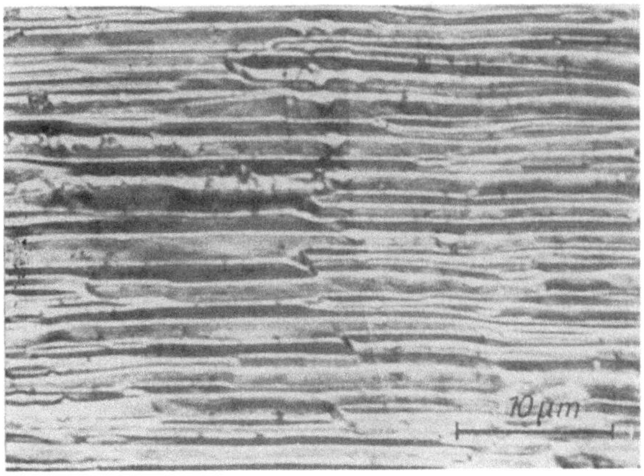

Abb. 5:
Lichtoptische Gleitlinienbilder der verformten Proben, $\eta = 50\%$.

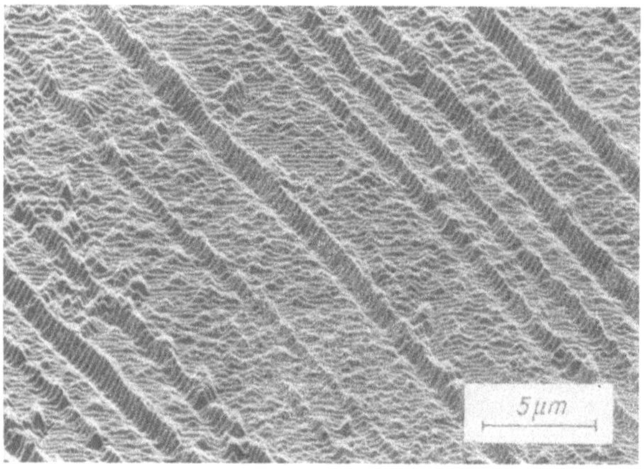

Abb. 6:
REM-Aufnahmen der Oberflächenprofile (Scanning-Mikroskop JSM 5DA, Fa. Joel, Y-Mod. 200L).

Abb. 7: Siehe Abb. 6 $\eta = 10\%$.

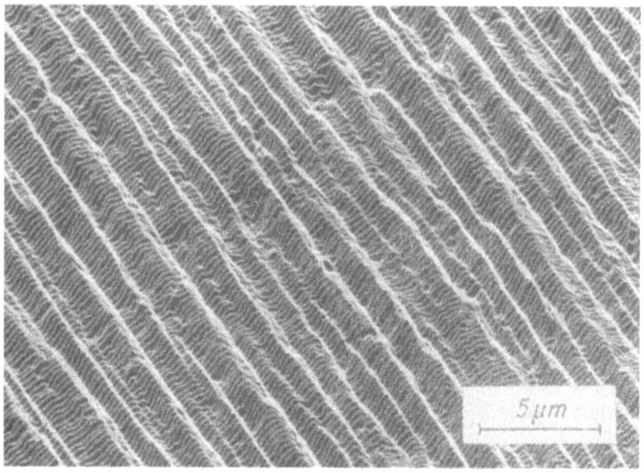

Abb. 8: Siehe Abb. 6, $\eta = 50\%$.

Wie Abb. 9 zeigt, nimmt die mittlere Gleitlinienlänge zwischen 3 und 10% Verformung besonders stark ab, was auf die mit zunehmender Versetzungsdichte häufiger werdende Wechselwirkung der Versetzungen im Bereich der Verfestigung zurückzuführen ist.

Die mittlere Gleitliniendichte als Kriterium des verformten Gefüges steigt proportional der Verformung an und kann daher als eine dem makroskopischen Verformungsgrad analoge Größe angesehen werden.

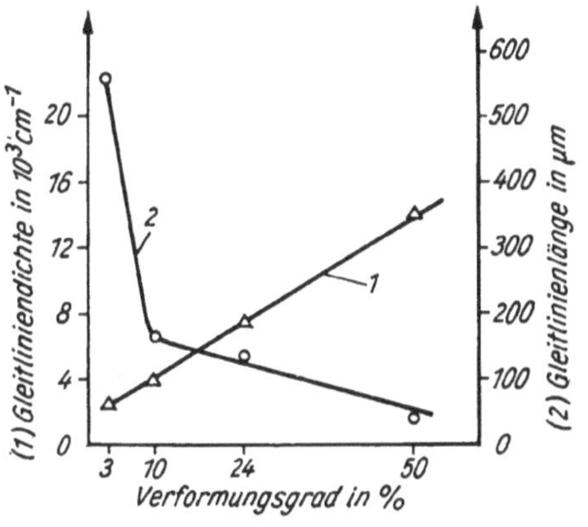

Abb. 9: Mittlere Gleitlinienlängen und -dichten.

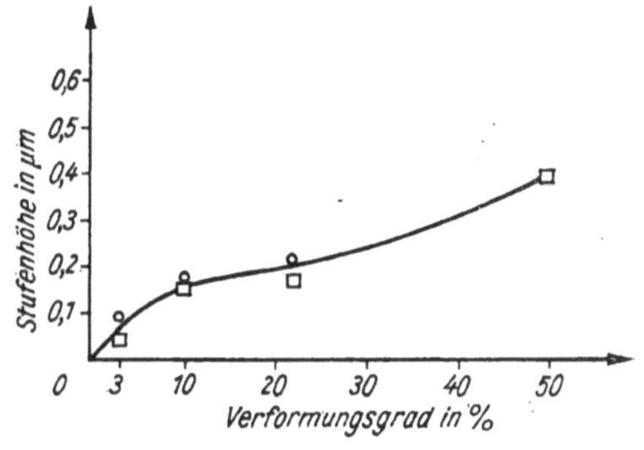

Abb. 10: Gleitstufenhöhen der verformten Proben.
— □ — REM; — ○ — m. Latexkugeln.

Die in Abb. 10 gezeigten Gleitstufenhöhen wurden einmal aus der Schattenlänge von Latexkugeln auf elektronenmikroskopischen Abdruckfolien [18, 19] und später aus stereomikrografischen REM-Aufnahmen ermittelt [20]. Auch die Gleitstufenhöhen steigen mit dem Verformungsgrad an, wobei zwischen 10 und 22% Verformung relativ kleine Änderungen auftreten.

Das anodische Verhalten der Proben wurde in O_2-freier 1 *n* H_2SO_4 bei 25 °C sowohl potentiodynamisch mit einer Polarisationsgeschwindigkeit von 15 mV/min als auch durch potentiostatische Wechselversuche ab 400 mV/GKE, d. h. im Passivbereich, untersucht [19]. Es wurde mit elektrolytisch polierten Proben und kurzer kathodischer Vorpolarisation (1 min, —800 mV) gearbeitet. Wie Abb. 11 zeigt, sinken die Passivströme bei Verformungsgraden bis 10% merklich ab, was in Verbindung mit

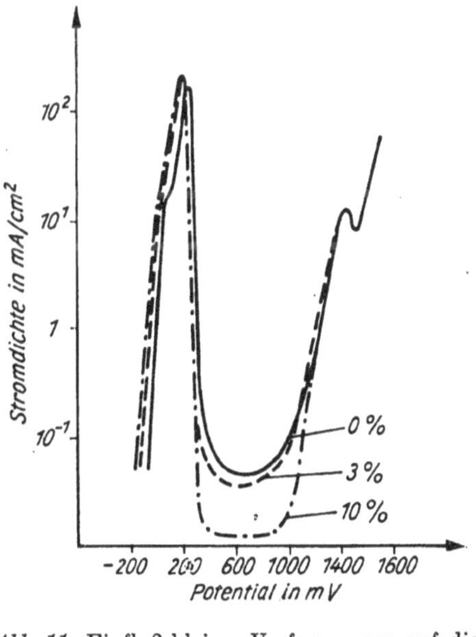

Abb. 11: Einfluß kleiner Verformungen auf die Stromdichte-Potential-Kurve in 1 N H_2SO_4 (Polarisationsgeschwindigkeit 15 mV/min).

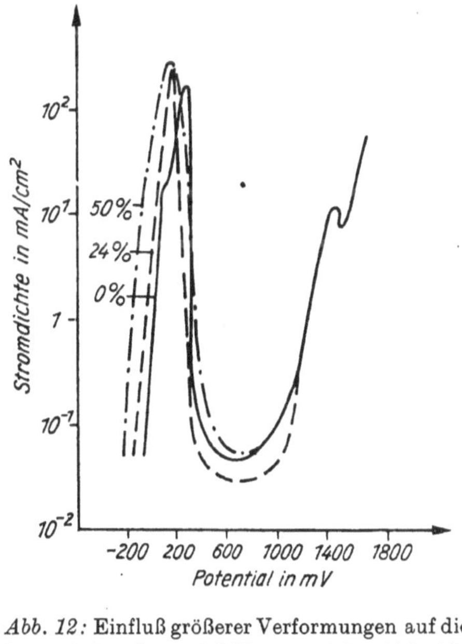

Abb. 12: Einfluß größerer Verformungen auf die Stromdichte-Potential-Kurve (siehe Abb. 11).

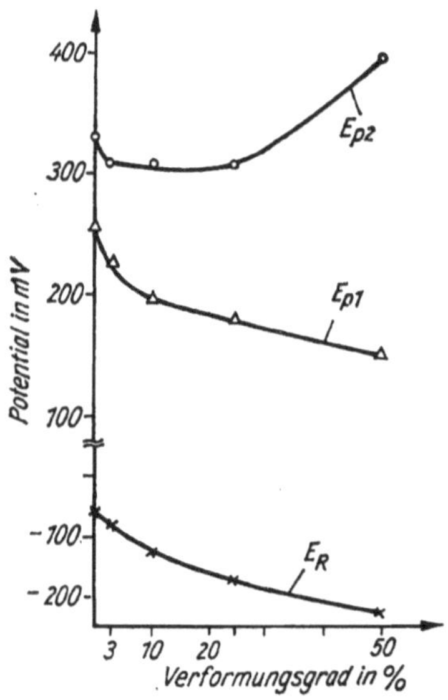

Abb. 13: Einfluß des Verformungsgrades auf charakteristische Potentiale.

E_{p1} — Potential des Stromdichtemaximums;
E_{p2} — Potential des Steilabfalls bei i = 0,1 mA/cm²;
E_R — Ruhepotential.

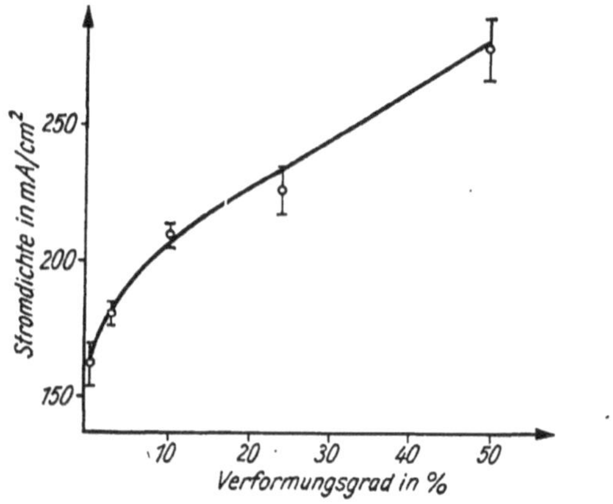

Abb. 14: Einfluß des Verformungsgrades auf die Maximalstromdichte.

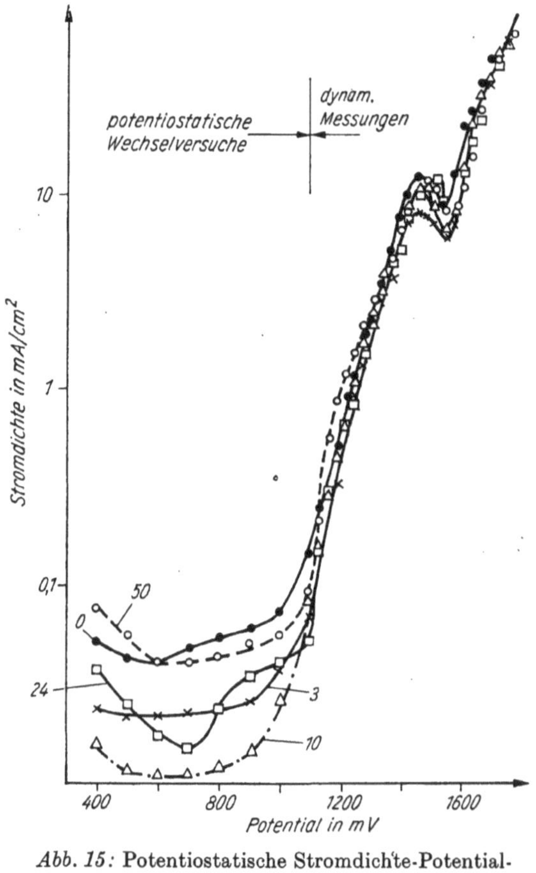

Abb. 15: Potentiostatische Stromdichte-Potential-Kurven in H_2SO_4 (Beginn der Polarisation bei 400 mV/GKE).

—• — η = 0%; —× — η = 3%; —△— η = 10%;
—□ — η = 24%; —○ — η = 50%.

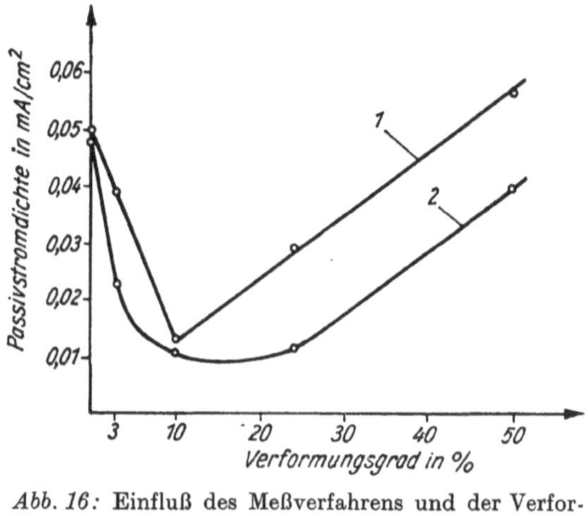

Abb. 16: Einfluß des Meßverfahrens und der Verformung auf den Minimalwert der Passivstromdichte.

1 — dynamische Messung; *2* — statische Messung.

einer Verschiebung des Passivierungspotentials in kathodischer Richtung für eine Verbesserung der Passivierbarkeit durch geringe Zugverformungen spricht. Die Aktivstromdichten und i_{max} werden dagegen erhöht. Im transpassiven Bereich ist bei dynamischer Messung kein merklicher Einfluß der Verformung festzustellen.

Abb. 12 läßt für die höheren Verformungsgrade noch deutlicher die Steigerung der aktiven Auflösung und von i_{max}, verbunden mit dem Verschwinden des ohnehin nicht sehr gut ausgebildeten Vormaximums, erkennen. Trotz weiterer Verringerung des Passivierungspotentials deutet sich beim Steilabfall und durch Wiederansteigen der Passivstromdichte eine Erschwerung der Passivierung durch höhere Verformungsgrade an.

Die z. T. gegenläufige Wirkung der Verformung auf die Passivierung ist in Abb. 13 an Hand der Werte von E_{p1} und E_{p2} (hier definiert als das Potential des Stromdichteabfalls auf 0,1 mA/cm²) dargestellt. E_{p1} sinkt mit steigender Verformung ab, wobei ab $\eta = 10\%$ ein linearer Verlauf und damit eine Analogie zur Zunahme der mittleren Gleitliniendichte festzustellen ist. Während sich die Verformung damit auf den Beginn der Passivierung günstig auswirkt, wird ihr Abschluß nur bei kleinen Verformungen begünstigt, bei großen jedoch verzögert. Die Breite des Übergangsbereichs, charakterisiert durch die Differenz $E_{p2} - E_{p1}$, nimmt deutlich mit η zu. Die Maximalstromdichten (Abb. 14) steigen ebenfalls ab $\eta = 10\%$ linear mit der Verformung und damit auch proportional der Gleitliniendichte an.

Die zur Vermeidung einer unterschiedlichen Aufrauhung der Proben bei der aktiven Auflösung durchgeführten potentiostatischen Messungen im Passivbereich (Abb. 15) bestätigen die günstige Wirkung kleiner Verformungen auf i_p. Abb. 16 zeigt an Hand der Minimalwerte der Stromdichten, daß sich die Ergebnisse beider Meßverfahren nicht grundlegend unterscheiden.

3. Diskussion

3.1. Die Passivierungsstromdichte

Durch eine plastische Deformation erhöht sich die Versetzungsdichte und die Konzentration von Punktdefekten, besonders von Gitterleerstellen, um mehrere Größenordnungen. Damit sinkt die Besetzungsdichte der betroffenen Gitterbereiche und ihre freie Oberflächenenergie wird erhöht, wozu auch die elastischen Verspannungen in der Umgebung der Versetzungskerne beitragen. Adsorptionsvorgänge werden durch Senkung der notwendigen Aktivierungsenergien erleichtert und die adsorbierten Teilchen infolge hoher Adsorptionsenergien sehr fest gebunden.

Die (primäre) Passivierung unter Adsorption des Sauerstoffes des Wassers nach der Bruttoreaktion

$$Ni + H_2O \rightarrow NiO + 2H^+ + 2e$$

wird dadurch prinzipiell bei negativeren Potentialen möglich. In der gefundenen Abhängigkeit des Passivierungspotentials vom Verformungsgrad kommt zum Ausdruck, daß bei sehr kleinen Verformungen die größten Aktivitätsänderungen in bezug auf den unverformten Einkristallzustand auftreten, während für die weitere Begünstigung der Passivierung Versetzungsanhäufungen maßgebend sind, für die die Gleitliniendichte eine repräsentative Größe darstellt.

Die Höhe der Passivierungsstromdichte i_{max} resultiert aus der Konkurrenz von Passivierungsreaktion und aktiver Metallauflösung, deren erster Teilvorgang ebenso wie der der Passivierung in einer Adsorption von H_2O besteht und deren Geschwindigkeit mit dem Verformungsgrad ansteigt. Begünstigt wird aber auch die Adsorption von Anionen aus der Lösung, deren zunehmend aktivierende Wirkung unter den vorliegenden Bedingungen auch durch das Verschwinden des Vormaximums deutlich wird, das allgemein der Bildung einer die Passivierung begünstigenden NiO-Schicht zugeschrieben wird. Das Ansteigen der Passivierungsstromdichte ergibt sich aus der Primäradsorption der Anionen an den aktiven Oberflächenstellen, die durch den passivierenden Sauerstoff erst verdrängt werden müssen.

Die hinsichtlich i_{max} bestehenden Differenzen zwischen Arbeiten verschiedener Autoren dürften im wesentlichen auf die Spezifik der mit der Verformung verbundenen Strukturänderungen zurückzuführen sein. Bei polykristallinem Nickel erfolgt die Sauerstoffbedeckung zunächst überwiegend an den Korngrenzen, während an den Kornflächen weiterhin eine Auflösung stattfindet. An verformten Proben bewirkt die Durchsetzung der Kornflächen mit Aktivstellen eine Homogenisierung der Oxidbildung, was die von MOVCHAN und JAGUPOLSKAJA [9] sowie von CROWE und FISHMAN [7] festgestellte Senkung von i_{max} mit zunehmender Verformung erklärt. Für den Abfall von i_{max} an Ni [*110*]-Proben, den GINDIN, JAGUPOLSKAJA u. a. [11] unter sonst ähnlichen Korrosionsbedingungen fanden, dürfte im wesentlichen die durch Walzen bei tiefen Temperaturen entstandene besonders homogene Versetzungsstruktur und die unverändert bleibende, wenig atomar gestufte [*110*]-Fläche mit ihrer vergleichsweise geringen aktiven Auflösungsgeschwindigkeit verantwortlich sein.

In diesem Zusammenhang ist auch zu erwähnen, daß in der Arbeit von MAZZA u. a. [12], die den Verformungseffekt als sehr gering einschätzen, auf die Abwesenheit von Gleitlinien bei Ni besonders hingewiesen wird.

3.2. Die Passivstromdichte

Für das im Passivbereich bestehende Phasenoxid hat schon SCHWABE [21] darauf hingewiesen, daß Störungen der Sauerstoffchemisorptionsschicht an Korngrenzen und anderen Defekten der Unterlage den Durchtritt von Metallionen und damit die Schichtbildung erleichtern, was durch das Absinken der Passivstromdichten bei Verformungen bis 10% bestätigt wird.

Der Passivstrom ist nach SIEJKA u. a. [22] als Ionenstrom zu betrachten, der aus einem dem direkten Übergang von Kationen in die Lösung und einem der chemischen Auflösung des Oxids entsprechenden Anteil für die Schichtbildung besteht. Beide Anteile können sich mit der Fehlstellenkonzentration ändern.

Unter der Annahme einer epitaktisch aufgewachsenen NiO-Schicht nach den Vorstellungen von PARK und FARNSWORTH [23] sowie von LATANISION und OPPENHAUER [24] ist für den direkten Übergang von Kationen entsprechend den für die Oxydation aus der Gasphase bei niedrigen Temperaturen entwickelten Vorstellungen [25—27] die Ausbildung sogenannter schneller Diffusionswege entscheidend. Die Versetzungsstruktur einer Ebene ist bei verformten Proben in erster Linie durch ihre Lage zur Verformungsrichtung und zu den Gleitebenen bestimmt. Die bei geringen Verformungen vorzugsweise in den Gleitebenen liegenden Versetzungen werden unter einem bestimmten Winkel von der Oberfläche geschnitten. Die so entstandenen Versetzungsanordnungen sind offensichtlich auch bei strukturierter Feingleitung im Bereich *II* der Verfestigungskurve noch so regelmäßig, daß sich eine relativ geordnete Oxidstruktur ohne bzw. mit nur wenigen schnellen Diffusionswegen bilden kann, so daß bei Verformungen bis 10% die erhöhte Sauerstoffadsorption auch zu einer verstärkt schützend wirkenden Passivschicht führt.

Große geometrische Störungen des Substrats, wie sie im Bereich III der τ-a-Kurve auftreten, führen mit hoher Wahrscheinlichkeit zu einer Fortsetzung von Versetzungslinien in und durch die Oxidschicht, zur Entstehung von Misfitgrenzen schräg zur Unterlage zwischen verschieden orientiert aufwachsenden Oxidbereichen und möglicherweise auch zur frühzeitigen Bildung von Großwinkelkorngrenzen, d. h. zu polykristallinen Schichten, was das Wiederansteigen von i_p bei hohen Verformungen begründet.

3.3. Der Aktiv-Passiv-Übergang

Wie Abb. 13 zeigt, wird der Aktiv-Passiv-Übergangsbereich mit steigender Verformung größer. Nach FRANKENTHAL [28] wird der Steilabfall der Stromdichte durch die Blockierung der Halbkristallagen verursacht, die im unverformten Zustand nur einen geringen Anteil der Oberfläche ausmachen. Die Verzögerung des Stromdichteabfalls bei hohen Verformungen dürfte daher im wesentlichen in dem zunehmenden Flächenanteil der *HK*-Lagen in Verbindung mit der wachsenden Schwierigkeit der Verdrängung der voradsorbierten Anionen, gegebenenfalls auch schon in der abnehmenden Schutzwirkung des sich bildenden Phasenoxids zu suchen sein.

3.4. Das Lochfraßpotential von Eisen-Nickel-Einkristallen

Im Zusammenhang mit der Verformungsabhängigkeit der Passivierung ist der Einfluß der Kaltverformung auf die Lochfraßkorrosion, besonders auf das Lochfraßpotial von Interesse, da Eisen und Nickel im Gegensatz zu rost- und säurebeständigen Stählen im gesamten Passivbereich Lochfraß aufweisen können. Bei potentiostatischer, vom Ruhepotential ausgehender anodischer Polarisation ist daher zu erwarten, daß die Lochfraßkorrosion am Passivierungspotential einsetzt. Für unverformte Nickeleinkristalle wurde zudem nachgewiesen, daß die Lochfraßpotentiale in bestimmten Chloridlösungen dem Passivierungspotential in einer chloridfreien Sulfatlösung gleichen pH-Wertes entsprechen [29]. Danach war ein Absinken der Lochfraßpotentiale mit zunehmender Kaltverformung zu erwarten, wie dies auch von Fujii und Kumada [16] für Cr—Ni-Stahl und Verformungsgrade bis 20% gefunden wurde.

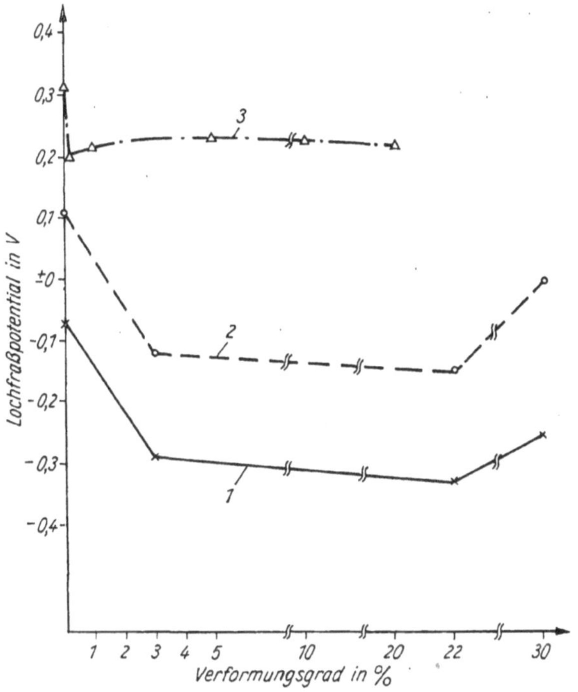

Abb. 17: Einfluß der Kaltverformung auf das Lochfraßpotential.

1 — FeNi(210) in 0,1 N-NaCl pH 3,5; *2* — dgl. unter Zusatz von 10^{-4} m Benzyl-n-propylsulfid als Inhibitor; *3* — 18.8CrNi-Stahl in 1n H_2SO_4 + 0,5n NaCl nach Fujii und Kumada [16].

Unsere Untersuchungen [15, 30, 31] erstreckten sich auf einen Fe—Ni-Einkristall mit $\sim$ 39% Ni und kfz-Struktur. An Arbeitsflächen der Orientierung [210] und einer Lage senkrecht zur Zugachse wurde durch potentiostatische Polarisation in 0,1 *n* NaCl bei kleinen Verformungen ebenfalls die erwartete Abnahme des Lochfraßpotentials erhalten (Abb. 17). Bei $\eta > 22\%$ steigt E_L jedoch wieder an. Das bedeutet, daß die Passivierung bei höheren Verformungsgraden durch die Voradsorption der Cl-Ionen offensichtlich so stark behindert wird, daß eine über größere Bereiche der Oberfläche wirksame Blockierung der Auflösung erst bei höheren Potentialen möglich wird. Dies steht in Einklang mit Ergebnissen von Kolotyrkin, Jagupolskaja und Besoljuk [32], nach denen an molybdänhaltigem, um 70% verformten Nickel eine Erschwerung des endgültigen Eintritts der Passivierung durch KJ-Zusätze in Schwefelsäure nachgewiesen wurde.

Die Ausbildung eines Minimums des Lochfraßpotentials von unlegiertem Stahl in chloridhaltiger Schwefelsäure bei einem Verformungsgrad von 10% [33] zeigt, daß die gefundene Verformungsabhängigkeit kein für Einkristalle spezifisches Ergebnis ist.

Herrn Dr.-Ing. G. Pusch, TH Magdeburg, und Herrn Dr. rer. nat. M. Schaper, Zentralinstitut für Festkörperphysik und Werkstofforschung Dresden, danke ich für die Mithilfe beim metallkundlichen Teil der Arbeit.

Literatur

[1] H. E. Bühler und W. Schwenk, Z. Metallkde, **56**, 24 (1965)
[2] E. Kunze, „*Einfluß der Verformung auf das Korrosionsverhalten in Schwefelsäure*", Vortrag, Freiberg 1970
[3] I. Garz und U. Höfke, Corrs. Sci., **11**, 329 (1971)
[4] W. J. Lorenz und G. Eichkorn, Ber. Bunsenges. Phys. Chem., **70**, 91 (1966)
[5] G. Eichkorn und W. J. Lorenz, Metalloberfl., **22**, 102 (1968)
[6] E. Kunze, Habilitationsschrift, Bergakademie Freiberg, 1970
[7] T. C. Finley und J. R. Myers, Corrosion, **26**, 150 (1970)
[8] K. Elayaperumal, P. K. De und J. Balachandra, Corrosion, **28**, 269 (1972)
[9] B. A. Movchan und L. N. Jagupolskaja, Zaschita metall., **5**, 511 (1969)
[10] C. R. Crowe und S. G. Fishman, Corros. Sci., **13**, 567 (1973)
[11] J. A. Gindin, L. N. Jagupolskaja, W. K. Aksenow und Ja. D. Starodubow, Corros. Sci., **13**, 967 (1973)
[12] B. Mazza, P. Pedeferri, D. Sinigaglia, U. Della Salla und J. Lazzari, Werkstoffe und Korr., **25**, 239 (1974)
[13] W. D. France, Corrosion, **26**, 189 (1970)
[14] R. P. Frankenthal, Corrosion Sci., **8**, 491 (1968)
[15] I. Garz, B. Glaser und G. Pusch, Korrosion (Rossendorf), **4**, 10 (1975)
[16] M. Fujii und M. Kumada, Nippon kinzoku Gakkai-shi, **34**, 1001 (1970)
[17] A. Seeger, „*Moderne Probleme der Metallphysik*", Bd. 1, Berlin—Heidelberg—New York, 1965
[18] H. Schmidtchen, Diplomarbeit, TH Magdeburg 1973
[19] K. Denkewitz, Diplomarbeit, TH Magdeburg 1973
[20] H. Bock, I. Garz und B. Suthoff, Wiss. Z. d. TH Magdeburg, **20**, 449 (1976)
[21] K. Schwabe, Z. Angew. Chem., **78**, 253 (1966)
[22] J. Siejka, C. Cherki und J. Yahalom, J. Electrochem. Soc., **119**, 927 (1972)
[23] L. Park und H. E. Farnsworth, J. Appl. Phys., **35**, 2220 (1954)
[24] R. M. Latanision und H. Oppenhauer, Corrosion, **27**, 509 (1971)
[25] J. V. Cathcart, G. F. Petersen und C. J. Sparks, J. Electrochem. Soc., **116**, 664 (1969)
[26] M. J. Caplan, M. J. Graham und M. Cohen, J. Electrochem. Soc., **119**, 1205 (1972)
[27] M. J. Graham, R. I. Hussey und M. Cohen, J. Electrochem. Soc., **120**, 1523 (1973)
[28] R. P. Frankenthal, J. Electrochem. Soc., **116**, 580 (1969)
[29] I. Garz, Wiss. Zt. d. TH Magdeburg, **14**, 779 (1970)
[30] B. Glaser, Dissertation TH Magdeburg, 1974
[31] I. Garz und B. Glaser, Vortrag „*II. Internationale RGW-Konferenz über Korrosion und Korrosionsschutz*", Prag 1975
[32] Ja. M. Kolotyrtin, L. N. Jagupolskaja und E. S. Besoljuk, Zaschita metallov, **9**, 551 (1973)
[33] H. Worch und E. Fritzsche, Arbeiten im Forschungsinstitut Meinsberg 1974
E. Fritzsche, Ingenieurarbeit TH Magdeburg, 1974

L. Franke und W. Forker

Technische Hochschule Karl-Marx-Stadt und Technische Universität Dresden

ÜBER DAS TRANSPASSIVE VERHALTEN VON EISEN UND STÄHLEN IN SALZLÖSUNGEN BEI HOHEN STROMDICHTEN

Der elektrochemische Abtrag von Eisen, Stählen und anderen metallischen Werkstoffen mit hoher Festigkeit ist in das Interesse der Techniker und Naturwissenschaftler gerückt, nachdem es gelungen war, Abtraggeschwindigkeiten bis zu etwa 1 cm/min zu erreichen und auf diese Weise Werkstücke durch gezielte anodische Auflösung wirtschaftlich zu bearbeiten. Die Auflösung der Metalle erfolgt dabei mit Stromdichten bis zu etwa 500 A/cm². Es ist leicht vorstellbar, daß derartige Stromdichten bei „aktiver" Metallauflösung möglich sind, wenn durch hinreichend starke Strömung der Elektrolytlösung in der Nähe der Metalloberfläche für eine hohe Geschwindigkeit der Transportvorgänge gesorgt wird. Bei gleichbleibendem Mechanismus könnte die Reaktionsstromdichte durch Vergrößerung der anodischen Überspannung um einige Zehntel Volt ohne weiteres vom Milliamperebereich bis auf einige 100 A/cm² ansteigen.

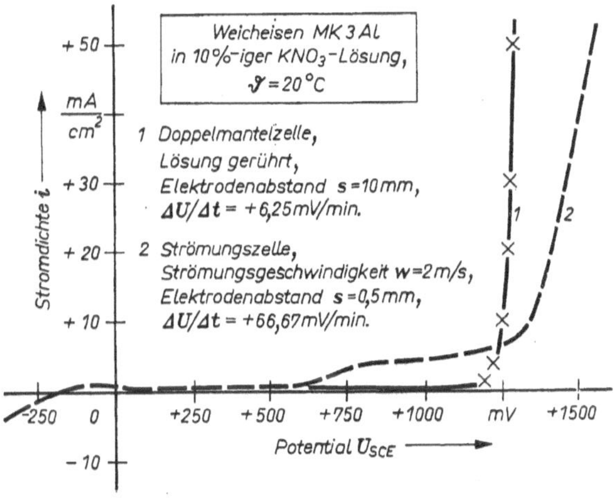

Abb. 1: Potentiodynamisch aufgenommene Stromdichte-Potential-Kurven.

1 — Doppelmantelzelle, Lösung gerührt, Elektrodenabstand $s = 10$ mm, $\Delta U/\Delta t = +6{,}25$ mV/min; *2* — Strömungszelle, Strömungsgeschwindigkeit $w = 2\ \mathrm{m \cdot s^1}$, Elektrodenabstand $s = 0{,}5$ mm, $\Delta U/\Delta t = +66{,}67$ mV/min.

Nicht so einfach deutbar ist das Auftreten derartig hoher Auflösungsgeschwindigkeiten bei Metall/Elektrolyt-Systemen, bei denen schon geringere anodische Polarisationen zur Passivität des Metalls führen und die Auflösung entsprechend dem bekannten Verlauf der Stromdichte-Potential-Kurven offensichtlich im Transpassivbereich durch eine oxidische Deckschicht hindurch erfolgt.

Die Stromdichte-Potential-Kurven von Eisen und Stählen in Nitrat- und Chloratlösungen stimmen unter den Bedingungen der elektrochemischen Metallbearbeitung, d. h. bei einem Elektrodenabstand von einigen Zehntel Millimetern und Strömungsgeschwindigkeiten der Elektrolytlösung von etwa 1 bis 50 m/s, nach Literaturangaben [1, 2] und eigenen Untersuchungen [3], mit denen bei niedrigen Strömungsgeschwindigkeiten im prinzipiellen Verlauf überein (Abb. 1).

Dem Bereich der aktiven Auflösung schließt sich der Passivbereich an. Danach folgt ein Anstieg der Stromdichte, der zunächst weitgehend der Sauerstoffentwicklung zuzuschreiben ist. Bei weiterer Potentialpositivierung nimmt der Anteil der Eisenauflösung ständig zu. Abb. 2 zeigt die Stromausbeute α_1 der Auflösung von Weicheisen MK3Al (Zusammensetzung nach TGL 101-18499;

≦ 0,25% Mn; 0,02 bis 0,15% Al; ≦ 0,1% Si; ≦ 0,04% C) in Abhängigkeit von der Gesamtstromdichte in 20%iger Kaliumnitrat- bzw. 10%iger Natriumchloratlösung bei 35°C.

In der Regel nähert sich die Stromausbeute mit wachsender Stromdichte dem Wert 1, wenn man die Auflösung zum zweiwertigen Eisen zugrunde legt. Durch diese Beobachtung und durch analytische Untersuchungen ist gesichert, daß die Bruttoreaktionsladungszahl der transpassiven Eisenauflösung gewöhnlich $Z_r = +2$ beträgt [4, 5, 6]. Das ist insofern bemerkenswert, als im Passivbereich das Eisen als Fe^{3+}-Ion in die Lösung übertritt.

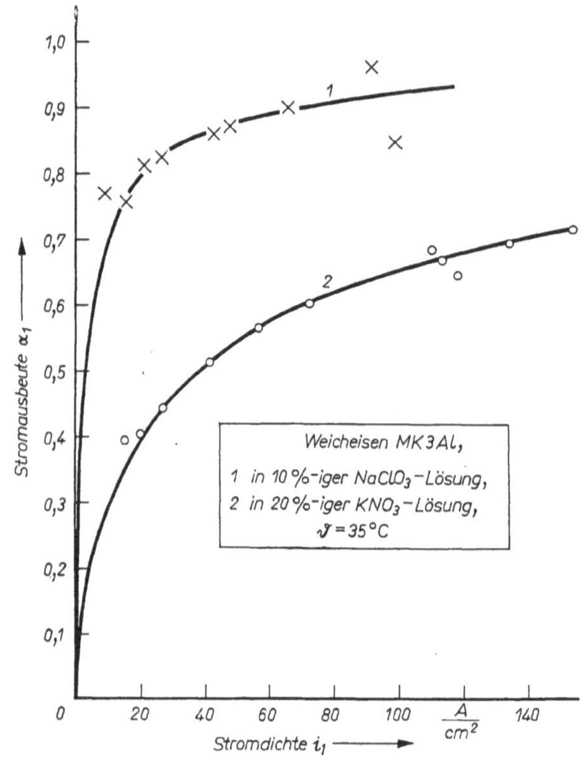

Abb. 2: Stromausbeute α_1 des Metallabtrages als Funktion der Reaktionsstromdichte i_1. Weicheisen MK3Al, 35°C, o – 20% KNO_3, × – 10% $NaClO_3$.

Bei anderen Metallen ist die Potentialpositivierung erwartungsgemäß mit einer Erhöhung der Bruttoreaktionsladungszahl verbunden.

So erklärt sich z. B. die Auflösung des Chroms im transpassiven Gebiet durch eine Erhöhung seiner Oxydationszahl [7]. In der Passivschicht liegt das Chrom 3wertig vor, bei der Auflösung im Transpassivbereich ist die Brutto-Reaktionsladungszahl +6. Die Erhöhung der Oxydationszahl von 3 auf 6 ist mit einer Veränderung der Löslichkeit der Chromverbindungen verbunden. In neutralen wäßrigen Lösungen ist das Dichromtrioxid Cr_2O_3 unlöslich, Chromtrioxid CrO_3 dagegen als Anhydrid der Chromsäure äußerst gut löslich, so daß damit der beschleunigte Schichtabbau verständlich wird.

Für das Eisen könnte zwar eine ähnliche Möglichkeit, nämlich die Bildung von Ferrat-Ionen, diskutiert werden. Die Messungen weisen unter den Arbeitsbedingungen der elektrochemischen Metallbearbeitung aber eine Brutto-Reaktionsladungszahl von $z_r = +2$ aus. Damit scheidet die Annahme einer beschleunigten transpassiven Eisenauflösung infolge Aufoxydation oberflächlicher Oxidschichten zum leichter löslichen Ferrat-Ion aus.

Die transpassive Auflösung des Eisens kann modellmäßig erfaßt werden, wenn man berücksichtigt, daß an der Grenzfläche Oxid/Lösung gleichzeitig 2 Reaktionen ablaufen, die Eisenauflösung und die Sauerstoffentwicklung, und wenn man annimmt, daß deren Stromdichte-Potential-Abhängigkeiten durch Tafelgeraden beschreibbar sind. Für die Abhängigkeit der Stromausbeute der Eisenauflösung von der Gesamtstromdichte lassen sich dann 3 Möglichkeiten diskutieren (Abb. 3), [8]:

Sind die TAFEL-Steigungen beider Teilreaktionen gleich, so ist die Stromausbeute unabhängig von der Stromdichte. Je nach Größe der Austauschstromdichte für die Simultanreaktionen liegt die Stromausbeute der Eisenauflösung zwischen 0 und 100%. Fallen beide TAFEL-Geraden zusammen, so ergibt sich eine 50%ige Stromausbeute. Ist die TAFEL-Steigung der Eisenauflösung größer als die der Nebenreaktion, so nimmt die Stromausbeute α_1 mit steigender Stromdichte ab, ist sie kleiner,

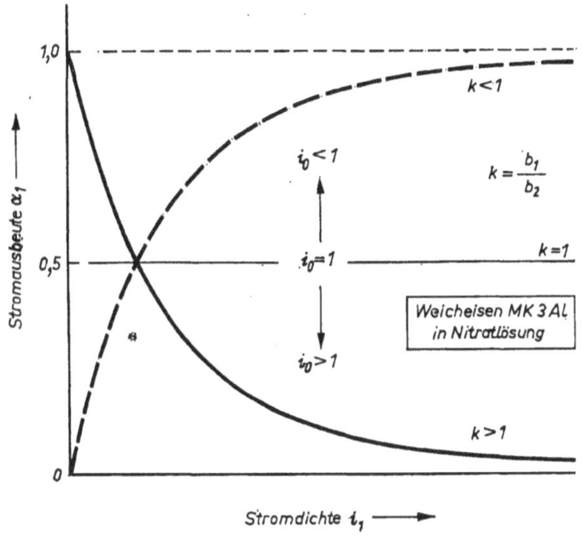

Abb. 3: Abhängigkeit der Stromausbeute α_1 der Metallauflösung von der Reaktionsstromdichte i_1 bei verschiedenen Werten für $k = \frac{b_1}{b_2}$.

dann steigt sie an. Ihre Abhängigkeit von der Reaktionsstromdichte i_1 der Eisenauflösung läßt sich für die mit steigender Stromdichte wachsende Stromausbeute durch die Beziehung

$$\frac{1}{\alpha_1} = 1 + i_0 \cdot i_1^{k-1} \tag{1}$$

ausdrücken [8]. Die Kurvenformen hängen von einem Exponenten $k = \frac{b_1}{b_2}$, dem Quotienten der TAFEL-Steigungen der Metallauflösung b_1 und der Nebenreaktion b_2, und einem Faktor i_0 ab, der wesentlich von den Austauschstromdichten der beiden Teilreaktionen beeinflußt wird. Experimentell konnte bei der transpassiven Metallauflösung bisher nur eine mit steigender Stromdichte wachsende Stromausbeute der Metallauflösung nachgewiesen werden [6, 8—14].

Abb. 2 zeigt Kurven für die transpassive Auflösung von Weicheisen MK 3 Al in Nitrat- und Chloratlösungen. Die Meßpunkte sind in die Abbildung eingezeichnet, die Kurvenzüge wurden nach Gl. (1) berechnet, nachdem aus den Meßwerten die Parameter k und i_0 ermittelt worden waren. Auf der Basis der Annahme von TAFEL-Beziehungen für beide Simultanreaktionen lassen sich schließlich auch die Konzentrations- und Temperaturabhängigkeit der Stromausbeute der Eisenauflösung beschreiben. Unter der Voraussetzung eines gleichbleibenden Reaktionsmechanismus hängt nur der Parameter i_0 von Temperatur und Konzentration der Elektrolytlösung ab. Es sind dann folgende Korrelationen zu erwarten [13]:

$$\lg i_0 = \lg k - \nu \cdot \lg c \tag{2}$$

$$\lg i_0 = A - \frac{B}{T}. \tag{3}$$

Sie werden durch die Meßergebnisse (Abb. 4 bis 7) bestätigt.

Alle abgeleiteten Beziehungen basieren auf der Annahme, daß die Galvanispannung zwischen Oxidschicht und Elektrolytlösung für die Geschwindigkeiten beider Simultanreaktionen bestimmend ist. Sehr wahrscheinlich sind Durchtrittsvorgänge für die Größe der Überspannung an der Grenzfläche Oxid/Lösung verantwortlich, da bei transportbestimmten Vorgängen andere als die experimentell ermittelten Abhängigkeiten von den Strömungsverhältnissen und auch von Temperatur und Konzentration zu erwarten wären. Die an den Durchtrittsreaktionen beteiligten Spezies müssen daher hin-

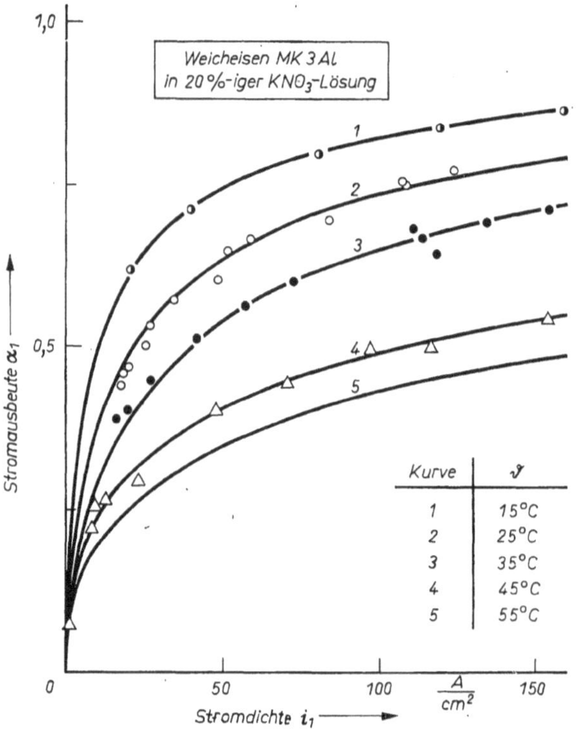

Abb. 4: Stromausbeute α_1 als Funktion der Reaktionsstromdichte i_1 der Auflösung von Eisen MK3Al in Nitratlösung (20% KNO_3) bei verschiedenen Temperaturen. (Kurven berechnet nach Gl. (1).)

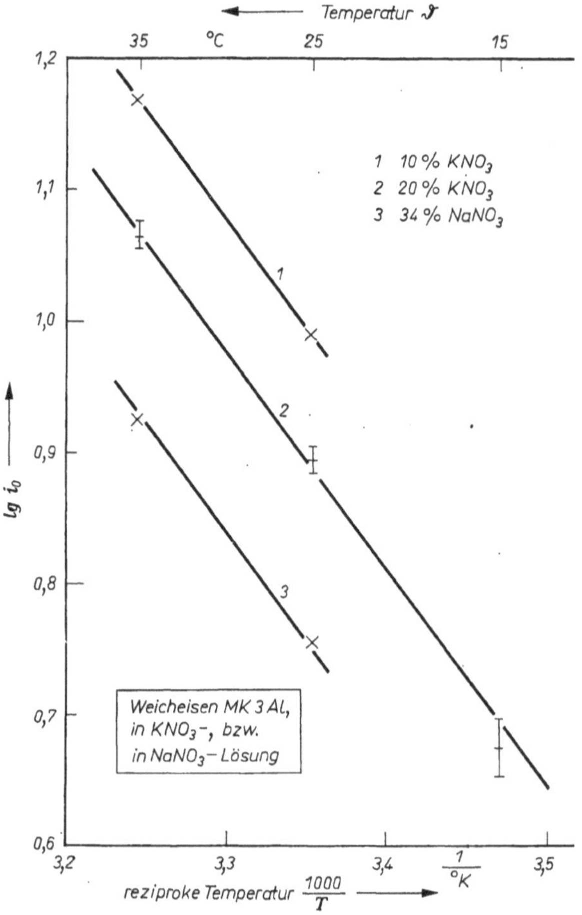

Abb. 5: Logarithmus i_0 als Funktion des Reziprokwertes der absoluten Temperatur für die Auflösung von Eisen MK3Al in Nitratlösung.

reichend schnell an die Phasengrenze herantransportiert oder von ihr weggeführt werden. Auf der Lösungsseite der Phasengrenzschicht kann dies durch eine hinreichend hohe Strömungsgeschwindigkeit der Elektrolytlösung gewährleistet werden. Der rasche Antransport der durchtretenden Eisenionen an die Grenzfläche muß hingegen im Oxidgitter erfolgen. Er setzt insbesondere bei niedrigen Temperaturen verhältnismäßig hohe Feldstärken voraus, deren Realisierbarkeit durch entsprechend hohe Elektrodenpotentiale wesentlich von der Dicke, der Zusammensetzung und den elektrischen Leiteigenschaften der transpassiven oxydischen Deckschicht abhängt.

Wir haben deshalb einige Untersuchungen durchgeführt, um zunächst die Veränderungen der Oxidschicht an den Elektrodenoberflächen beim Übergang vom passiven in den transpassiven Bereich zu erfassen. Die Charakterisierung der Schichten erfolgte elektrochemisch durch kathodische Reduktion und durch den zeitlichen Verlauf der Selbstaktivierung. Zunächst wurden Eisenelektroden in Boratpufferlösungen anodisch bis in den Bereich des transpassiven Stromanstieges polarisiert. Die

Untersuchungen erfolgten in Elektrolysezellen mit gerührter Elektrolytlösung, also bei relativ kleinen Strömungsgeschwindigkeiten. Die Abhängigkeit der kathodischen Reduktionskurven vom Potential der anodischen Schichtbildung zeigt Abb. 8 [3]. Die Reduktion wurde ohne Rührung der Elektrolytlösung durchgeführt. Die Reproduzierbarkeit der Meßergebnisse im Transpassivbereich ist begrenzt. So muß insbesondere dem Sauerstoff eine Beeinflussung der Meßergebnisse zugeschrieben werden, die sich z. B. in Potentialschwankungen auswirkte (Kurve 7). Dennoch erkennt man deutlich ein ausgeprägtes Anwachsen der Oxidschichten im Transpassivbereich, wozu die Fe_3O_4-Stufe stärker beiträgt als die γ-Fe_2O_3-Stufe.

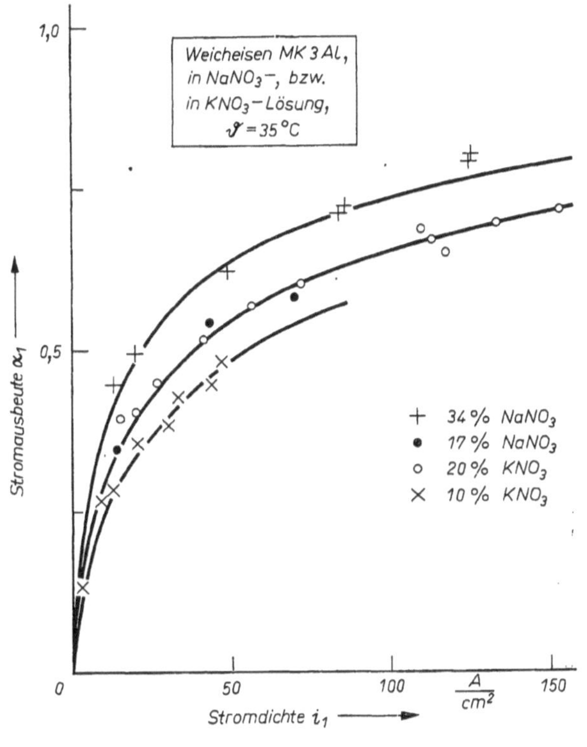

Abb. 6: Stromausbeute α_1 als Funktion der Reaktionsstromdichte i_1 der Auflösung von Eisen MK3Al in Nitratlösungen verschiedener Zusammensetzung. (Kurven berechnet nach Gl. (1).)
+ – 34% $NaNO_3$; • – 17% $NaNO_3$; ○ – 20% KNO_3; × – 10% KNO_3.

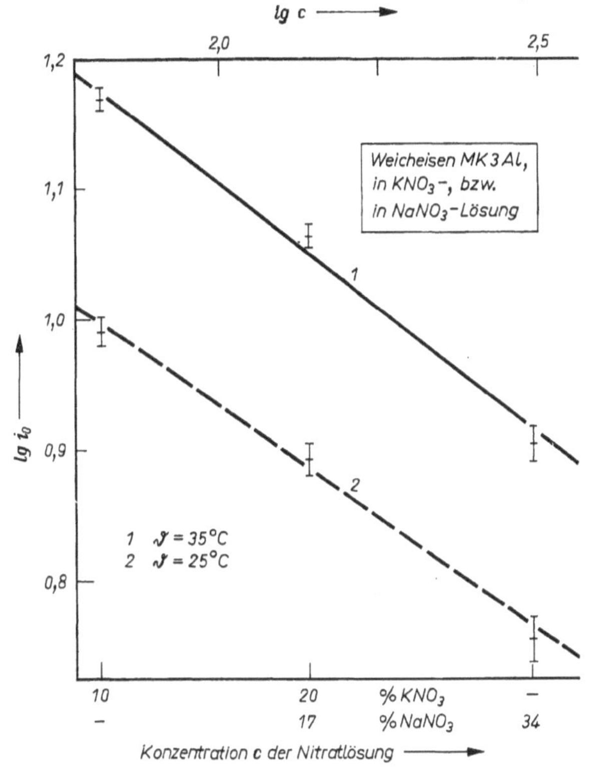

Abb. 7: Logarithmus des Kurvenparameters i_0 als Funktion des Logarithmus der Nitrationenkonzentration bei 25°C und 35°C.

Aus Abb. 9, in dem die für die Reduktion erforderlichen Ladungsmengen aufgetragen sind, ist auch das unterschiedliche Verhalten gegenüber dem Passivbereich ersichtlich und die Kurve für den Sauerstoffgehalt im Spülgas angegeben. Bemerkenswert ist, daß die Anstiege der Stromdichte-Potential-Kurve und der Schichtdickenkurve gegeneinander verschoben sind. Daraus kann gefolgert werden, daß die Schichtdickenbestimmung in diesem Bereich nicht maßgeblich durch die verhältnismäßig schwache Sauerstoffentwicklung beeinflußt wird. Die coulometrische Analyse von Oxidschichten, die sich bei Stromdichten bis zu 140 A/cm² in Nitratlösungen ausbilden, stieß auf erhebliche Schwierigkeiten. Man kann aber abschätzen, daß die Gesamtstärke der Schichten zunächst auf etwa 1000 Å anwächst, bei Stromdichten über 20 A/cm² dann aber rasch wieder abnimmt. Der äußere Teil der Schichten läßt sich abwischen, der metallseitige Teil ist fest haftend und offenbar dicht [16].

Zur Untersuchung der γ-Fe_2O_3-Teilschicht wurde als weitere Methode die Selbstaktivierung der Elektroden herangezogen. In einer Zelle mit gerührter oder ruhender Elektrolytlösung wurden Potential-Zeit-Kurven aufgenommen [3], wie sie in Abb. 10 dargestellt sind. Der steile Potential-

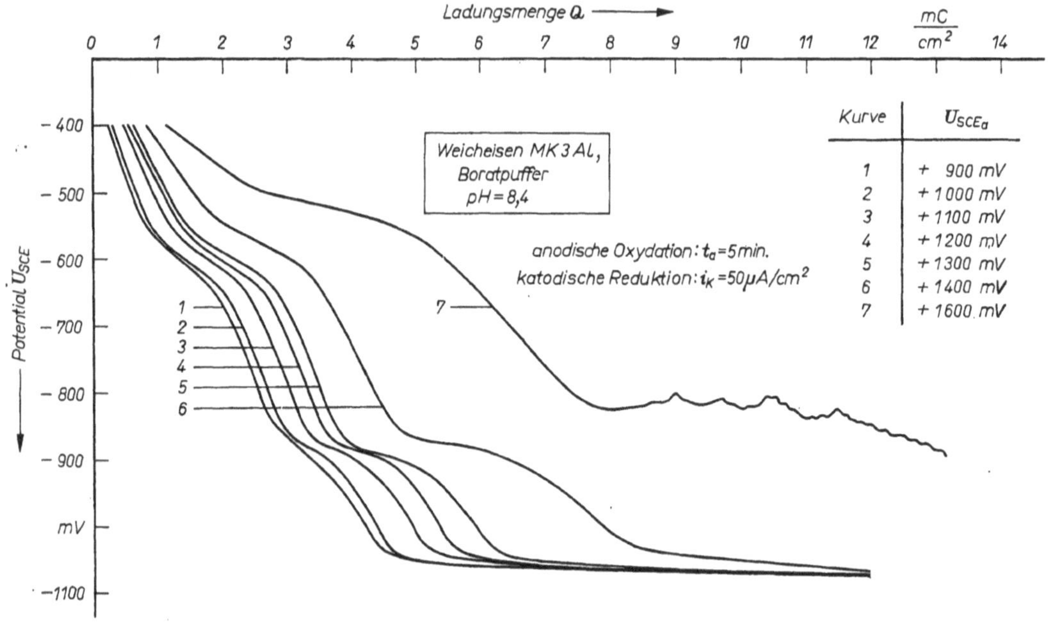

Abb. 8: Einfluß des Potentials U_{SCE_a} bei der Bildung der Oxidschichten auf die Potential-Zeit-Kurven ihrer kathodischen Reduktion.

i_k = 50 µA cm^{-2}, anodische Oxydation t_a = 5 min.

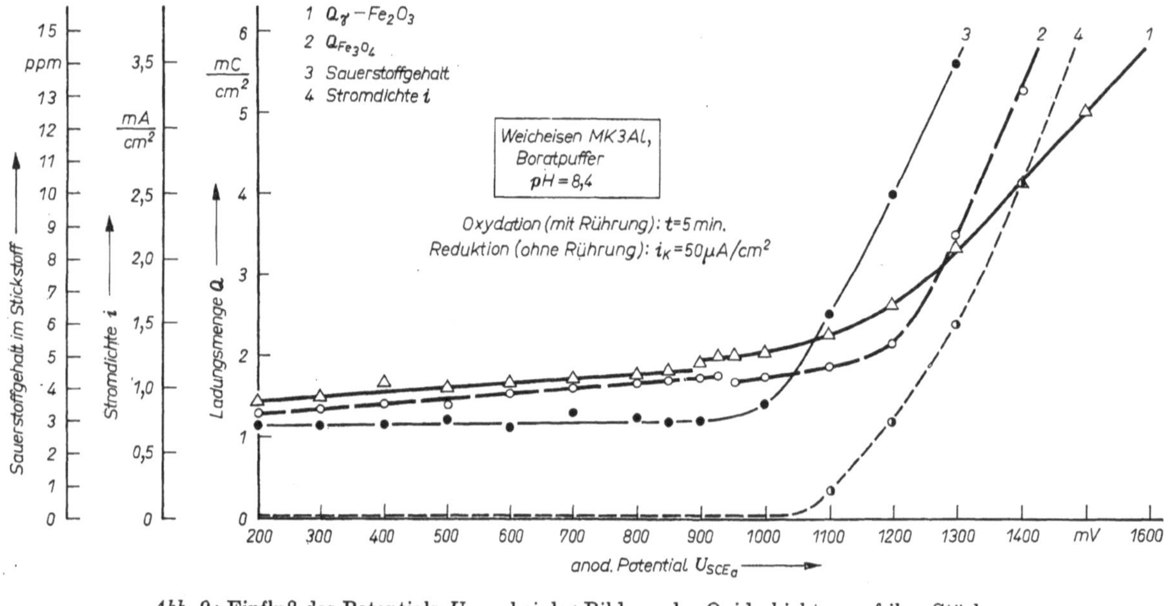

Abb. 9: Einfluß des Potentials U_{SCE_a} bei der Bildung der Oxidschichten auf ihre Stärke.

Oxydation: 5 min, mit Rührung; Reduktion: i_k = 50 µA cm^{-2}, ohne Rührung.

abfall entspricht der Aktivierung der Elektroden, d. h. einem Abbau bzw. Durchbruch der Fe_2O_3-Schicht. Die Zeit bis zur Aktivierung ist vom Potential der Schichtbildung abhängig. Sie wächst zunächst mit positiver werdendem Potential, um dann zwischen 1450 und 1900 mV (SCE) wieder abzunehmen (Abb. 11). Die Steilheit dieses Abfalles hängt von den Versuchsbedingungen ab, die bei der Selbstaktivierung vorlagen. Man muß deshalb in Übereinstimmung mit kinetischen Überlegungen berücksichtigen, daß die Selbstaktivierungszeit nur ein qualitatives Maß für die Stärke der Fe_2O_3-Schicht sein kann.

Potential-Zeit-Kurven der Ausschaltaktivierung wurden auch in Versuchszellen aufgenommen, die hinsichtlich des Elektrodenabstandes und der Strömungsgeschwindigkeit der Elektrolytlösung den Arbeitsbedingungen der elektrochemischen Metallbearbeitung besser angepaßt sind. Das Potential für die Bildung der Oxidschicht wurde dabei im Bereich von 1100 bis 1800 mV (SCE) verändert. Diese Potentiale der Schichtbildung liegen am Ende des Passivbereiches und im Transpassivgebiet. Mit positiver werdendem Potential nimmt die Aktivierungszeit ab. Es handelt sich dabei offenbar um das Übergangsgebiet vom passiven zum transpassiven Zustand, in dem die korrosionsfestere Fe_2O_3-

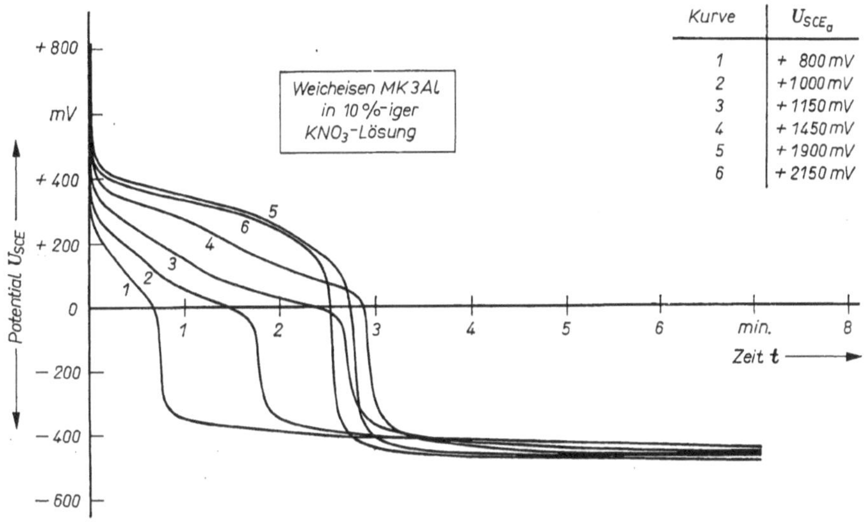

Abb. 10: Ausschaltaktivierungskurven von Eisen MK3Al in KNO_3-Lösung, 10%ig, in Abhängigkeit von dem Potential U_{SCE_a} der Oxidschichtbildung.

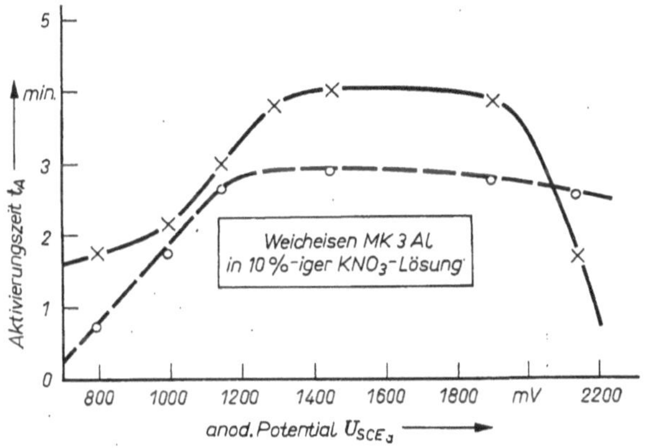

Abb. 11: Abhängigkeit der Aktivierungszeit vom Potential U_{SCE_a} der Oxidschichtbildung.

Schicht dünner wird (Abb. 12). Der nahezu vollständige Abbau der Schicht erfolgt unter diesen Arbeitsbedingungen bei ca. +1,800 V, was einer Stromdichte von etwa 1 A/cm² entspricht. Das Verschwinden der γ-Fe_2O_3-Schicht im Transpassivgebiet war auch von Foley u. Mitarb. [16] unter anderen Arbeitsbedingungen durch Elektronenbeugungsaufnahmen nachgewiesen worden.

Geht man davon aus, daß das Passivoxid auf Eisen lösungsseitig der Zusammensetzung Fe_2O_3 nahekommt, dann läßt sich entsprechend den Modellvorstellungen von Vetter [17] und Wagner [18] der sehr geringe stationäre Passivstrom auf der Basis eines stark gehemmten Fe^{3+}-Ionen-Durchtritts durch die Grenzfläche Oxid/Lösung bei der Gleichgewichts-Galvanispannung der Reaktion $H_2O \rightleftharpoons O^{2-}_{oxid} + 2H^+$ verstehen. Man sollte erwarten, daß auch bei weiter zunehmender anodischer

Polarisation nur Fe^{3+}-Ionen bzw. höherwertige Ionen in die Lösung übertreten könnten. Die im Transpassivgebiet auftretende Abnahme des Eisendefizits im lösungsseitigen Teil der Oxidschicht und die damit verbundene Änderung der Reaktionsladungszahl von $+3$ auf $+2$ steht offenbar in engem Zusammenhang mit der stark zunehmenden Sauerstoffentwicklung.

So ist denkbar, daß die intensive Sauerstoffentwicklung trotz hoher Strömungsgeschwindigkeiten der Elektrolytlösung zu einer Abnahme des pH-Wertes in Elektrodennähe führt. Die Erhöhung der Wasserstoffionenaktivität aber würde eine Verschiebung der Gleichgewichtsgalvanispannung der

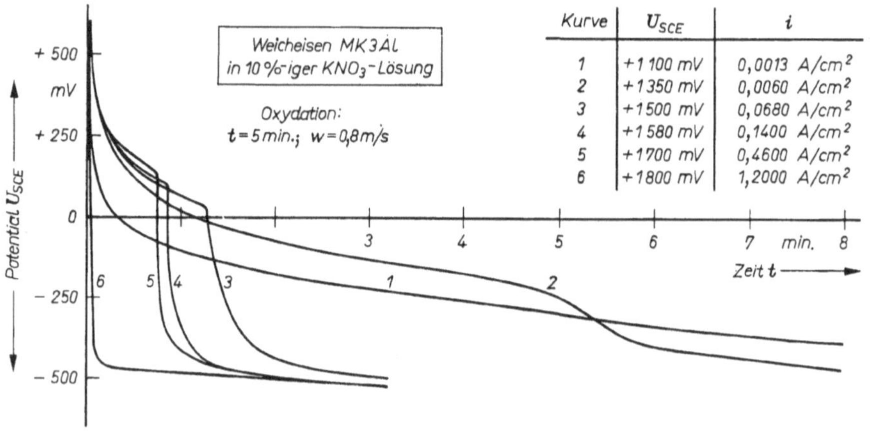

Abb. 12: **Einfluß des anodischen Potentials der Oxidschichtbildung auf die Potential-Zeit-Kurven bei der Ausschaltaktivierung; $U-t$-Kurven mit ruhender Lösung.**

Reaktion $H_2O \rightleftharpoons O^{2-}_{oxid} + 2H^+$ nach positiveren Werten mit sich bringen, die im stationären Zustand bei vorgegebenem Gesamt-Elektrodenpotential durch einen gegenläufigen Einfluß kompensiert werden muß. Als solcher würde die Erniedrigung der O^{2-}-Ionenaktivität im Oxid wirken, wodurch die Umwandlung der lösungsseitigen Fe_2O_3-Schicht verständlich wird. Der damit verbundene starke Anstieg des Fe^{2+}-Ionen-Durchtrittstromes ist einerseits auf die Zunahme der Konzentration der Fe^{2+}-Ionen in der Oxidgrenzschicht, andererseits aber auch darauf zurückzuführen, daß der Durchtritt von Fe^{2+}-Ionen offensichtlich weniger gehemmt ist als der von Fe^{3+}-Ionen. Letzteres ist z. B. auch daraus zu schließen, daß die stationären Stromdichten beim Übergang vom passiven in den vorpassiven Bereich trotz negativer werdender Galvanispannung Oxid/Lösung zunehmen — ein Effekt, der auf den einsetzenden Durchtritt von Fe^{2+} — an Stelle von Fe^{3+}-Ionen zurückzuführen ist.

Unabhängig von der Annahme einer pH-Wert-Erniedrigung in der lösungsseitigen Phasengrenzschicht muß der Gesichtspunkt in Betracht gezogen werden, daß die mit dem Elektrodenpotential zunehmende Sauerstoffentwicklung bzw. der damit ansteigende Elektronenstrom in der Oxidschicht Veränderungen des elektrochemischen Potentialgefälles $\Delta\tilde{\mu}_e = \Delta\mu_e - F \cdot \Delta\varphi$ zur Voraussetzung hat. Unter Berücksichtigung der Richtung des Elektronenstromes kann diese Änderung auf eine Vergrößerung von $\Delta\varphi$ und/oder eine Verkleinerung von $\Delta\mu_e$ zurückzuführen sein. Letzteres wäre bei Annahme eines Gleichgewichtes der Elektronen mit dem Ionengitter gleichbedeutend mit einer Einengung des Stöchiometriebereiches in dem Sinne, daß der Oxydationsgrad des lösungsseitigen Oxids von Fe_2O_3 in Richtung Fe_3O_4 abnimmt.

Die veränderte Oxidzusammensetzung im Transpassivbereich des Eisens begründet damit einen Auflösungsmechanismus, der sich von dem der transpassiven Chromauflösung merklich unterscheidet. Während beim Chrom die transpassive Auflösung durch eine Aufoxydation der Chromionen der Passivschicht ausgelöst wird, leitet beim Eisen eine Simultanreaktion, nämlich die Sauerstoffentwicklung, Vorgänge ein, die über eine Verringerung des Oxydationsgrades der lösungsseitigen Oxidschicht zu einer transpassiven Metallauflösung mit kleinerer Reaktionsladungszahl als im Passivbereich führen.

Literatur

[1] A. D. Davydov, V. D. Kaschejev und B. N. Kabanov, Elektrochimija, **5**, 221 (1969); **6**, 1760 (1970)
[2] Chin, D.-T. J. electrochem. Soc., **119**, 1181 (1972)
[3] H. Lüdenbach, Dissertation TH Karl-Marx-Stadt (in Vorbereitung)
[4] K.-W. Mao, J. electrochem. Soc., **118**, 1870 (1971); **118**, 1876 (1971)
[5] E. Rössner, Dissertation TH Karl-Marx-Stadt, 1972
[6] L. Franke, W. Forker und P. Schultheiss, Electrochim. Acta, **18**, 877 (1973)
[7] W. J. Plieth und K. J. Vetter, Ber. Bunsenges., **73**, 1077 (1969)
[8] L. Franke und W. Forker, Electrochim. Acta, **19**, 27 (1974)
[9] K. W. Mao und D.-T. Chin, J. electrochem. Soc., **121**, 191 (1974)
[10] D.-T. Chin und A. J. Wallace, Jr., J. electrochem. Soc., **120**, 1487 (1973)
[11] D. Pahl, VDI-Berichte, Düsseldorf 1970, S. 83
[12] D. Pahl, Dissertation TH Aachen, 1969
[13] L. Franke, Dissertation B TU Dresden, 1974
[14] G. Treffer, Dissertation TH Karl-Marx-Stadt, 1974
[15] W. Krysmann, Diplomarbeit TU Dresden, 1975
[16] C.-L. Foley, J. Kruger und C. J. Bechtoldt, J. Electrochem-Soc., **114**, 994 (1967)
[17] K.-J. Vetter, Electrochim. Acta, **16**, 1923 (1971)
[18] C. Wagner, Ber. Bunsenges., **77**, 1090 (1973)

L. Kiss und M. L. Varsányi

Lehrstuhl für physikalische Chemie der Loránd-Eötvös-Universität, Budapest

DIE PASSIVIERUNG DES EISENS IN NICHTWÄSSRIGEN LÖSUNGSMITTELN

Zusammenfassung

Mit Hilfe der potentiostatischen Methode wurde die anodische Auflösung von Eisen in Lösungen von Natriumacetat, Perchlorsäure und Natriumperchlorat in wasserfreier und in wasserhaltiger Essigsäure untersucht. Es wurde festgestellt, daß in Natriumacetatlösungen eine Passivierung des Eisens eintritt, wenn Wasser zugegen ist, beim Fehlen von Wasser in der Lösung sich jedoch keine Passivierung zeigt. In Perchlorsäure- und Natriumperchloratlösungen ergibt sich das entgegengesetzte Bild: im wasserfreien Medium kommt es zu einer Passivierung, in Gegenwart des Wassers jedoch nicht.

In Formamidlösungen von Essigsäure und Natriumacetat findet eine Passivierung nur in Gegenwart von Wasser statt.

Die Untersuchung der Kinetik von Elektrodenprozessen in nichtwäßrigen Lösungsmitteln, darunter des anodischen Verhaltens der Metalle, ist sowohl für die Theorie als auch für die Praxis sehr interessant. Die in nichtwäßrigen Lösungen gewonnenen Ergebnisse ermöglichen es, den Einfluß der Lösungszusammensetzung auf die Kinetik des Elektrodenprozesses zu beurteilen sowie den Mechanismus der in den wäßrigen Lösungen ablaufenden Prozesse besser zu verstehen. Außerdem ist bekannt, daß in nichtwäßrigen Lösungen unter industriellen Bedingungen bisweilen eine äußerst schnelle Metallkorrosion stattfinden kann.

Aus diesem Grunde erschienen in den letzten zehn Jahren viele Arbeiten zu diesem Thema. Vom praktischen und theoretischen Standpunkt aus sind jene sehr wichtig, die sich mit den Erscheinungsformen der Metallpassivierung und dem Einfluß der Zusammensetzung der Lösung, darunter mit dem Einfluß des Wassers, auf die Passivierung beschäftigen.

Aus diesen Arbeiten geht hervor, daß die Passivierung der Metalle aller Wahrscheinlichkeit nach von einer Adsorption des Sauerstoffs des Wassers mit Bildung der Metalloxide herrührt.

So untersuchten Schwabe und Schmidt [1, 2] die Passivierung des Nickels in einer schwefelsauren Lösung von Dimethylformamid und Acetonitril. Sie stellten fest, daß Nickel nur dann passiviert wird, wenn in der Lösung auch Wasser vorhanden ist. Die Auflösungsgeschwindigkeit des Nickels in Dimethylformamid hängt im passiven Zustand nichtlinear von der Konzentration des Wassers ab: bei einer Vergrößerung der Wasserkonzentration verringert sich anfangs die Geschwindigkeit, später wächst sie. Aus den Ergebnissen geht hervor, daß in Acetonitril der passive Zustand schwerer erreicht wird als in Dimethylformamid.

Kolotyrkin und Kossyj [3] untersuchten die Passivierung von Chrom und Eisen in methanolischer Lösung. Sie stellten fest, daß eine Passivierung dieser Metalle nur in Gegenwart von Wasser beobachtet wird und daß die minimale Wassermenge, die für die Passivierung des Chroms in Chloridlösungen erforderlich ist, mit der Konzentration des Chlorids wächst.

Für die Wichtigkeit des Wassers bei der Passivierung der Metalle sprechen auch die experimentellen Ergebnisse von Cinman u. Mitarb. [4, 5, 6]. Bei der Untersuchung von Titan und verschiedenen Stählen in Dimethylformamid und in Essigsäure kam es nur in Gegenwart von Wasser zu einer Passivierung, wobei die für die Passivierung des Metalls erforderliche minimale Wassermenge von einer Reihe von Faktoren (Temperatur, Lösungszusammensetzung, Lösungsmittel) abhängt. Deswegen kann in Abwesenheit von Wasser, aber in Gegenwart eines Depolarisators eine sehr rasche Korrosion auftreten. In diesem Fall vergrößert sich die Korrosionsstabilität von Stahl bei Wasserzugabe.

Um die Passivierung des Eisens zu untersuchen, experimentierten wir auf der Basis von wasserfreier Essigsäure [7, 8] und von wasserfreiem Formamid [9] (Methodik und Versuchsbedingungen siehe [7, 8, 10]). Wir bestimmten stationäre potentiostatische Kurven und versuchten, auf der Grundlage dieser Kurven Schlüsse bez. der ablaufenden Prozesse zu ziehen. Die Messungen in Essigsäure wurden hauptsächlich in Lösungen von Natriumacetat und Perchlorsäure durchgeführt.

Abb. 1 zeigt charakteristische potentiostatische Polarisationskurven, aufgenommen in 0,5 *n* Natriumacetatlösung bei verschiedenen Zusammensetzungen des Lösungsmittels. Wie ersichtlich, wird das Eisen in der wasserfreien Lösung nicht passiviert (gesättigte wäßrige Kalomel-Elektrode). In Gegenwart von Wasser wird das Eisen passiviert, und eine Vergrößerung des Wassergehaltes verringert die Auflösungsgeschwindigkeit im passiven Zustand.

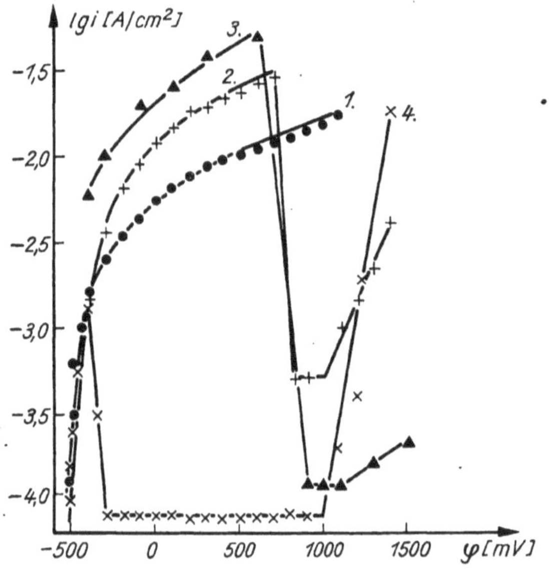

Abb. 1: Potentiostatische Polarisationskurven von Eisen, aufgenommen in 0,5 N CH_3COONa-Lösungen.
Zusammensetzung des Lösungsmittels: *1* – wasserfreie Essigsäure; *2* – 90% CH_3COOH + 10% H_2O; *3* – 50% CH_3COOH + 50% H_2O; *4* – 100% H_2O

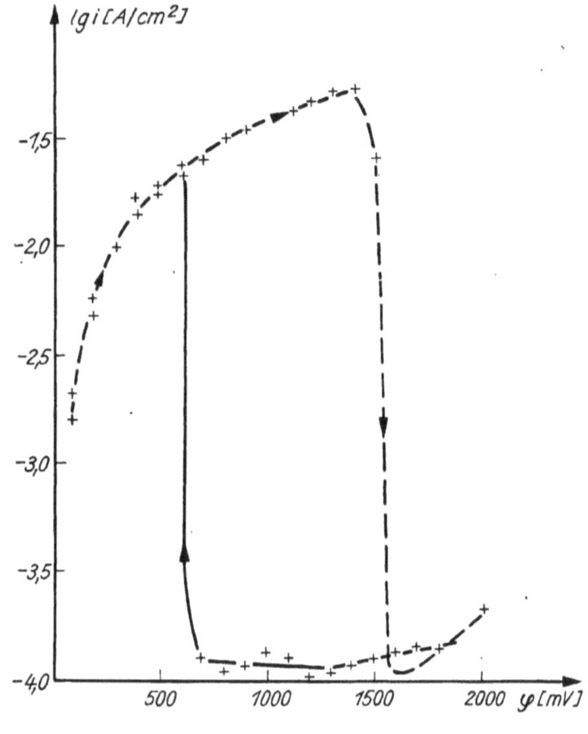

Abb. 2: Potentiostatische Polarisationskurven, gemessen an einer Eisenelektrode in 0,5 N CH_3COONa-Lösung.
Zusammensetzung des Lösungsmittels: 50% CH_3COOH + 50% H_2O; *1* – direkter Verlauf; *2* – umgekehrter Verlauf

Die in Abb. 1 dargestellten Kurven wurden bei einer von negativen zu positiven Werten verlaufenden Potentialänderung aufgenommen.

Wenn die potentiostatische Kurve in umgekehrter Richtung aufgenommen wird, wie das oft geschieht, ist eine Hysterese zu beobachten (siehe Abb. 2).

Auf Grund dieser experimentellen Ergebnisse kann der Schluß gezogen werden, daß in Übereinstimmung mit den oben genannten Literaturwerten eine Passivierung des Eisens in essigsauren Natriumacetatlösungen nur in Gegenwart von Wasser auftritt.

In Perchlorsäurelösungen werden allerdings andere Ergebnisse erhalten. Abb. 3 zeigt potentiostatische Kurven, die an einer Eisenelektrode in 0,5 *n* Perchlorsäurelösung gemessen wurden. In der wasserfreien Lösung zeigt die Polarisationskurve dasselbe Minimum, das bei der Passivierung erhalten wird. Das Minimum, das die Passivierung charakterisiert, verschwindet jedoch durch den Zusatz von Wasser. Ähnliche Ergebnisse werden in Natriumperchloratlösungen erhalten.

Um zusätzliche Informationen über diese Systeme zu erhalten, bestimmten wir die potentiostatische Polarisationskurven mit Hilfe einer rotierenden Scheibenelektrode. Bei Natriumacetatlösungen beeinflußte die Rotation der Elektrode den Kurvenverlauf nicht. Bei Perchlorsäure jedoch verschwindet das für die Passivierung charakteristische Minimum, wenn die Rotationsgeschwindigkeit der Elektrode erhöht wird. Das geht aus Abb. 4 hervor. Diese Ergebnisse zeigen, daß die hier behandelte Passivierung nur in Natriumacetatlösungen auftritt, wenn Wasser zugegen ist, und daß in Perchloratlösungen nur eine Diffusionshemmung zu beobachten ist.

Daß es in Gegenwart von Perchlorat nicht zu einer Passivierung kommt, hat seine Ursache offensichtlich in einer unter bestimmten Bedingungen in der wäßrigen Lösung auftretenden aktivierenden Wirkung des Perchlorats. Außerdem ist vermutlich auch wichtig, daß im Falle des Natriumacetats beim Zusetzen von Wasser infolge der Hydrolyse Hydroxylionen gebildet werden, die bekanntlich die Passivierung fördern.

In Formamidlösungen wurden die Ergebnisse hauptsächlich in 0,5 N Natriumacetat- und in 0,5 N Essigsäurelösung gewonnen. Abb. 5 zeigt für 0,5 N Natriumacetatlösung typische Kurven.

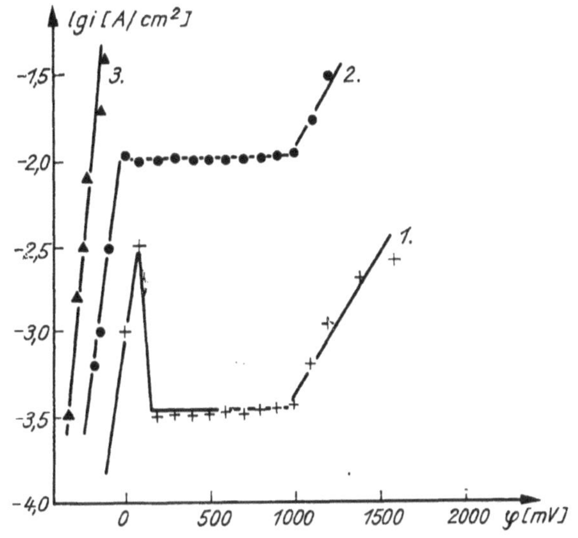

Abb. 3: Potentiostatische Polarisationskurven von Eisen, aufgenommen in 0,5 N $HClO_4$-Lösungen.

Zusammensetzung des Lösungsmittels: *1* – wasserfreie Essigsäure; *2* – 98% CH_3COOH + 2% H_2O; *3* – 90% CH_3COOH + 10% H_2O.

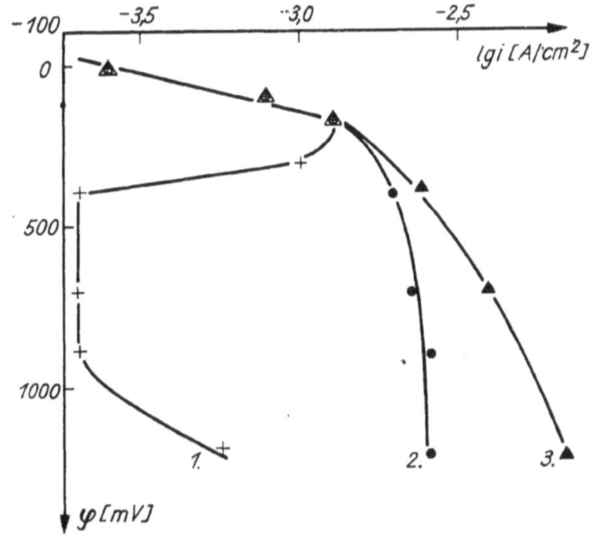

Abb. 4: Einfluß der Rotationsgeschwindigkeit der Elektrode auf den Verlauf der potentiostatischen Kurve in wasserfreier 0,5 N essigsaurer $HClO_4$-Lösung.

Zahl der Umdrehungen der Elektrode: *1* – 0 min^{-1}; *2* – 230 min^{-1}; *3* – 1320 min^{-1}.

Nach diesen Resultaten wird in wasserfreier Lösung keine Passivierung beobachtet. In Gegenwart von 10% Wasser wird das Eisen passiviert. Wenn jedoch der Wassergehalt in der Lösung erhöht wird, steigt die Geschwindigkeit der passiven Auflösung. Eine ähnliche Erscheinung beobachteten Schwabe und Schmidt [1, 2] bei der Auflösung von Nickel in schwefelsaurer Dimethylformamidlösung.

Kurven, die in wasserfreier Lösung aufgenommen wurden, besitzen einen horizontalen Abschnitt, der an den Grenzstrom erinnert. Dieser Abschnitt rührt offensichtlich von einer Salzabscheidung an der Eisenoberfläche her. Oft ist dieser Prozeß der erste Schritt der Passivierung eines Metalls. Die hier behandelte Passivierung tritt jedoch nicht auf, wenn infolge der Lösungszusammensetzung die Reaktivierung ein sehr schneller Prozeß ist. Das Metall wird natürlich, wie im oben erwähnten Fall, auch dann nicht passiv, wenn in der Lösung solche Komponenten nicht oder nicht in ausreichender Menge vorhanden sind, die über eine Reaktion zu einer Passivierung der Metalloberfläche führen.

Die in Abb. 6 dargestellten Ergebnisse bestätigten, daß dem horizontalen Abschnitt eine Salzausscheidung auf der Metalloberfläche entspricht und daß die Reaktionsgeschwindigkeit hauptsächlich durch die Abdiffusion der Auflösungsprodukte von der Elektrodenoberfläche bestimmt wird. Wird die Polarisationskurve in wasserfreier Formamidlösung mit einer rotierenden Eisenelektrode gemessen, so steigt bei einer Erhöhung der Rotationsgeschwindigkeit der Elektrode auch die Stromdichte.

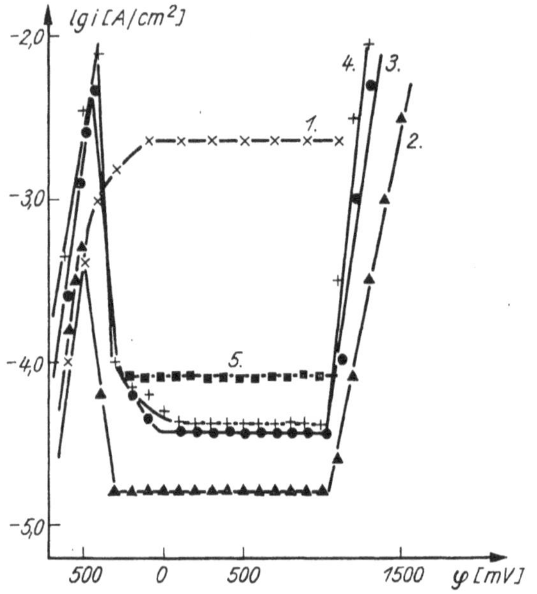

Abb. 5: Potentiostatische Polarisationskurven, gemessen an einer Eisenelektrode in 0,5 N CH_3COONa-Lösungen.

Zusammensetzung des Lösungsmittels: *1* — wasserfreies Formamid; *2* — 90% Formamid + 10% Wasser; *3* — 80% Formamid + 20% Wasser; *4* — 50% Formamid + 50% Wasser; *5* — 100% Wasser.

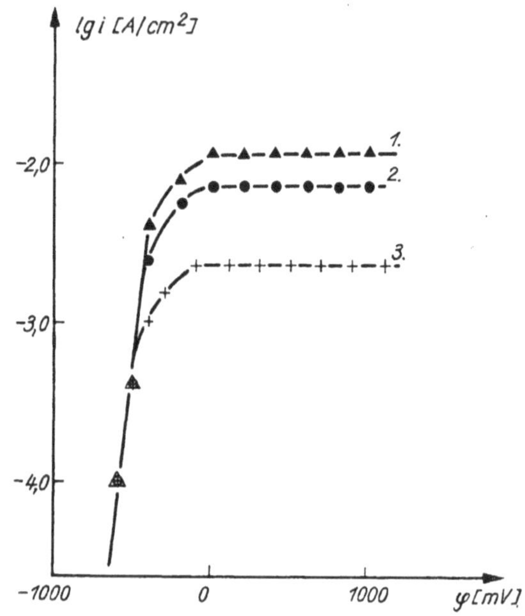

Abb. 6: Einfluß der Rotationsgeschwindigkeit der Elektrode auf den Verlauf der potentiostatischen Kurve in 0,5 N Lösung von CH_3COONa in wasserfreiem Formamid.

Zahl der Umdrehungen der Elektrode: *1* — 0 min^{-1}; *2* — 2120 min^{-1}; *3* — 3820 min^{-1}.

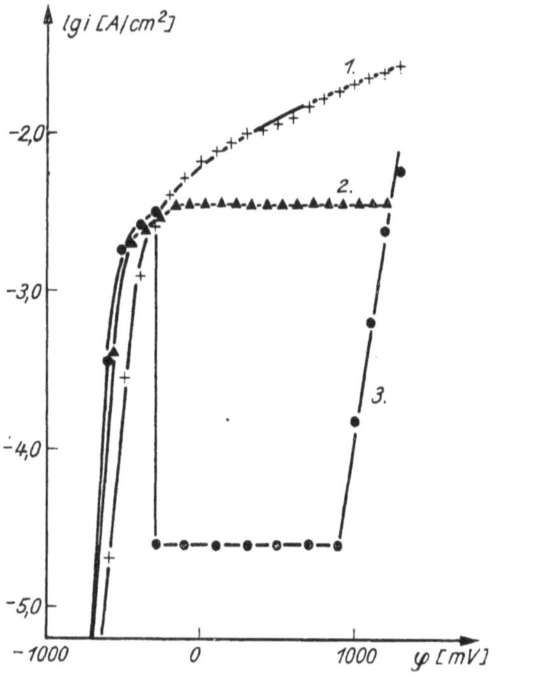

Abb. 7: Potentiostatische Polarisationskurven von Eisen in folgenden Lösungen: *1* und *2* — 0,5 N CH_3COOH, *3* — 0,5 N CH_3COOH + 0,5 N CH_3COONa.

Zusammensetzung des Lösungsmittels: *1* — wasserfreies Formamid; *2* und *3* — 95% Formamid + 5% Wasser.

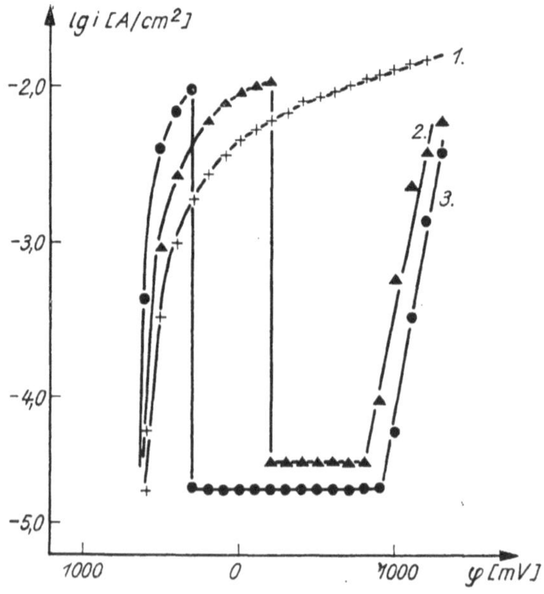

Abb. 8: Potentiostatische Polarisationskurven von Eisen.

Zusammensetzung der Lösungen: *1* — 0,5 N CH_3COOH in wasserfreiem Formamid; *2* — 0,5 N CH_3COOH in 95% Formamid + 5% Wasser; *3* — 0,5 N CH_3COOH + 0,5 N CH_3COONa in 90% Formamid + 10% Wasser.

In Formamid, das 0,5 N Essigsäure enthält, tritt eine Passivierung nur ein, wenn große Mengen Wasser zugegen sind. Auch das Hinzufügen von Natriumacetat fördert die Passivierung, wie aus den Abb. 7 und 8 zu sehen ist. Abb. 7 zeigt, daß bei einem Anteil von 5% Wasser keine Passivierung beobachtet wird, jedoch beim Zusetzen von Natriumacetat Passivierung des Eisens erfolgt.

Aus den in Abb. 8 dargestellten Ergebnissen folgt, daß das Eisen in Gegenwart von 10% Wasser passiviert wird, sofern das Formamid 0,5 N an Essigsäure ist. Allerdings zeigt die Kurve *3* der Abb. 8, daß das Passivierungspotential beim Zusatz von Natriumacetat in negativer Richtung verschoben wird.

Aus unseren Ergebnissen folgt, daß bei den von uns untersuchten Systemen die Anwesenheit von Wasser eine notwendige, aber nicht hinreichende Bedingung für eine Passivierung ist. Ähnlich der in wäßrigen Lösungen beobachteten Tatsache findet in Gegenwart aktivierender Ionen oder Moleküle keine Passivierung von Metallen statt.

Literatur

[1] W. SCHMIDT, Zaščita metallov, **5**, 608 (1969)
[2] K. SCHWABE und W. SCHMIDT, Corrosion Sci., **10**, 143 (1970)
[3] Ja. M. KOLOTYRKIN und G. G. KOSSYJ, Zaščita metallov, **1**, 272 (1965)
[4] A. I. CINMAN und A. N. KATREVIČ, Zaščita metallov, **5**, 517 (1969)
[5] A. I. CINMAN, L. M. PISČIK, V. M. BRUSENCOVA und L. E. ZACHARENKOVA, Zaščita metallov, **7**, 700 (1971)
[6] A. I. CINMAN, L. M. PISČIK, V. M. BRUSENCOVA und L. E. ZACHARENKOVA, Zaščita metallov, **8**, 567 (1972)
[7] L. KISS, Do Ngoc LIEN und M. L. VARSÁNYI, Acta Chim. Acad. Sci. Hung., **70**, 313 (1971)
[8] L. KISS, Do Hgoc LIEN und M. L. VARSÁNYI, Acta Chim. Acad. Sci. Hung., **74**, 135 (1972)
[9] L. KISS, Do Ngoc LIEN und M. L. VARSÁNYI, Collect. Czechoslovak. Chem. Commun., **36**, 914 (1971)
[10] L. KISS, Do Ngoc LIEN und M. L. VARSÁNYI, Ann. Univ. Sci. Bp. Sec. Chim., **12**, 145 (1971)

FSC
www.fsc.org
MIX
Papier | Fördert gute Waldnutzung
FSC® C083411

Kolibri 360 GmbH
Ferdinand-Jühlke-Straße 7
99095 Erfurt, Deutschland
produktsicherheit@kolibri360.de